ELECTRICAL POWER SYSTEMS

VOLUME 1

Second Edition (SI/Metric Units)

Other Titles of Related Interest

BADEN-FULLER
Microwaves, 2nd Edition

CALVER
Power Systems: Modelling and Control Applications

CEGB
Advances in Power Station Construction

DAVIES
Protection of Industrial Power Systems

DUMMER & WINTON
Elementary Guide to Reliability, 4th Edition

DUNN & REAY
Heat Pipes, 3rd Edition

GANDHI
Microwave Engineering and Applications

HOLLAND
Integrated Circuits and Microprocessors
Microprocessors and their Operating Systems

HORLOCK
Cogeneration—Combined Heat and Power

MURPHY & TURNBULL
Power Electronic Control of AC Motors

MURRAY
Nuclear Energy, 3rd Edition

REAY & MACMICHAEL
Heat Pumps, 2nd Edition

YORKE
Electric Circuit Theory, 2nd Edition

Pergamon Related Journals (free sample copies gladly sent on request)

Computers and Electrical Engineering
International Journal of Engineering Science
Mechatronics
Microelectronics and Reliability
Solid-State Electronics

ELECTRICAL POWER SYSTEMS

In two volumes

A. E. GUILE

D.Sc.(Eng.), Ph.D., B.Sc.(Eng.), C.Eng., F.I.E.E.

Reader in Electrical Engineering,
University of Leeds

the late W. PATERSON

B.Sc., B.Sc.(Eng.), C.Eng., F.I.E.E.

one-time Principal Lecturer in Electrical Engineering,
Leeds Polytechnic

VOLUME 1

Second Edition (SI/Metric Units)

PERGAMON PRESS

Member of Maxwell Macmillan Pergamon Publishing Corporation

OXFORD · NEW YORK · BEIJING · FRANKFURT
SÃO PAULO · SYDNEY · TOKYO · TORONTO

U.K.	Pergamon Press plc, Headington Hill Hall, Oxford OX3 0BW, England
U.S.A.	Pergamon Press Inc., Maxwell House, Fairview Park, Elmsford, NY 10523, U.S.A.
PEOPLE'S REPUBLIC OF CHINA	Pergamon Press, Room 4037, Qianmen Hotel, Beijing, People's Republic of China
FEDERAL REPUBLIC OF GERMANY	Pergamon Press GmbH, Hammerweg 6, D-6242 Kronberg, Federal Republic of Germany
BRAZIL	Pergamon Editora Ltda, Rua Eça de Queiros, 346, CEP 04011, Paraiso, São Paulo, Brazil
AUSTRALIA	Pergamon Press Australia Pty Ltd, P.O. Box 544, Potts Point, N.S.W. 2011, Australia
JAPAN	Pergamon Press, 5th Floor, Matsuoka Central Building, 1-7-1 Nishishinjuku, Shinjuku-ku, Tokyo 160, Japan
CANADA	Pergamon Press Canada Ltd., Suite No. 271, 253 College Street, Toronto, Ontario, Canada M5T 1R5

First edition 1969
Second edition 1977
Reprinted 1978, 1979, 1981, 1982, 1985, 1986, 1991

Library of Congress Cataloging in Publication Data

Guile, Alan Elliott.
Electrical power systems.
Includes bibliographical references and indexes.
1. Electrical power systems. 2. Electric power transmission. I. Paterson, William, B.Sc., joint author. II. Title.
TK 1001.G84 1977 621.319 77.1789

ISBN 0-08-021728-1 (Hardcover)
ISBN 0-08-021729-X (Flexicover)

Printed in Great Britain by BPCC Wheatons Ltd, Exeter

CONTENTS

CONTENTS OF VOLUME 2

Chapter 3 STABILITY OF POWER SYSTEMS

Chapter 4 TRAVELLING WAVES IN TRANSMISSION LINES

Chapter 5 INSULATION CO-ORDINATION

Chapter 6 RECTIFICATION, INVERSION AND HIGH-VOLTAGE D.C. SYSTEMS

Chapter 7 3-PHASE NETWORK MATRICES AND CO-ORDINATE SYSTEM TRANSFORMATIONS

Chapter 8 POWER SYSTEM ANALYSIS

Appendix A.2

ECONOMICS OF ELECTRICAL POWER GENERATION

PREFACE TO FIRST EDITION

This textbook is the first of two volumes written to suit the needs of students studying electrical power systems at universities, polytechnics and colleges of technology. There is considerable variation in the syllabus content of this subject in different institutions, but as a general guide it may be said that this first volume treats the subject to about the level of the second year of a university Honours Course having some specialisation in electrical power engineering. At the same time the breadth and treatment of the topics chosen are such as to suit also Ordinary Degree, C.E.I. part II, H.N.D. or H.N.C. students. For example, although matrices have been used where appropriate, the text has been designed to be readable by a student not familiar with them. The second volume will extend the subject to final year Honours Degree level. The division of material between the two volumes has been carefully considered by the authors. Where possible, a topic has been developed to about Honours Degree level in this volume, so that it need not be dealt with further in Volume 2. In the case of unbalanced faults, for example, this has involved including some material usually dealt with in the final year of an Honours Degree course. In the case of protection, however, only the more elementary topics are dealt with in Volume 1, while Volume 2 will cover more advanced topics together with the associated subject of circuit interruption.

For each chapter the text provides an introduction, gives definitions and develops the basic theory. Worked examples are given only where it is felt that they supplement and clarify the text in a really useful way, and they are omitted where they would only involve mere substitution in formulae. At the end of each chapter a variety of examples are given with answers, and at the end of the book there are miscellaneous examples. The latter are given because it is felt that students tend to be willing only to think and apply knowledge within artificial watertight compartments, and need practice in integrating information from different aspects of a subject.

It is hoped that lecturers will find the book useful in providing material which students can be asked to study for themselves. The lecturer will then have more time, both to extend the coverage of this textbook to suit his particular syllabus, and to include tutorial work on the numerous examples and points suggested in the text for the student to develop further. All too often, students spend nearly all the lecture periods copying rapidly with little opportunity to think, to understand, to discuss, and to practice oral presentation of technical information.

The authors will welcome any constructive criticism of this textbook and will be grateful if they are informed of any errata.

The authors wish to thank Dr C. H. Gosling and Messrs D. J. Bell, W. Bowman and F. R. Haigh who have read particular chapters; John Wiley and Sons for permission to reprint Charts 1, 2 and 3 from *Principles of Electric Power Transmission* by L. F. Woodruff; and the English Electric Company Limited for permission to reproduce Fig. 9.5. Thanks are also due to the University of London, University of Leeds, Leeds College of Technology, and the Institution of Electrical Engineers for permission to reproduce questions from past examination papers, the answers to which are the responsibility of the authors alone. Those taken from College Diploma examination papers of the Leeds College of Technology are denoted by L.C.T., those from Higher National Certificate or Diploma papers by H.N.C. or H.N.D. and those from the Part III Examination of the Institution of Electrical Engineers by I.E.E. It should be noted that to save space descriptive and theoretical sections of these questions have generally been omitted, so that some of the examples are no longer full-length questions.

PREFACE TO SECOND EDITION

This edition has been revised to use SI units throughout. The opportunity has also been taken to include references to some recent developments, e.g. the growth of alternator ratings and the possible further growth in the future, and the present position in high-voltage cables of large ratings.

Chapter 1

TRANSMISSION

1.1 Introduction

This chapter will discuss the theory of power transmission lines operating under steady-state conditions, but first the general features of a transmission system as organised in the United Kingdom will be outlined. Throughout this chapter (and generally throughout the book) all the data and comment on typical examples and current practice, refer to the 50Hz power system in Great Britain.

Transmission lines transmit electrical power from a sending-station (S) to a receiving-station (R) without supplying any consumers en route: by contrast, a distribution line or distributor supplies consumers directly at short intervals along the line. Thus a distributor is subject to the legal requirement that power should be supplied at a voltage within $\pm 6\%$ of a declared (service or rated) voltage, whereas a transmission line is not subject to that restriction and its voltage can vary as much as 10 or even 15% due to variations in load. Any restriction on transmission voltage is technical not legal. In practice the distinction between transmission and distribution is not so clear-cut as several large consumers are now being supplied at 132 kV and above, and single-phase loads are being taken from two phases of a 132-kV line. The transmission system is referred to collectively as the grid, the 275 and 400-kV systems being referred to as the super-grid. Although the present 400–420-kV network of some 7000-km length is expected to meet the main transmission requirements until about the end of the century, research continues at higher voltages, e.g. on a 1300-kV transmission tower.

The two ends of a transmission line can terminate at generating-stations, grid-switching stations or grid-supply points. The busbars (terminals) of a modern generating station are now normally outdoor. Grid-switching stations have similar outdoor busbars but are not associated with a particular generating station. They comprise transformers for interconnecting the various voltage levels of the grid and circuit breakers for clearing faults and controlling the system. A grid-supply point is a point on the system at which the Central Electricity Generating Board (wholesaler) supplies an Electricity Board (retailer) and comprises transformers to step down to 33 kV and circuit breakers to control the outgoing circuits: it is from this point outwards that the system voltage should be held within the legal limits.

The primary parameters (so-called constants) of a transmission line are its series and shunt resistance, its series inductance (self and mutual) and its shunt capacitance. Inductance is due to the magnetic fields set up by the line currents and is a series inductance carrying a line current (see Chapter 2). Capacitance, and its charging current, are due to the electric field set up by the system voltage and appear as a shunt circuit across the system voltage, usually the line-to-neutral, or per-phase-star, voltage (see Chapter 3).

The insulation of a line is seldom perfect and leakage currents flow over the surface of the insulators especially during foul weather (see Chapter 4). This leakage is simulated by a shunt resistance of high ohmic value across the system line-to-neutral (phase) voltage and so is in parallel with the system capacitance. Leakage causes a power loss which is a function of the system voltage (not the line current). Corona (see Chapter 3) causes a power loss from the system which, being dependent on the system voltage, is lumped together with the leakage effect. Normally leakage and corona effects are ignored (unless the relevant data is given) since the effective shunt resistance is generally many times greater than the shunt capacitive reactance. The analyses in this chapter will assume that these primary parameters are given.

Power transmission lines (as distinct from communication lines) may be classified as short or long. A short line is defined as one for which all shunting effects can be ignored. A short line therefore has a simple series equivalent circuit (or model). This approximation is usually justified for lines up to about 80 km route length, but the real criterion is the accuracy required in the answer to a problem.

An approximate solution of a long line (sometimes referred to as a medium line) assumes that all shunting effects can be lumped at a few selected points along the line: for example, all at the middle giving the nominal T (tee) circuit, or half at each end giving the nominal Π (pi) circuit. These two circuits tend to give some quantities on opposite sides of the truth (rigorous solution of a uniform long line) but it usually requires careful work with four significant figures to show up any difference in their answers. The word nominal is used to emphasise that their configurations are merely convenient in that the two circuits both have the same total series impedance and total shunt admittance as the line itself, based upon the formulae in Chapters 2 and 3. The nominal circuits should not be confused with the equivalent T and Π circuits which are accurate representations of a long line, so far as the terminal conditions are concerned. Now that digital computers are available to perform the tedious arithmetical calculations involved, one or more Π circuits in cascade are being used to represent long lines.

The main assumption made in the rigorous solution of a long line is that the line is uniform, so that the primary parameters can be given per km. The formulae derived in Chapters 2 and 3 assume that the conductors are straight, parallel and at a constant height above ground. Due to line sag, these conditions are not strictly valid and there is some doubt as to the precise meaning of route length.

The modern tendency is to generalise: that is, to develop a theory which can be applied to many types of circuits, and is not restricted to a special case. The theory of the 4-terminal (or 2-port) network with its **ABCD** constants, discussed later in this chapter, is important in its application to all types of transmission lines (and other items of plant).

Transmission systems can consist of overhead lines (see Chapter 4) or underground cables (see Chapter 5), or a mixture, since many lines terminate in cable ends. To avoid tedious repetition, in this chapter, only lines will be mentioned, but the theories apply equally to cables when due allowance is made for their different primary parameters. Cables tend to have a lower series impedance and a higher shunt capacitance and are usually only a few km long.

The theory of power transmission lines will here be treated for steady-state conditions only: their operation under transient conditions is discussed in Volume 2.

1.2 Short Lines

1.2.1 ANALYTICAL METHODS

A very large number of power lines in Great Britain have route lengths of less than 80 km, so each can be represented approximately as a series equivalent circuit as in Fig. 1.1(a). The same circuit can also represent an alternator (see Chapter 7), and a transformer, if the exciting circuit can be neglected (see Chapter 6). Thus short-line theory can be applied to a large part of the grid network.

The student is reminded that a balanced 3-phase circuit, with a star-connected supply and load, can be represented by a single-phase (single-line) diagram. Since the phasor sum of the three line currents

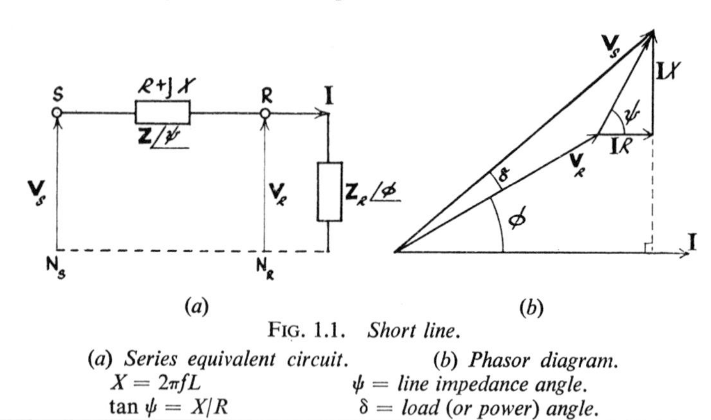

(*a*) (*b*)

FIG. 1.1. *Short line.*

(*a*) *Series equivalent circuit.* $X = 2\pi fL$, $\tan\psi = X/R$

(*b*) *Phasor diagram.* ψ = *line impedance angle.* δ = *load (or power) angle.*

is zero, no current would flow between the neutral points N_S and N_R even if they were connected by a fourth wire. Thus there can be no potential difference between N_S and N_R; to preserve this feature in the single-phase circuit, the return neutral conductor (dotted) must be deemed to have no impedance. Usually N_S is at earth potential (except possibly during fault conditions) and therefore so is N_R. Throughout this book all loads will be assumed to be balanced 3-phase loads (unless otherwise stated).

When working with balanced 3-phase circuits, it is usual to assume that all voltages stated are line-to-line values, that all currents are line currents, that all voltamperes, power and reactive power are 3-phase, and that all power factors are phase power factors (cosine of angle between phase voltage and corresponding phase current). Answers to problems are normally given in similar form, but in all calculations the student is strongly advised to use phase-to-neutral (star) voltage, phase current and phase impedance (impedance data is always per phase). In this book bold-face type (**V**) is used to denote an r.m.s. quantity which varies sinusoidally in time (d.c. quantities appearing only in section 5.9), when its phasor nature is involved. Italicised symbols (V) or $|\mathbf{V}|$ again refer to the r.m.s. value, but only when magnitude and not phase is involved. The peak value or amplitude of the alternating quantity is $\sqrt{2}V$. Complex quantities such as impedance **Z** and line constants **A**, **B**, **C** and **D**, although they do not vary in time, are also shown bold-face where their complex values rather than their magnitudes must be used. When they are italicised (Z) only the magnitude is required. Since quantities such as inductance L, capacitance C, resistance R, reactance X are not complex, they are italicised in all cases.

The sending- and receiving-end voltages $\mathbf{V}_S$ and $\mathbf{V}_R$ are both line-to-neutral (per phase star) values equal to the corresponding line-to-line values divided by $\sqrt{3}$. It follows from Fig. 1.1(b) given the load-end conditions, that

$$V_S = [(V_R \cos\phi + IR)^2 + (V_R \sin\phi + IX)^2]^{\frac{1}{2}} \text{ volts} \qquad [1.1]$$

and sending-end power factor is

$$\cos(\phi+\delta) = (V_R \cos\phi + IR)/V_S \qquad [1.2]$$

from which the transmission efficiency and all sending-end data can be found.

Fig. 1.1(b) can be redrawn as Fig. 1.2 to illustrate the calculation of $\mathbf{V}_S$ using j-notation with $\mathbf{V}_R$ as the reference phasor. Thus

$$\begin{aligned} \mathbf{V}_S &= \mathbf{V}_R + \mathbf{IZ} \qquad [1.3] \\ &= V_R\underline{/0^\circ} + I\underline{/-\phi} \times Z\underline{/\psi} \\ &= V_R + IZ\underline{/\psi-\phi} \\ &= (\overline{oa}) + (\overline{ab} + \mathrm{j}\,\overline{bc}) \\ &= (\overline{ob}) + \mathrm{j}\,(\overline{bc}) \quad : \text{ Cartesian form} \\ &= V_S\underline{/\delta} \quad : \text{ polar form} \end{aligned}$$

where

$$V_S = (\overline{ob}^2 + \overline{bc}^2)^{\frac{1}{2}}$$

and

$$\tan\delta = \overline{bc}/\overline{ob}.$$

Equation [*1.3*] is valid in the usual system of units and also in the per-unit system (see Appendix 1).

The impedance voltage is defined as the numerical product IZ and, unless otherwise qualified, it is assumed that I is rated current. It follows that the impedance voltage at rated current is a constant for a given circuit and, in particular, is independent of power factor. If the impedance voltage is divided by the rated (base) voltage, it becomes the circuit impedance in per-unit form corresponding to rated MVA (see Appendix 1).

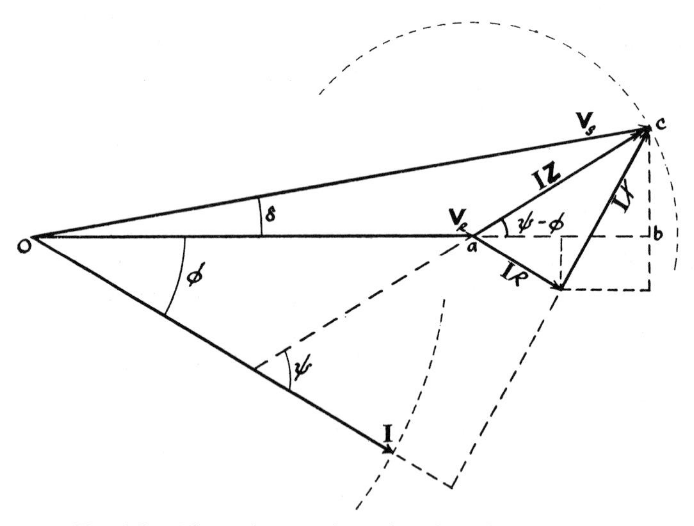

FIG. 1.2. *Phasor diagram for a short line, showing locus of* $\mathbf{V}_S$ *for constant I at varying load power factor.*

The voltage drop along a transmission line is defined as the numerical difference between the sending-, and receiving-end voltages and, unless otherwise qualified, is assumed to refer to rated current and voltage at the receiving-end. It should be clear from Fig. 1.2 that voltage drop is a function of the load power factor, so the power factor should be stated. Thus

$$\text{voltage drop} = V_S - V_R \text{ volts.} \qquad [1.4]$$

If the voltage drop is divided by rated (base) phase voltage the result is in per-unit form and if further multiplied by 100 the result is in percentage form.

The inherent voltage regulation of a line is defined as the numerical rise in receiving-end voltage when rated load current, at rated voltage, at a stated power factor, is switched off, with the input voltage and frequency held constant (hence the qualification inherent—the rise in voltage is due to the line alone and not to the source of supply). Since for a short line (and only for a short line), the no-load receiving-end voltage $V_{Ro} = V_S$, it follows that

$$\text{voltage regulation} = V_{Ro} - V_R = V_S - V_R \qquad [1.5]$$

so that the voltage regulation is equal to the voltage drop. The regulation can be given in volts, per-unit or percentage (usually percentage) form. The student should check that the regulation is a maximum when $\psi - \phi = 0$ or ϕ (lagging) $= \psi$, and that this maximum regulation is equal to the impedance voltage: and further, that there is a particular load power factor (usually leading) when $V_S = V_R$, i.e. when the regulation is zero. If the load power factor falls to a still lower (leading) value then $V_S < V_R$ and the regulation is negative. Regulation is a measure of the fluctuations which occur at a consumer's terminals as the system load fluctuates. A low regulation is achieved by having a low system impedance up to the consumer's terminals and by not allowing the power factor to fall to a low lagging value.

In some textbooks it is shown that (for ϕ lagging)

$$\text{voltage regulation} \simeq IR \cos \phi + IX \sin \phi$$

$$\simeq IZ \cos (\psi - \phi) \text{ volts} \qquad [1.6]$$

(see Fig. 1.2). From this it is clear that regulation is proportional to load current (the student should not confuse this remark with the previous one that regulation is normally quoted at rated current). Equation [*1.6*] is reasonably true for the normal lagging power factors but it should not be used for leading power factors. It follows from [*1.6*] that the regulation is a maximum $= IZ$ when ϕ (lagging) $= \psi$: the student should deduce from Fig. 1.2 that the statement is exact and not approximate (except in so far as short lines are approximations). By dividing [*1.6*] by rated voltage it follows that

$$\text{voltage regulation} \simeq R_{pu} \cos \phi + X_{pu} \sin \phi \quad \text{pu}$$

$$\simeq Z_{pu} \cos (\psi - \phi) \quad \text{pu.} \qquad [1.7]$$

Students are recommended not to memorise the approximations [*1.6*] or [*1.7*], nor the construction lines in Fig. 1.2 from which

[*1.6*] was deduced, but to determine the correct value of V_S from [*1.1*] or [*1.3*] and to substitute the value in [*1.5*].

For most practical transmission lines, IZ at rated load is not likely to exceed 10% of the rated voltage V_R and the load (or power) angle δ is only a few degrees. Most diagrams in this chapter show IZ and δ greatly exaggerated in order to make the diagrams clearer.

Worked example 1.1

A short 3-phase, 33-kV line delivers a load of 7 MW, power factor 0·85 lagging, 33 kV. If the series impedance of the line is 20+j 30 ohms/phase, calculate the sending-end voltage and the load angle.

SOLUTION

Reference voltage $\mathbf{V}_R = (33/\sqrt{3})\underline{/0°} = 19{\cdot}07\underline{/0°}$ kV/phase

$$\mathbf{I} = (7\times 10^6/\sqrt{3}\times 33\,000\times 0{\cdot}85)(0{\cdot}85-\mathrm{j}\,0{\cdot}527)$$
$$= 144{\cdot}1(0{\cdot}85-\mathrm{j}\,0{\cdot}527) = 122{\cdot}3-\mathrm{j}\,75{\cdot}9 \text{ A}$$
$$\mathbf{IZ} = (122{\cdot}3-\mathrm{j}\,75{\cdot}9)(20+\mathrm{j}\,30) = 4723+\mathrm{j}\,2151 \text{ V/phase}$$
$$\mathbf{V}_S = \mathbf{V}_R+\mathbf{IZ} = 23\,793+\mathrm{j}\,2151 = 23\,880\underline{/5{\cdot}18°} \text{ V/phase}$$

or 41·4 kV (line).

Load angle = 5·18°.

Alternatively:

$$V_R \cos\phi_R + IR = 19\,070\times 0{\cdot}85+144{\cdot}1\times 20 = 19\,082$$
$$V_R \sin\phi_R + IX = 19\,070\times 0{\cdot}527+144{\cdot}1\times 30 = 14\,353$$
$$V_S = (19{\cdot}08^2+14{\cdot}35^2)^{\frac{1}{2}} = 23{\cdot}88 \text{ kV/phase}$$

or 41·4 kV (line)

$$\phi_S = \phi_R+\delta = \tan^{-1}(14{\cdot}35/19{\cdot}08) = 36{\cdot}98°$$
$$\delta = \phi_S-\phi_R = 36{\cdot}98-31{\cdot}8 = 5{\cdot}18°.$$

Alternatively:

Power loss in line $= 3I^2R = 3\times 144{\cdot}1^2\times 20\times 10^{-6} = 1{\cdot}252$ MW

Input power to line = 8·252 MW

VAr loss in line $= 3I^2X = 3\times 144{\cdot}1^2\times 30\times 10^{-6} = 1{\cdot}878$ MVAr lagging

VAr input to line $= (7\times 0{\cdot}527/0{\cdot}85)+1{\cdot}878 = 6{\cdot}218$ MVAr lagging

$$\text{VA input to line} = (8{\cdot}25^2+6{\cdot}22^2)^{\frac{1}{2}} = 10{\cdot}32 \text{ MVA}$$

$$V_S = 10{\cdot}32/\sqrt{3}\times 144{\cdot}1 = 41{\cdot}4 \text{ kV (line)}$$

$$\cos\phi_S = 8{\cdot}25/10{\cdot}32 = 0{\cdot}799 \text{ lagging}$$

$$\phi_S = 36{\cdot}98^\circ$$

$$\delta = 36{\cdot}98 - 31{\cdot}8 = 5{\cdot}18^\circ.$$

1.2.2 GRAPHICAL METHODS; PERFORMANCE CHARTS; OPERATION OF A POWER SYSTEM

A graphical method for dealing with a short line will now be discussed. The student is reminded that this discussion could apply also to an alternator. Although the performance charts to be discussed are seldom drawn to scale, they are very useful in giving a visual picture of the behaviour of a power system as the load varies in magnitude and power factor. We shall discuss first the receiving-end charts which assume that the receiving-end voltage is constant at the remote end of a line of fixed impedance.

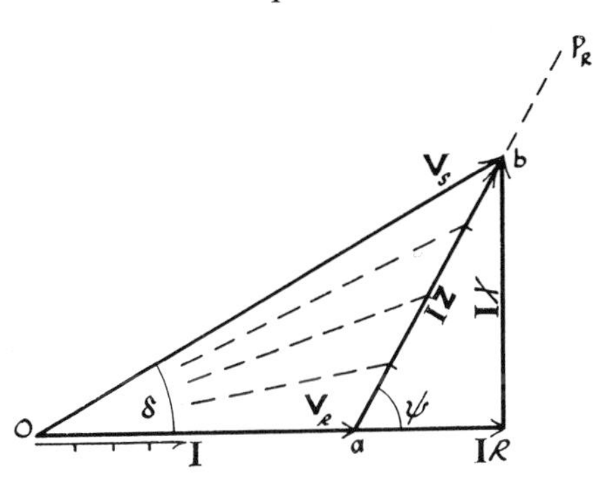

FIG. 1.3. *Short line on unity power factor load, V_R = constant.*

Fig. 1.3 shows the phasor diagram of a given short line delivering power at a fixed voltage and at unity power factor measured at the receiving-end of the line. On no-load $\mathbf{V}_S$ would coincide with $\mathbf{V}_R$. If now the load were to rise in equal increments, the end of the phasor $\mathbf{V}_S$ would move along $\overline{ab}$ in equal increments. The line *ab* is called the power axis as it could be scaled off in units of delivered power P_R (MW). The power axis leads $\mathbf{V}_R$ by the fixed angle ψ.

If the load power factor were to change to zero lagging, both **I** and **IZ** would move 90° clockwise from their positions shown in Fig. 1.3.

Also, if the current were to increase in the same equal increments as before, the end of the phasor $\mathbf{V}_S$ would move in a straight line, perpendicular to the power axis, and by equal increments each equal to those along the power axis. This new axis is called the reactive power (var or reactive voltamperes) axis (Q_R) as it can be scaled off in MVAr lagging (see Appendix 4).

Fig. 1.4 shows the receiving-end chart for the general case of a load of lagging power factor cos ϕ. This chart can be obtained by a process of superposition of power and reactive power. The scales for MW along $\overline{ab}$, for MVAr lagging along $\overline{ac}$ and for MVA along $\overline{ad}$ are all numerically equal. Since V_R and Z are both constants,

$$IZ \propto I \propto V_R I \propto (\text{load VA/phase})$$

and this argument is true also for the two components of the load current.

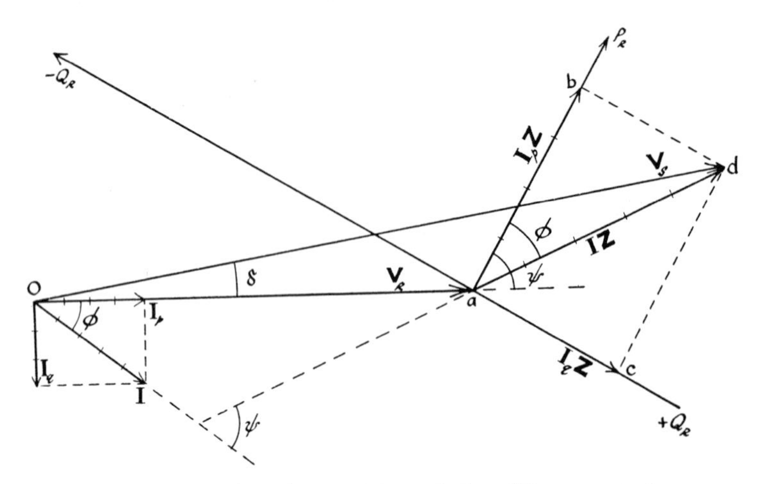

FIG. 1.4. *Short line receiving-end chart (V_R = constant).*
P_R = load power (locus of $\mathbf{V}_S$ for unity power factor load).
Q_R = load vars (locus of $\mathbf{V}_S$ for zero power factor load).

The receiving-end performance chart for a short line is given in Fig. 1.5 which shows the loci of $\mathbf{V}_S$ for several varying operating conditions (but all with constant V_R). Consider a load taking a constant power at a varying power factor. For low lagging power factors $V_S > V_R$, whilst for low leading power factors V_S may be less than V_R, and there is a particular leading power factor for which

$V_S = V_R$. The chart gives the value of $\mathbf{V}_S$ which is required to satisfy a given load demand at the receiving end of a given line.

Consider now the alternator at the sending end of the line, which has to supply this demand. Fig. 1.5 is valid for an alternator alone if $\mathbf{V}_S$ is changed to the open-circuit e.m.f. $\mathbf{E}_o$ and $\mathbf{V}_R$ is changed to

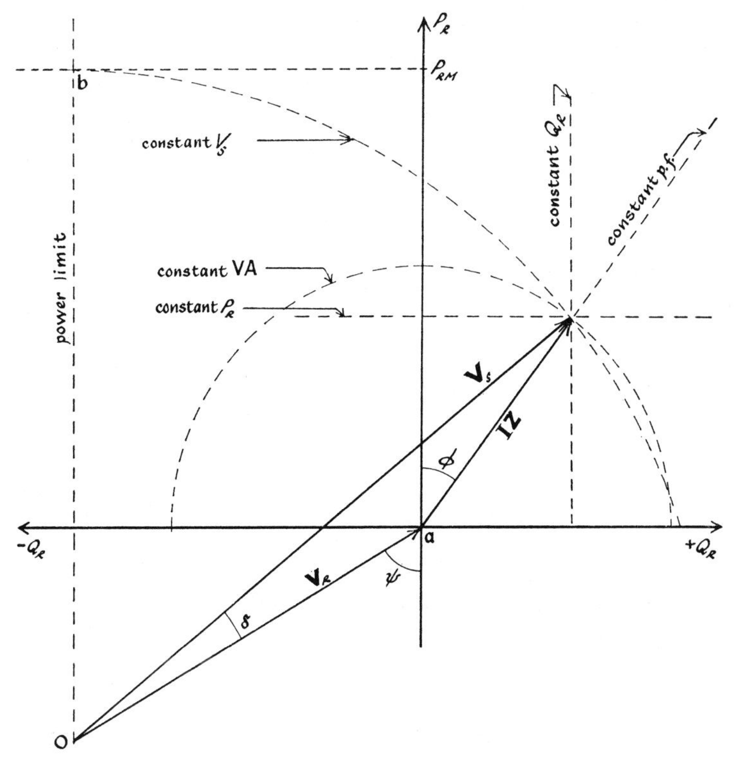

FIG. 1.5. *Loci of* $\mathbf{V}_S$ *for a short line* (V_R = *constant*).

the terminal voltage $\mathbf{V}$ and $\mathbf{Z}$ is the (constant) synchronous impedance of the alternator. The theory of the alternator is discussed in some detail in Chapter 7, but the following is sufficient for our present purpose. E_o is controlled by the excitation of the alternator. If an alternator is floating on live busbars (i.e. connected to the busbars but not supplying any electrical load) the steam supply to the turbine is just sufficient to supply all the losses in the set. If now the steam supply is increased, the power balance is upset and the additional power is absorbed temporarily in the form of kinetic energy by the alternator increasing speed slightly. Thus $\mathbf{E}_o$ advances in phase relative to $\mathbf{V}$ and the alternator generates at a leading power factor. When the electrical output of the alternator is such that the power balance is restored (input = output+losses) the alternator synchronises and runs at synchronous speed. If now the excitation is increased the end of the phasor $\mathbf{E}_o$ moves horizontally to the right and the alternator generates at a lagging power factor. Thus by suitably adjusting the excitation and the steam supply, an alternator can supply the necessary sending-end conditions to enable a given line to supply a given load. It is important to note that Q is positive when the alternator is supplying lagging vars to an inductive load (see Appendix 4).

Considering the line alone, Fig. 1.5 shows the locus of V_S = constant. As $\mathbf{V}_S$ advances in phase (due to an increasing steam supply) the power delivered at the load end of the line increases until $\mathbf{V}_S$ coincides with $\overline{ob}$. If $\mathbf{V}_S$ advances further in phase the delivered power decreases. There is therefore a limit to the amount of power which can be delivered by a given line operating with fixed values of V_S and V_R. The limit (P_{RM}) is called the power limit of the line: it is a function of V_S and increases as V_S increases. At the power limit, $\delta = \psi$, the line current is proportional to $\overline{ab}$ and greatly exceeds the rated current of the line. Also the load power factor is a relatively low leading value (cosine of angle between $\overline{ab}$ and the power axis). The power limit is therefore not a practical operating condition and, in any case, the line cannot be considered alone without reference to its supply alternator.

Fig. 1.5 will now be considered as applying to an alternator alone. If, with constant excitation, the steam supply is slowly increased, $\mathbf{E}_o$ (replacing $\mathbf{V}_S$) advances in phase and the power output of the alternator increases until $\mathbf{E}_o$ coincides with $\overline{ob}$. Up to this condition, the alternator set is stable in that it could operate at synchronous speed for a given steam supply. This neglects transient conditions due to any sudden changes in the system. If, however, $\mathbf{E}_o$ advances in phase beyond $\overline{ob}$, due to an increase in the steam supply, the electrical output of the alternator decreases. The difference between the input power to, and the output power from the set must be absorbed inside the set in the form of kinetic energy. Thus the set increases speed and is no longer running at synchronous speed. The set is now unstable since any further increase in input power would result in a decrease in output power. For a given set, operating with fixed values of E_o and V, there is a limit to the power available from that set and this limit is called the steady-state stability limit. The phrase 'steady state' implies that the steam supply is increased slowly. The limits within which an alternator can remain stable are dealt with in Chapter 7, while Volume 2 discusses the transient stability of a power system.

Since one of the assumptions made in developing performance charts was that the impedance was a constant, it follows that the impedance of the alternator has been assumed to be a constant (see Chapter 7). Another assumption implicit in the above discussion is that the alternator is running in parallel with the grid which fixes the synchronous speed.

If the alternator, line and load are considered as a separate system not connected to the grid, any increase in steam supply will increase the system frequency and force the load to accept more power since its a.c. motors will be running faster. Any increase in alternator excitation will tend to raise the lagging var output of the alternator (see Fig. 1.4) but it will also tend to raise the system voltage and force the load to take a greater exciting current for its transformers and induction motors. If the load voltage rises, the power taken by any resistance (heating) load will rise. Thus a rise in system frequency and voltage indicates that the alternator is generating more than the load demands. Conversely, if the load demand rises, the system frequency and voltage tend to fall. The fall in frequency will operate the steam governors which will raise the steam supply and frequency. The fall in voltage will operate the automatic voltage regulators (AVR) which will increase the excitation and voltage. This argument can be applied to the grid system considered as one power system. Thus if the grid frequency and voltage are constant, the generation and load demand are equal (the losses in the system being part of the demand). (See Chapter 7 for further details.)

For academic purposes, the student is strongly advised to draw performance charts on a per-phase basis and to a voltage scale.

It is sometimes convenient to draw a performance chart to a volt-ampere scale rather than to a voltage scale. In Fig. 1.5 IZ is proportional to the load voltamperes since V_R and Z are both constant. The proportionality can be changed to an equality by multiplying

IZ, and hence also V_R and V_S, by V_R/Z (see Fig. 1.6). Since $V_S V_R/Z$ and V_R^2/Z both have the dimensions of voltamperes, the whole diagram can be drawn to a voltampere scale. It should be noted that Fig. 1.6 contains complex quantities (voltamperes) not phasors (see Appendix 4).

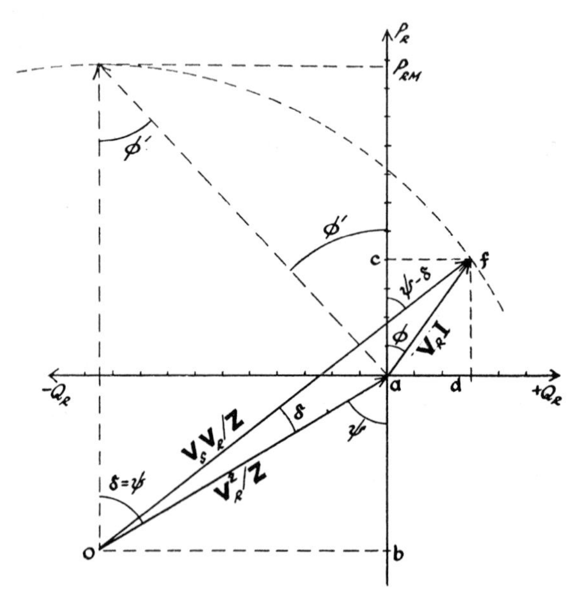

FIG. 1.6. *Short line performance chart showing conditions at the power limit.*

If the three vectors in Fig. 1.6 are each multiplied by 3 (i.e. the scale is multiplied by 3), $\overline{af}$ will now represent 3-phase load voltamperes. The factor of 3 is obtained automatically by using line-to-line values for V_S and V_R instead of line-to-neutral values.

In Fig. 1.6 $\overline{af}$ was derived from IZ. If I is multiplied by any factor K and Z is divided by the same factor (i.e. the route length is divided by K), then $\overline{af}$ is unchanged, i.e. $\overline{af}$ is a constant independent of K. Thus $\overline{af}$ can represent the product of the load voltamperes and the length of the line providing these are inversely proportional to each other. In practice, line performance charts may be drawn to a scale of 3-phase MVA-km. If on such a chart, $\overline{af}$ = 200 MVA-km, it could be interpreted as 20 MVA delivered by a 10 km line or 10

MVA delivered by a 20 km line. The charts are drawn for a given voltage and a given type of line construction but they are independent of the route length. For convenience such charts are usually drawn with the power axis along the $+x$ axis and reactive power axis (lagging vars to the load) along the $+y$ axis. It should be noted that such charts will show the power angle δ increasing in the clockwise direction which is the opposite of the normal convention.

Performance charts can also be drawn on the basis of constant sending-end voltage. These are called sending-end charts and their development is left to the student as an exercise.

Worked example 1.2

Draw the receiving-end performance chart for a receiving-end voltage of 33 kV for the line and load given in worked example 1.1 to the following scales: (*a*) voltage/phase, (*b*) 3-phase MVA, (*c*) 3-phase MVA-km. Hence estimate (i) the sending-end voltage, (ii) the power factor of the total load which will limit the sending-end voltage to 36 kV, assuming the power factor improvement equipment adds 500 kW to the original load, and (iii) the power limit of the line for a voltage drop of 10%.

SOLUTION

(*a*) Voltage/phase.

To obtain reasonable accuracy, charts should occupy at least half a page. Remembering that the power limit is required, a suitable scale will be taken as 1 cm = 2kV/phase. Thus the length of V_R is

$$(33/\sqrt{3})/2 = 19{\cdot}07/2 = 9{\cdot}53 \text{ cm}$$

The angle between $\mathbf{V}_R$ and the negative power axis is

$$= 90° - \tan^{-1}(20/30) = 56{\cdot}3°.$$

It is usual to draw the power axis along the $+y$ axis and lagging var axis (to the load) along the $+x$ axis. For the given load

$$IZ = 144{\cdot}1 \times 36{\cdot}05 = 5190 \text{ V} = 2{\cdot}59 \text{ cm}$$

and is drawn at an angle, clockwise from the power axis, equal to

$$\phi_R = \cos^{-1} 0{\cdot}85 = 31{\cdot}8°.$$

(i) To scale, $V_S = 11{\cdot}95$ cm $= 23{\cdot}9$ kV/phase

or $41{\cdot}4$ kV (line).

(ii) To scale, $V_S = 36/\sqrt{3} = 20{\cdot}8$ kV/phase $= 10{\cdot}4$ cm

The in-phase current due to the total load is

$$I_p = 7.5\times 10^6/\sqrt{3}\times 33\,000 = 131{\cdot}3 \text{ A.}$$

$$I_pZ = 131{\cdot}3\times 36{\cdot}05 = 4730 \text{ V/phase} = 2{\cdot}36 \text{ cm}$$

This length is measured along the power axis, and then a line drawn perpendicular to the power axis will cut a circle of radius 10.4 cm at the answer to the problem. The angle between the new IZ and the power axis is measured as 22·5° and the total power factor is 0·92 leading.

(iii) To scale, $V_S = 1{\cdot}1\times 19{\cdot}07 = 20{\cdot}99$ kV/phase = 10·5 cm. A line is drawn, perpendicular to the power axis, and tangential to a circle of radius 10·5 cm. The length, measured along the power axis is 5 cm or 10 kV/phase. The in-phase component of current at the power limit is

$$10\,000/36{\cdot}05 = 277{\cdot}5 \text{ A}$$

and the power limit is

$$\sqrt{3}\times 33\times 0{\cdot}2775 = 15{\cdot}83 \text{ MW (3 phase).}$$

(*b*) 3-phase MVA.

The voltage scale used above can be converted into a 3-phase MVA scale by noting that the original 3-phase load MVA = 7/0·85 = 8·23 MVA was represented by a length of 2·6 cm. Thus the 3-phase MVA scale is 8·23/2·6 = 3·16 MVA/cm, and the new load of 7·5 MW is represented by 7·5/3·16 = 2·36 cm (as before).

It is more usual to choose a suitable MVA scale and deduce the voltage scale. Take a scale of 1 cm = 2 MVA (3-phase). Referring to Fig. 1.6, the original V_R (in line kV) is now

$$V_R^2/Z = 33^2/36{\cdot}05 = 30{\cdot}2 \text{ MVA} = 15{\cdot}1 \text{ cm}$$

and the line kV scale is 33/15·1 = 2·18 line kV/cm. The original load of 7 MW = 7/2 = 3·5 cm along the power axis. The original V_S (which is now converted to V_SV_R/Z) is measured as 19 cm = 41·4 kV. The student should now be able to obtain the remaining answers.

(*c*) 3-phase, MVA-km.

Assume the line is 50 km long. Then the load MW × route km = 7 × 50 = 350 3-phase MW-km and this is represented by a length of 3·5 cm. Thus the scale is 350/3·5 = 100 3-phase MW (or MVA)-km/cm. The diagram is the same as for (*b*) above: the voltage scale is the same. Circles can be drawn representing fixed voltage regulations and radial lines representing fixed power factors. For any given sending-end voltage and given load power factor, the operating point on the diagram is fixed, and the load MW-km can be scaled off: if one is known the other is determined. This illustrates that the voltage drop along a line depends on the product of the load and the route length.

The student should deduce for himself that Fig. 1.6 can be drawn directly to a 3-phase, MVA-km scale by replacing Z, the total impedance of the line, by Z_1 the impedance per km. In this case $Z_1 = 36{\cdot}05/50 = 0{\cdot}721$ ohms/km, and $V_R^2\ Z_1 = 33^2/0{\cdot}721 = 1510$ MVA-km. Thus V_R is represented to scale by $1510/100 = 15{\cdot}1$ cm (as before).

1.2.3 POWER FORMULAE: STEADY-STATE POWER LIMIT

Referring to Fig. 1.6, the power delivered to the receiving-end of the line of impedance $\mathbf{Z} = Z\underline{/\psi}$ is given by

$$P_R = \overline{ac} = \overline{bc} - \overline{ba}$$

$$= (V_S V_R/Z)\cos(\psi-\delta) - (V_R^2/Z)\cos\psi. \qquad [1.8]$$

Similarly the delivered reactive power

$$Q_R = (V_S V_R/Z)\sin(\psi-\delta) - (V_R^2/Z)\sin\psi. \qquad [1.9]$$

If V_S and V_R are the line-to-neutral (per phase star) values then P_R and Q_R are the values per phase. If P_R and Q_R are multiplied by 3, by using the line-to-line values of V_S and V_R, then the formulae give the 3-phase power and reactive power delivered to a balanced load. These formulae can also be obtained by an analytical method. In Appendix 4 it is shown that voltamperes, treated as a complex number, is equal to the product of phasor voltage and the conjugate of phasor current, when lagging vars to an inductive load are defined as positive.

In Fig. 1.7, $\mathbf{V}_R$ is taken as the reference phasor, i.e.

$$\mathbf{V}_R = V_R\underline{/0^\circ}$$

$$\mathbf{V}_S = \mathbf{V}_R + \mathbf{IZ}$$

$$\mathbf{I} = \mathbf{V}_S/\mathbf{Z} - \mathbf{V}_R/\mathbf{Z}$$

$$= (V_S/Z)\underline{/\delta-\psi} - (V_R/Z)\underline{/-\psi}.$$

The conjugate of **I** is

$$\mathbf{I}^* = (V_S/Z)\underline{/\psi-\delta}-(V_R/Z)\underline{/\psi},$$

The voltamperes delivered to the load is given by

$$\begin{aligned}\mathbf{S}_R &= \mathbf{V}_R\mathbf{I}^* \\ &= (V_S V_R/Z)\underline{/\psi-\delta}-(V_R^2/Z)\underline{/\psi} \qquad [1.10] \\ &= P_R + \mathrm{j}\, Q_R,\end{aligned}$$

where P_R and Q_R are the in-phase and quadrature components respectively of $\mathbf{S}_R$. Equations [*1.8*] and [*1.9*] follow. These equations are useful when the line current is not known (otherwise [*1.1*], [*1.2*] and [*1.3*] would be used). For example, given P, V_S, V_R and $\mathbf{Z}$, [*1.8*] will give δ which, when substituted in [*1.9*], gives Q_R.

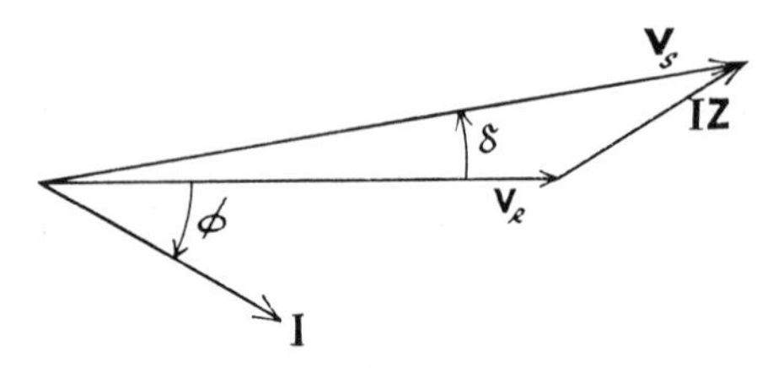

FIG. 1.7.

Fig. 1.6 shows the locus of $V_S V_R/Z$ = constant (i.e. V_S = constant). The equation of this circle can be obtained by eliminating δ between [*1.8*] and [*1.9*] using

$$\cos^2(\psi-\delta)+\sin^2(\psi-\delta) = 1.$$

Hence

$$[P_R+(V_R^2/Z)\cos\psi]^2+[Q_R+(V_R^2/Z)\sin\psi]^2 = (V_S V_R/Z)^2. \qquad [1.11]$$

This is the equation of a circle if P_R and Q_R are the only variables. The problem posed above can be solved using [*1.11*].

The fact that a given short line, operating with fixed voltages at each end, has an upper limit to the power it can deliver has already been discussed in Section 1.2.2 (Fig. 1.6). This conclusion can be derived from [*1.8*]. If P_R is a function of δ only (all other parameters being constant) then P_R is a maximum when $\delta = \psi$ and the maximum value of P_R is given by

$$\begin{aligned}P_{RM} &= (V_S V_R/Z)-(V_R^2/Z)\cos\psi \\ &= (V_S V_R/Z)-(V_R^2 R/Z^2). \qquad [1.12]\end{aligned}$$

This equation shows the parameters upon which the power limit of a line depends. If $R = 0$ then the second term is zero. If $Z = KR$ and $V_S = V_R$ then the two terms are in the ratio 1 : $1/K$. The value of K varies from just over 1 for light 11-kV lines to about 8 for heavy 400-kV lines.

If $V_S = V_R$ then

$$P_{RM} = (V_R^2/Z)(1-\cos\psi)$$
$$= (V_R/Z)^2(Z-R). \qquad [1.13]$$

Equations [*1.12*] and [*1.13*] show that if V_S and V_R can both be raised 10%, the power limit is raised 20%.

At the power limit condition the reactive power delivered to the load is given by

$$Q_R = -(V_R^2/Z)\sin\psi \qquad [1.14]$$

and is independent of V_S (see Fig. 1.6). The negative sign indicates that the load is a sink of leading vars (going to the load) or a source of lagging vars (going from the load to the supply). (See Appendix 4.)

Using Fig. 1.7 the student should now derive the following formulae for the power and vars into the sending-end of a short line (lagging vars into the sending-end being defined as positive):

$$P_S = (V_S^2/Z)\cos\psi-(V_SV_R/Z)\cos(\psi+\delta) \qquad [1.15]$$

$$Q_S = (V_S^2/Z)\sin\psi-(V_SV_R/Z)\sin(\psi+\delta). \qquad [1.16]$$

If P_S is a function of δ only, then P_S is a maximum (P_{SM}) when $(\psi+\delta) = 180°$: since a typical value of ψ is about 60°, the corresponding value of δ is about 120° ($\mathbf{V}_S$ leading $\mathbf{V}_R$). The maximum input power is

$$P_{SM} = (V_S^2/Z)\cos\psi+(V_SV_R/Z)$$
$$= V_S^2R/Z^2+V_SV_R/Z. \qquad [1.17]$$

If $V_S = V_R$ then

$$P_{SM} = (V_S/Z)^2(R+Z). \qquad [1.18]$$

The input vars corresponding to maximum input power are given by

$$Q_S = (V_S^2/Z)\sin\psi \qquad [1.19]$$

and are positive and independent of V_R.

The student should carefully correlate the above formulae with the corresponding sending-end charts (V_S = constant). An indication of the power factor at the load end of the line can be obtained by visualising the current **I** lagging **IZ** by the angle ψ.

1.2.4 VOLTAGE CONTROL OF A RADIAL LINE

There is a legal requirement that the voltage at consumers' terminals should not vary by more than $\pm 6\%$ from the declared value. It is desirable that the voltage at all generating station and substation busbars should not vary by more than about $\pm 10\%$ so that all the auxiliary equipment in the stations can operate satisfactorily at or near their rated voltages and not have overstressed insulation due to excessive voltages. The main causes of voltage drop along a line are the impedance of the line and a load of low lagging power factor (the latter being due to the magnetising currents taken by the load).

The resistance of the line can be reduced by installing a heavier line but this adds to the cost so that there is an economic limit to this method of reducing voltage drop. The series reactance of the line is proportional to the conductor self-inductance which decreases as the spacing between the conductors decreases (see Chapter 2). The minimum conductor spacing must be such as to prevent flash-over between the conductors when these are swinging (dancing) during foul weather conditions.

Consumers are encouraged to raise their lagging power factors nearer to unity power factor by being charged on a tariff which penalises a low power factor (usually by metering the kVA of maximum demand). The cost of installing and maintaining the power-factor improvement equipment is balanced against the reduction in the electricity accounts. The student should be able to show that if a consumer has a load demand of P kW and raises his power factor from $\cos \phi_1$ to $\cos \phi_2$, the most economic value of $\cos \phi_2$ is determined from $\sin \phi_2 = a/b$ where a is the annual cost (interest, depreciation and maintenance) per installed kVAr of power-factor improvement equipment and b is the annual charge per kVA of maximum demand. The current taken by the power-factor improvement equipment is the difference between the original and the new reactive component of current (see Fig. 1.8). The sending-end voltage before, and after installing the power-factor improvement equipment is $\mathbf{V}_{S1}$ and $\mathbf{V}_{S2}$ respectively.

Static capacitors connected in shunt (either star-, or delta-connected) across the load provide the normal method of improving the total load power factor. A capacitor is a sink of leading vars and hence a source of lagging vars (see Appendix 4) which are supplied to the load thus reducing the lagging vars taken from the supply. Voltage-sensitive relays control, in steps, the amount of capacitance in circuit to hold V_R approximately constant for a fixed V_S. If a fixed value of capacitance were used, it should be clear from Fig. 1.8(c) that at light load V_R might rise to a high value ($V_R > V_S$).

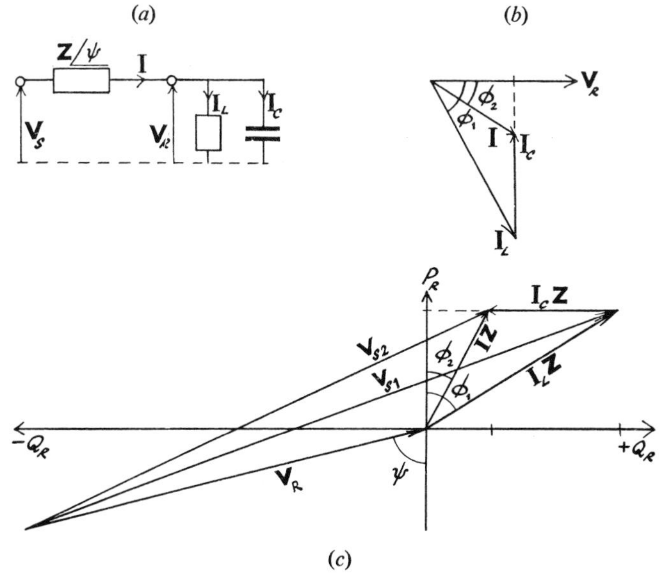

FIG. 1.8. *Power factor improvement.*

For large industrial motor loads, overall power factor improvement can be obtained by using synchronous motors as part of the installation (see Chapter 7). When over-excited these behave like capacitors in the sense that they take current from the supply at a leading power factor. It is an easy matter to provide fine control of the excitation. In this context such synchronous motors are referred to as synchronous compensators (synchronous capacitors or synchronous phase modifiers).

Most transmission lines have on-load, tap-changing transformers at one or both ends of the line (see Chapter 6). These give a numerical (or in-phase) control of the voltage. Assume, for example, that power is to be transferred from 11-kV to 33-kV busbars via a 132-kV line with tap-changing gear on the 132-kV windings, and that both busbars have to be maintained at their rated values. The h.v. turns on the sending-end transformer can be raised 10% so that the sending-end voltage is 10% above nominal. If the load is such that the voltage at the load end of the 132-kV line is 10% below nominal, the h.v. turns on the receiving-end transformer can be reduced by 10%, thus giving nominal volts/turn and 33 kV on the l.v. side. Tap-changing gear is operated by voltage-sensitive relay which detects any change in the system voltage. The relay is time-delayed to prevent it hunting with any transient changes in voltage.

1.2.5 CONTROL OF PARALLEL LINES AND RINGS: CONTROL OF THE GRID SYSTEM

Fig. 1.9(a) shows two short lines in parallel. Clearly

$$\mathbf{I}_A\mathbf{Z}_A = \mathbf{I}_B\mathbf{Z}_B$$

and hence

$$\mathbf{I}_A = \mathbf{I}\mathbf{Z}_B/(\mathbf{Z}_A+\mathbf{Z}_B). \qquad [1.20]$$

The sharing of the total load current in magnitude and phase by the two lines, can be chosen in the design stage by a suitable choice of $\mathbf{Z}_A$ and $\mathbf{Z}_B$, or after erection by adding reactance to one line. If $\mathbf{Z}_A$

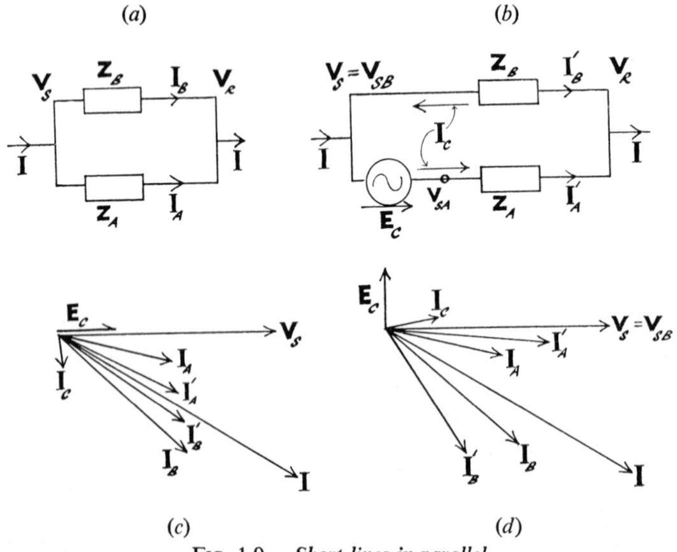

FIG. 1.9. *Short lines in parallel.*

and $\mathbf{Z}_B$ both have the same X/R ratio, all three currents will be in phase. This gives the most economical utilization of conductor carrying capacity.

Consider now the following problem. Assume that in Fig. 1.9(b) $\mathbf{I}$, $\mathbf{I}'_A$, $\mathbf{I}'_B$, $\mathbf{Z}_A$, $\mathbf{Z}_B$ and $\mathbf{V}_R$ are all given (i.e. the sharing of a given load by given lines is predetermined). Then $\mathbf{V}_{SA}$ and $\mathbf{V}_{SB}$, which are the sending-end voltages of lines A and B are fixed and are not (in general) equal. It is necessary therefore to make up the difference by a boost e.m.f. $\mathbf{E}_C = \mathbf{V}_{SA} - \mathbf{V}_{SB}$ (any impedance in the booster is treated as part of $\mathbf{Z}_A$). A simple method of solving this type of problem is to assume that $\mathbf{I}_A$ and $\mathbf{I}_B$ are given by the natural sharing by the two lines (Fig. 1.9(a) and [*1.20*]) and that to them is added a current

$$\mathbf{I}_C = \mathbf{E}_C/(\mathbf{Z}_A + \mathbf{Z}_B) \qquad [1.21]$$

which circulates round the two lines (it cannot flow in the load since the load current is fixed). Thus the actual current in line A is $\mathbf{I}'_A = \mathbf{I}_A + \mathbf{I}_C$ while that in line B is $\mathbf{I}'_B = \mathbf{I}_B - \mathbf{I}_C$ (flowing from sending- to receiving-end).

The booster is a source of voltamperes and is usually a pair of transformers whose primaries are fed from the sending-end busbars (assuming it is installed at the sending-end) at voltage $\mathbf{V}_{SB}$. One transformer is usually the normal tap-changing transformer which gives a boost e.m.f. in phase with $\mathbf{V}_{SB}$ while the other is a quadrature transformer which gives a boost e.m.f. in quadrature with $\mathbf{V}_{SB}$.

Fig. 1.9(c) shows an in-phase boost e.m.f. $\mathbf{E}_C$ and a circulating current $\mathbf{I}_C$ lagging it by nearly 90° (i.e. the lines have a high X/R ratio). The effect is to bring the naturally shared currents $\mathbf{I}_A$ and $\mathbf{I}_B$ more nearly into phase with each other i.e. $\mathbf{I}'_A$ and $\mathbf{I}'_B$. Clearly the in-phase boost has a marked effect on the var distribution between the lines, especially for high X/R ratios. Lagging vars have been taken from line B and injected into line A.

Fig. 1.9(d) shows a boost e.m.f. leading $\mathbf{V}_{SB}$ by 90°, and shows that the quadrature boost has a marked effect on the power distribution between the lines. Power has been transferred from line B to line A. The operation of a quadrature transformer depends on the fact that in a balanced 3-phase system $\mathbf{V}_{RN}$ and $\mathbf{V}_{YB}$ are in quadrature with each other. By means of an isolating transformer, fitted with on-load tap-changing and reversing gear, a proportion of $\mathbf{V}_{YB}$ is injected into the red line to add to $\mathbf{V}_{RN}$: and similarly for the other two phases.

This equipment is usually referred to as a quadrature booster since its rating is only that of the voltamperes boosted into the line ($E_C I'_A$). In the British grid system only a few quadrature boosters are in use; they can be used to force power along a line of high impedance (relative to alternative parallel paths).

If over a short interval of time it can be assumed that all consumers are supplied at a fixed voltage and frequency, then the load demand for power and vars is fixed in magnitude and location. By controlling the mechanical power supply to, and excitation of, all the alternator sets, the generation of power and vars is determined in magnitude and location. The grid transmission system is the link between generation and load demand. It is the duty of the electricity control engineers so to order the settings of (steam) power supply, alternator excitation and transformer taps that the generation satisfies the load demand (including losses) with minimum operating costs and without overloading any equipment or endangering system security or stability. It has been mentioned already that any discrepancy between supply and demand will result in changes in system voltage and/or frequency until supply and demand are equal.

1.3 Long Lines

The short line representation of a transmission line neglects the resistance and capacitance shunt circuits discussed in section 1.1 (power-factor improvement equipment is treated as part of the load). The long line representation allows for these shunt circuits which are shown as consisting of admittances (rather than impedances) since they are then proportional to route length, and can thus be given in siemens/km.

1.3.1 NOMINAL Π AND T CIRCUITS

In the nominal circuits, the shunt admittances are placed at certain specified points on the line as distinct from distributing them uniformly along the line as occurs in practice. In the nominal Π (pi) form, half the shunt admittance is placed at each end, with the series impedance between the ends as shown in Fig. 1.10(a). In the nominal-T (tee) form, the shunt admittance is concentrated at the centre of the line, while the series impedance is split into two equal parts, as shown in Fig. 1.10(b). The qualification 'nominal' infers that these are arbitrary but convenient decisions giving an accuracy which is sufficient for many practical purposes.

The main shunting effect is the capacitance of the line (see Chapter 3), and this will take a current leading 90° on the voltage across it. Allowance can be made for any shunt power loss from the line, due to leakage over the surface of the insulators under foul weather conditions, and corona power loss into the atmosphere (see

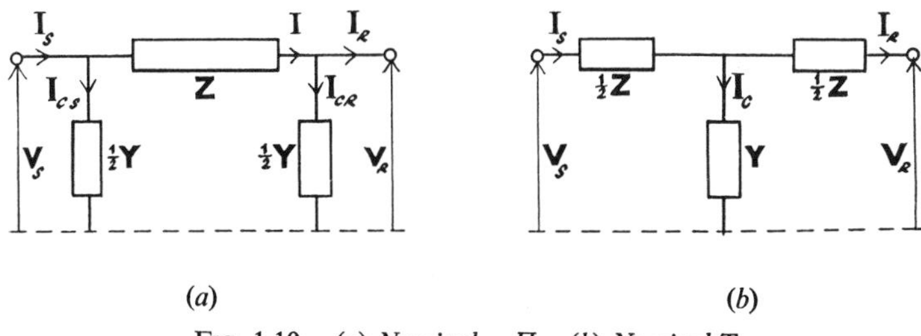

FIG. 1.10. (*a*) *Nominal* $-\Pi$. (*b*) *Nominal-T.*

Chapter 3). Both can be lumped together and treated as a shunt resistance, having a conductance G equal to the reciprocal of the shunt resistance. Thus the admittance of the parallel shunt circuit is given by

$$\mathbf{Y} = G + \mathrm{j}\,2\pi fC$$

and the corresponding shunt current by

$$\mathbf{I} = \mathbf{VY} = \mathbf{V}G + \mathrm{j}\,\mathbf{V}2\pi fC.$$

The shunt current is the sum of an in-phase current and a current leading the voltage by 90°. If leakage and corona can be neglected then the shunt resistance is infinite, and $G = 0$. C is the capacitance to neutral/phase and is calculated from the physical dimensions of the line (see Chapter 3). The shunting circuit is an equivalent star circuit using line-to-neutral values.

If the phasor diagram of the nominal Π line is now drawn, it will be seen that the effect of the shunt capacitance at the load end of the line is to give a form of power-factor improvement which reduces $\mathbf{V}_S$ for a given $\mathbf{V}_R$, compared with the corresponding short line. The series impedance of the line is

$$\mathbf{Z} = R + \mathrm{j}\,X = R + \mathrm{j}\,2\pi fL = Z/\underline{\psi}\ \Omega/\text{phase}.$$

R is the resistance per conductor and is usually obtained from data tables which allow for operating temperature, stranding and skin effect at 50Hz. L is the inductance per phase and is calculated from the physical dimensions of the line (see Chapter 2).

It is important to note that the voltage and current formulae to be derived for the nominal lines apply only to the terminal values and give no indication as to the values at any point along the line.

The nominal-Π circuit (Fig. 1.10(a)) will now be analysed, assuming that $\mathbf{V}_R$ and $\mathbf{I}_R$ are given.

$$\begin{aligned}
\mathbf{I}_{CR} &= \mathbf{V}_R\mathbf{Y}/2 \\
\mathbf{I} &= \mathbf{I}_R+\mathbf{I}_{CR} \\
\mathbf{V}_S &= \mathbf{V}_R+\mathbf{IZ} \\
&= \mathbf{V}_R+(\mathbf{I}_R+\mathbf{V}_R\mathbf{Y}/2)\mathbf{Z} \\
&= (1+\mathbf{YZ}/2)\mathbf{V}_R+\mathbf{ZI}_R \qquad [1.22a] \\
\mathbf{I}_{CS} &= \mathbf{V}_S\mathbf{Y}/2 \\
\mathbf{I}_S &= \mathbf{I}+\mathbf{I}_{CS} \\
&= \mathbf{I}_R+\mathbf{V}_R\mathbf{Y}/2+(1+\mathbf{YZ}/2)\mathbf{V}_R\mathbf{Y}/2+\mathbf{ZI}_R\mathbf{Y}/2 \\
&= \mathbf{Y}(1+\mathbf{YZ}/4)\mathbf{V}_R+(1+\mathbf{YZ}/2)\mathbf{I}_R. \qquad [1.22b]
\end{aligned}$$

The student should now analyse the nominal-T circuit Fig. 1.10(b) and show that

$$\begin{aligned}
\mathbf{V}_S &= (1+\mathbf{YZ}/2)\mathbf{V}_R+\mathbf{Z}(1+\mathbf{YZ}/4)\mathbf{I}_R \qquad [1.23a] \\
\mathbf{I}_S &= \mathbf{YV}_R+(1+\mathbf{YZ}/2)\mathbf{I}_R. \qquad [1.23b]
\end{aligned}$$

Some impression of the numerical significance of the above equations can be obtained by inserting typical values for 50Hz power lines. The inductance of a line is about 1·25mH/km giving a reactance of about 0·4Ω/km while the impedance angle is of the order of 70° ($X/R \simeq 3$) so $\mathbf{Z} \simeq 0{\cdot}42\underline{/70°}\,\Omega$/km. The capacitance of a line is about 0·01 μF to neutral/km, giving an admittance $Y \simeq 3\mu$ S/km. Leakage and corona are usually neglected, so the phase angle of $\mathbf{Y}$ is $+90°$. Thus for a line of route length 100 km, $\mathbf{YZ} \simeq 0{\cdot}013\underline{/160°}$. Sketching the vector addition of 1 to half or a quarter of $\mathbf{YZ}$, will show that $(1+\mathbf{YZ}/2)$ and $(1+\mathbf{YZ}/4)$ are each slightly less than 1 numerically, with a very small positive phase angle.

The nominal Π and T circuits are not equivalent to each other or to the real line: they are different approximations to a real line. Thus a star-delta transformation (see Appendix 2) cannot be used to derive one circuit from the other. The exact representations of a uniform long line, in Π and T form, are called the equivalent Π and T forms (see section 1.3.4).

In section 1.2.1 the terms 'voltage drop' and 'inherent voltage regulation' were defined and it was shown that for a short line the two were equal in value. The equality does not hold however for long lines because, under no-load conditions, the voltage at the load end of the line is no longer equal to the sending-end voltage. The difference is due to the line charging current flowing in the series impedance of the line; this point was discussed in section 1.2.4 with reference to power factor improvement (see Fig. 1.8). The rise in voltage from the sending end towards the load end, under no-load conditions (Ferranti effect), can be quite pronounced in the case of very long, high voltage lines (see section 1.3.2).

Although the nominal lines are mainly of academic interest, the nominal Π form is often used to represent long lines during a power system analysis using a digital computer.

1.3.2 UNIFORM LONG LINE

In the nominal Π and T representations of a transmission line, the parameters of resistance, inductance and capacitance were lumped at particular points along the line. In the analysis to follow, these parameters are assumed to be uniformly distributed along the whole length of the line; their values are usually given per km. Again, as for the nominal representations of the line, the values of inductance and capacitance are calculated from the physical dimensions of the line using the formulae in Chapters 2 and 3. The uniform line representation is used for very long, high voltage power lines as it gives a more accurate representation than does a lumped parameter form, but it should be noted that because of the sag of the lines between the towers (see Chapter 4), there is some doubt as to the length of the line and its height above the ground, which affect the calculation of inductance and capacitance. The analysis will give the steady-state response of the line to the application of a sinusoidal voltage of fixed frequency f.

In this section subscript 1 will denote a value per km e.g. $\mathbf{Y}_1$ and $\mathbf{Z}_1$ are respectively the shunt admittance in siemens/km, and the series impedance in ohms/km: the subscript 1 does not infer a positive sequence value (see Chapter 8), although in fact this chapter deals only with balanced, i.e. positive-sequence conditions. $\mathbf{Y}$ and $\mathbf{Z}$ are the corresponding total values of the whole length of the line, i.e. $\mathbf{Y} = \mathbf{Y}_1 l$ and $\mathbf{Z} = \mathbf{Z}_1 l$.

Fig. 1.11 shows a uniform line. For a very short section dx at a distance x from the receiving end, its series impedance is $\mathbf{Z}_1$ dx where $\mathbf{Z}_1 = R_1 + \mathrm{j}\, 2\pi f L_1$, and its shunt admittance is $\mathbf{Y}_1$ dx where $\mathbf{Y}_1 = G_1 + \mathrm{j}\, 2\pi f C_1$. The drop in voltage in the section (from left to right) is

$$(\mathbf{V}_x + \mathrm{d}\mathbf{V}_x) - \mathbf{V}_x = \mathrm{d}\mathbf{V}_x = (\mathbf{I}_x + \mathrm{d}\mathbf{I}_x)\mathbf{Z}_1\, \mathrm{d}x \simeq \mathbf{I}_x \mathbf{Z}_1\, \mathrm{d}x$$

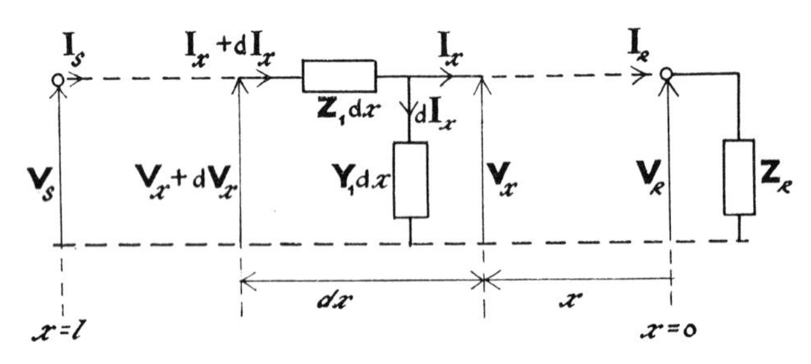

FIG. 1.11. *Uniform long line.*

when second order terms are neglected. Similarly,

$$\mathrm{d}\mathbf{I}_x \simeq \mathbf{V}_x \mathbf{Y}_1\, \mathrm{d}x.$$

Thus

$$\frac{\mathrm{d}\mathbf{V}_x}{\mathrm{d}x} = \mathbf{Z}_1 \mathbf{I}_x, \qquad [1.24a]$$

$$\frac{\mathrm{d}\mathbf{I}_x}{\mathrm{d}x} = \mathbf{Y}_1 \mathbf{V}_x. \qquad [1.24b]$$

The student should check that these same equations are obtained for any representation of the line within the infinitesimally small section, e.g. with the shunting circuit to the left of the series section. Differentiating one equation with respect to x and substituting from the other equation gives

$$\frac{\mathrm{d}^2\mathbf{V}_x}{\mathrm{d}x^2} = \mathbf{Y}_1 \mathbf{Z}_1 \mathbf{V}_x. \qquad [1.25a]$$

$$\frac{\mathrm{d}^2\mathbf{I}_x}{\mathrm{d}x^2} = \mathbf{Y}_1 \mathbf{Z}_1 \mathbf{I}_x. \qquad [1.25b]$$

These equations are standard second order, ordinary differential equations with constant coefficients.

It is convenient at this stage to introduce terms which will be discussed in more detail later (section 1.3.5). We define $\sqrt{(\mathbf{Y}_1\mathbf{Z}_1)} = \boldsymbol{\gamma} = \alpha + \mathrm{j}\,\beta$ where γ is the propagation coefficient/km, α is the attenuation coefficient in nepers/km and β is the phase-change coefficient in radians/km. We also define $\mathbf{Z}_c = \sqrt{(\mathbf{Z}_1/\mathbf{Y}_1)} = \sqrt{(\mathbf{Z}/\mathbf{Y})}$ ohms where $\mathbf{Z}_c$ is the characteristic (or surge or natural) impedance of the line. Note that $\mathbf{Z}_c$ is independent of the length of the line. (An alternative symbol for $\mathbf{Z}_c$ is $\mathbf{Z}_o$. The latter is not used here because of its possible confusion with zero-sequence impedance—see Chapter 8 and B.S.1991.)

The solution of [*1.25a*] can be written in exponential form as

$$\mathbf{V}_x = \mathbf{a}\,\mathrm{e}^{\gamma x} + \mathbf{b}\,\mathrm{e}^{-\gamma x}, \qquad [1.26a]$$

where **a** and **b** are the two constants of integration. By differentiating [*1.26a*] with respect to x and substituting from [*1.24a*], we obtain

$$\mathbf{I}_x = \frac{\mathbf{a}\gamma\,\mathrm{e}^{\gamma x} - \mathbf{b}\gamma\,\mathrm{e}^{-\gamma x}}{\mathbf{Z}_1}$$

$$= \frac{\mathbf{a}\,\mathrm{e}^{\gamma x} - \mathbf{b}\,\mathrm{e}^{-\gamma x}}{\mathbf{Z}_c}. \qquad [1.26b]$$

Values for the constants **a** and **b** can be obtained by inserting the known boundary conditions, namely, that for $x = 0$, $\mathbf{V}_x = \mathbf{V}_R$ and $\mathbf{I}_x = \mathbf{I}_R$. Thus

$$\begin{aligned} \mathbf{V}_R &= \mathbf{a} + \mathbf{b} \\ \mathbf{I}_R &= (\mathbf{a} - \mathbf{b})/\mathbf{Z}_c \\ \mathbf{a} &= \tfrac{1}{2}(\mathbf{V}_R + \mathbf{Z}_c\mathbf{I}_R) \\ \mathbf{b} &= \tfrac{1}{2}(\mathbf{V}_R - \mathbf{Z}_c\mathbf{I}_R). \end{aligned}$$

Formulae for the sending end voltage and current can now be obtained by making the substitution $x = l$ in [*1.26*] and by noting that

$$\begin{aligned} \mathrm{e}^{\gamma l} &= \mathrm{e}^{\alpha l + \mathrm{j}\beta l} = \mathrm{e}^{\alpha l}\,\mathrm{e}^{\mathrm{j}\beta l} \\ &= \mathrm{e}^{\alpha l}(\cos\beta l + \mathrm{j}\sin\beta l) \\ &= \mathrm{e}^{\alpha l}\underline{/\beta l}. \end{aligned}$$

$$\begin{aligned} \mathbf{V}_S &= \tfrac{1}{2}(\mathbf{V}_R + \mathbf{Z}_c\mathbf{I}_R)\mathrm{e}^{\gamma l} + \tfrac{1}{2}(\mathbf{V}_R - \mathbf{Z}_c\mathbf{I}_R)\mathrm{e}^{-\gamma l} \\ &= \tfrac{1}{2}(\mathbf{V}_R + \mathbf{Z}_c\mathbf{I}_R)\mathrm{e}^{\alpha l}\underline{/\beta l} + \tfrac{1}{2}(\mathbf{V}_R - \mathbf{Z}_c\mathbf{I}_R)\mathrm{e}^{-\alpha l}\underline{/-\beta l} \end{aligned} \qquad [1.27a]$$

$$\mathbf{I}_S = \tfrac{1}{2}\left(\frac{\mathbf{V}_R}{\mathbf{Z}_c}+\mathbf{I}_R\right)e^{\gamma l}-\tfrac{1}{2}\left(\frac{\mathbf{V}_R}{\mathbf{Z}_c}-\mathbf{I}_R\right)e^{-\gamma l}$$

$$= \tfrac{1}{2}\left(\frac{\mathbf{V}_R}{\mathbf{Z}_c}+\mathbf{I}_R\right)e^{\alpha l}\underline{/\beta l}-\tfrac{1}{2}\left(\frac{\mathbf{V}_R}{\mathbf{Z}_c}-\mathbf{I}_R\right)e^{-\alpha l}\underline{/-\beta l} \qquad [1.27b]$$

Some appreciation of the significance of these formulae can be obtained by assuming, for a long high voltage power line, that $\mathbf{Y}_1 = 3{\cdot}5\underline{/90^\circ}$ μS/km (leakage and corona power losses are assumed to be negligible) and that $\mathbf{Z}_1 = 0{\cdot}039+j\,0{\cdot}327 = 0{\cdot}33\underline{/83{\cdot}2^\circ}$ ohms/km. The student should check that

$$\gamma \simeq 1{\cdot}08\times 10^{-3}\underline{/86{\cdot}6^\circ} \text{ per km}$$

$$\alpha \simeq 6{\cdot}2\times 10^{-5} \text{ nepers/km}$$

$$\beta \simeq 1{\cdot}08\times 10^{-3} \text{ radians/km} \simeq 0{\cdot}06^\circ/\text{km}$$

$$\mathbf{Z}_c \simeq 306\underline{/-7^\circ} \text{ ohms}$$

$\mathbf{Z}_c$ represents a series circuit of a resistor and a large capacitor. Re-arranging [*1.27*] and making the substitutions

$$\tfrac{1}{2}(e^{\gamma l}+e^{-\gamma l}) = \cosh \gamma l$$

$$\tfrac{1}{2}(e^{\gamma l}-e^{-\gamma l}) = \sinh \gamma l$$

gives

$$\mathbf{V}_S = (\cosh \gamma l)\mathbf{V}_R+(\sinh \gamma l)\mathbf{Z}_c\mathbf{I}_R \qquad [1.28a]$$

$$\mathbf{I}_S = \left(\frac{\sinh \gamma l}{\mathbf{Z}_c}\right)\mathbf{V}_R+(\cosh \gamma l)\mathbf{I}_R. \qquad [1.28b]$$

Equations [*1.28a*] and [*1.28b*] are useful if tables of complex hyperbolic functions are available. Alternatively the following expansions can be used:

$$\sinh \gamma l = \sinh \alpha l \cos \beta l + j \cosh \alpha l \sin \beta l \qquad [1.28c]$$

$$\cosh \gamma l = \cosh \alpha l \cos \beta l + j \sinh \alpha l \sin \beta l. \qquad [1.28d]$$

Substituting for γl and $\mathbf{Z}_c$ in terms of $\mathbf{Y}$ and $\mathbf{Z}$, the total shunt admittance and series impedance respectively of the whole line, gives

$$\mathbf{V}_S = (\cosh \sqrt{(\mathbf{YZ})})\mathbf{V}_R+\sqrt{(\mathbf{Z}/\mathbf{Y})}(\sinh \sqrt{(\mathbf{YZ})})\mathbf{I}_R \qquad [1.29a]$$

$$\mathbf{I}_S = \sqrt{(\mathbf{Y}/\mathbf{Z})}(\sinh \sqrt{(\mathbf{YZ})})\mathbf{V}_R+(\cosh \sqrt{(\mathbf{YZ})})\mathbf{I}_R \qquad [1.29b]$$

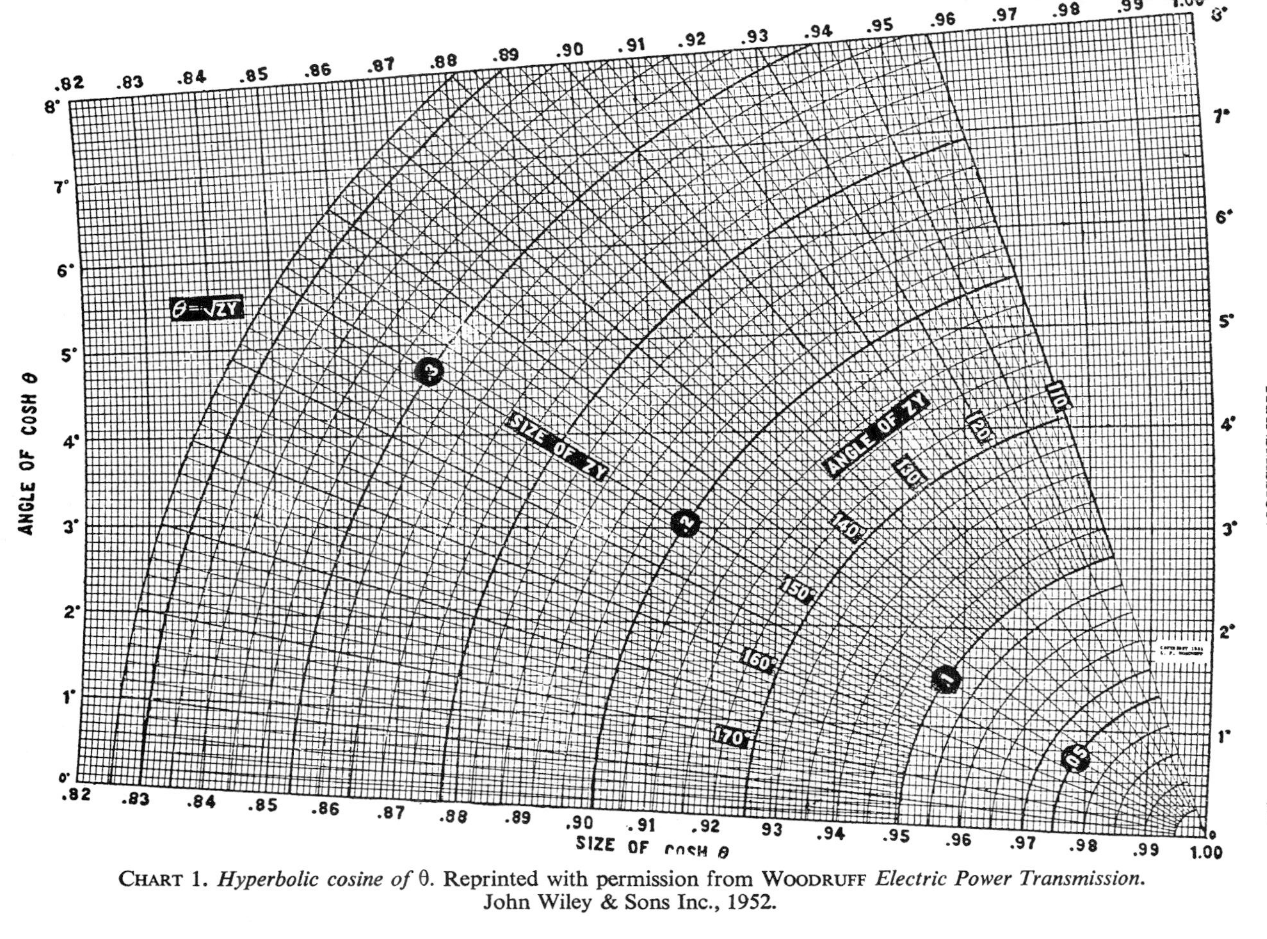

CHART 1. *Hyperbolic cosine of* θ. Reprinted with permission from WOODRUFF *Electric Power Transmission.* John Wiley & Sons Inc., 1952.

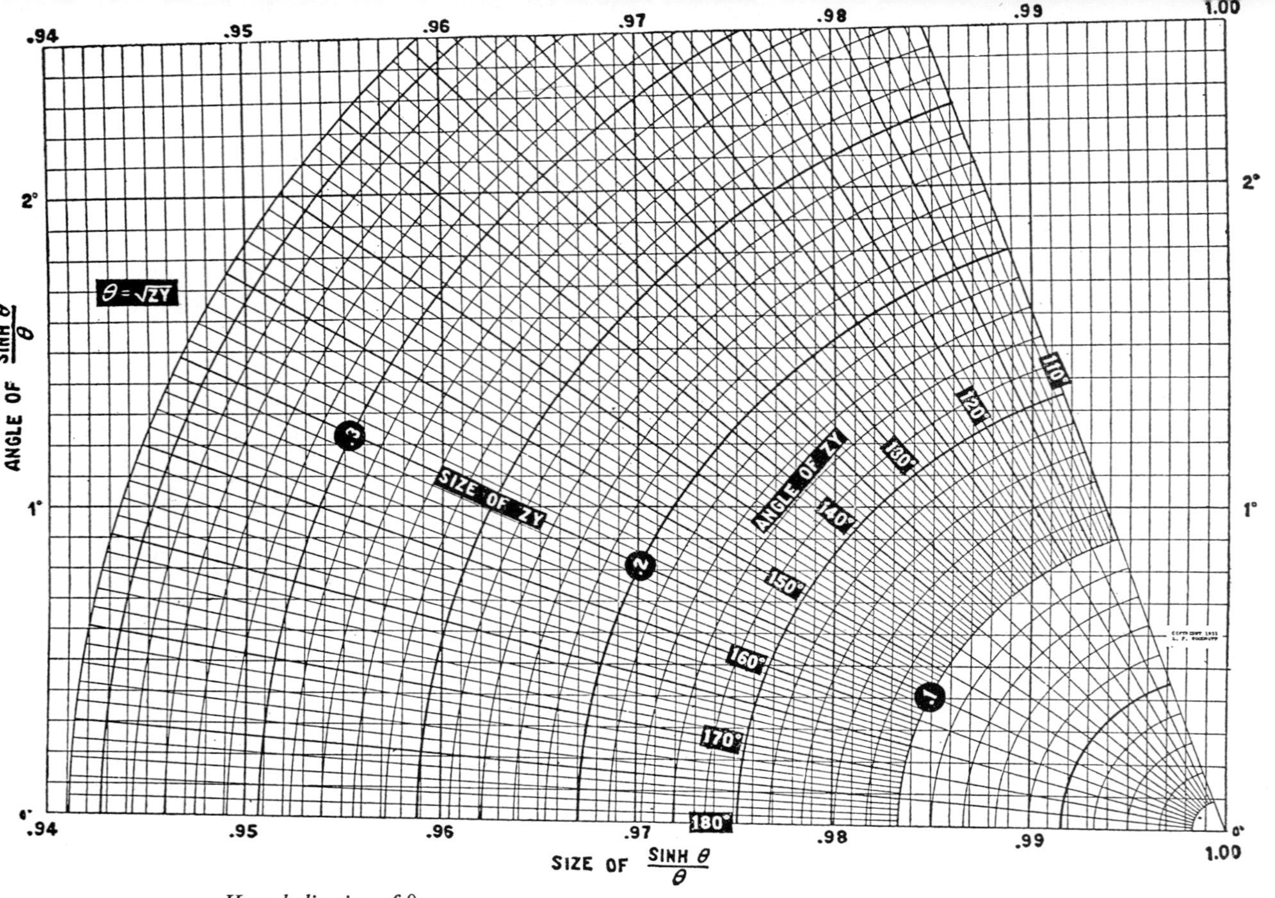

CHART 2.—$\frac{\textit{Hyperbolic sine of } \theta.}{\theta}$ Reprinted with permission from WOODRUFF *Electric Power Transmission*. John Wiley & Son Inc., 1952.

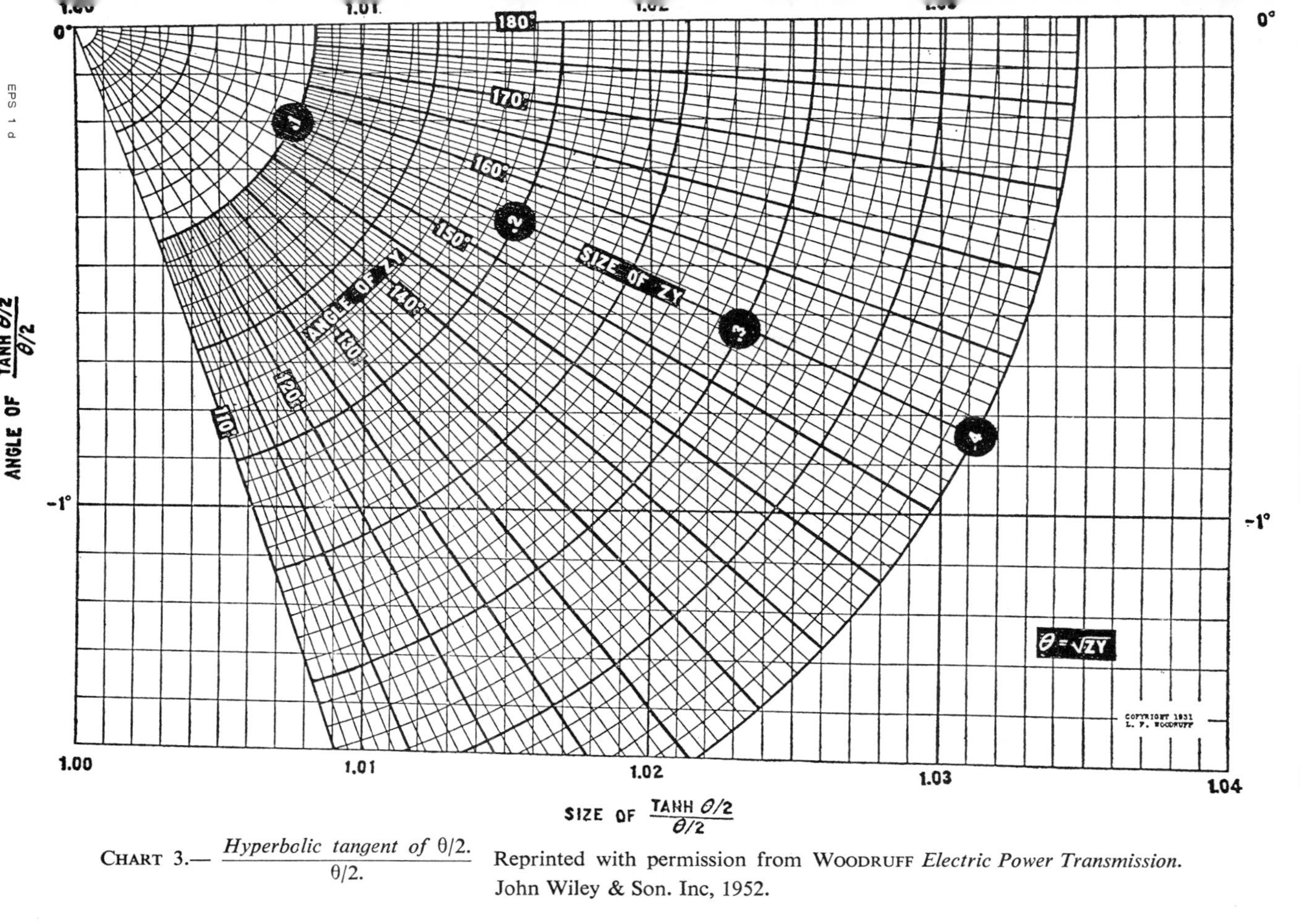

CHART 3.—*Hyperbolic tangent of* θ/2 / θ/2. Reprinted with permission from WOODRUFF *Electric Power Transmission.* John Wiley & Son. Inc, 1952.

or alternatively

$$\mathbf{V}_S = (\cosh \sqrt{(\mathbf{YZ})})\mathbf{V}_R + \left(\frac{\sinh \sqrt{(\mathbf{YZ})}}{\sqrt{(\mathbf{YZ})}}\right)\mathbf{ZI}_R \qquad [1.29c]$$

$$\mathbf{I}_S = \left(\frac{\sinh \sqrt{(\mathbf{YZ})}}{\sqrt{(\mathbf{YZ})}}\right)\mathbf{YV}_R + (\cosh \sqrt{(\mathbf{YZ})})\mathbf{I}_R \qquad [1.29d]$$

The advantages of [*1.29c*] and [*1.29d*] are that the factors in brackets can readily be evaluated using Woodruff's charts (see pages 31, 32, 33) and can also be expanded as convergent infinite series, since $\sqrt{(\mathbf{YZ})}$ is numerically very small, as follows:

$$\cosh \sqrt{(\mathbf{YZ})} = 1 + \frac{\mathbf{YZ}}{2!} + \frac{(\mathbf{YZ})^2}{4!} + \frac{(\mathbf{YZ})^3}{6!} + \ldots \qquad [1.30a]$$

$$\frac{\sinh \sqrt{(\mathbf{YZ})}}{\sqrt{(\mathbf{YZ})}} = 1 + \frac{\mathbf{YZ}}{3!} + \frac{(\mathbf{YZ})^2}{5!} + \frac{(\mathbf{YZ})^3}{7!} + \ldots \qquad [1.30b]$$

Equations [*1.27*], [*1.28*] and [*1.29*] give the sending-end voltage and current in terms of the receiving-end voltage and current. Clearly they can also give the voltage $\mathbf{V}_x$ and current $\mathbf{I}_x$ at any distance x from the receiving-end if l is replaced by x or if $\mathbf{Y}$ and $\mathbf{Z}$ are interpreted as the shunt admittance and series impedance respectively for the length x of the line. Thus the equations of the uniform long line can give the conditions at any point along the line: by contrast the other representations are only valid for the terminal conditions.

Equations [*1.28*] and [*1.29*] can be written in a generalised form as

$$\mathbf{V}_S = \mathbf{AV}_R + \mathbf{BI}_R \qquad [1.31a]$$

$$\mathbf{I}_S = \mathbf{CV}_R + \mathbf{DI}_R \qquad [1.31b]$$

(see section 1.4). Applying these general equations to the line representations so far discussed we obtain Table 1.1 (in which $\mathbf{Y}_c = 1/\mathbf{Z}_c$).

The infinite series formulae show clearly the errors involved in using the short and nominal representations. The expansion in terms of line length l shows that for very long lines, the convergence could be slow and the errors in using short and nominal representations relatively large.

TABLE 1.1. *Transmission line formulae*

Short line

$\mathbf{A} = 1$	$\mathbf{B} = \mathbf{Z}$
$\mathbf{C} = 0$	$\mathbf{D} = \mathbf{A}$

Nominal Π line

$\mathbf{A} = 1+\dfrac{\mathbf{YZ}}{2}$	$\mathbf{B} = \mathbf{Z}$
$\mathbf{C} = \mathbf{Y}\left(1+\dfrac{\mathbf{YZ}}{4}\right)$	$\mathbf{D} = \mathbf{A}$

Nominal T line

$\mathbf{A} = 1+\dfrac{\mathbf{YZ}}{2}$	$\mathbf{B} = \mathbf{Z}\left(1+\dfrac{\mathbf{YZ}}{4}\right)$
$\mathbf{C} = \mathbf{Y}$	$\mathbf{D} = \mathbf{A}$

Uniform long line

$\mathbf{A} = \cosh \gamma l$	$\mathbf{B} = \mathbf{Z}_c \sinh \gamma\, l$
$= \cosh \sqrt{\mathbf{YZ}}$	$= \mathbf{Z}\left(\dfrac{\sinh\sqrt{\mathbf{YZ}}}{\sqrt{\mathbf{YZ}}}\right)$
$= 1+\dfrac{\mathbf{YZ}}{2}+\dfrac{(\mathbf{YZ})^2}{24}+\ldots$	$= \mathbf{Z}\left(1+\dfrac{\mathbf{YZ}}{6}+\dfrac{(\mathbf{YZ})^2}{120}+\ldots\right)$
$= 1+\dfrac{\mathbf{Y}_1\mathbf{Z}_1}{2}l^2+\dfrac{(\mathbf{Y}_1\mathbf{Z}_1)^2}{24}l^4+\ldots$	$= \mathbf{Z}_1\left(l+\dfrac{\mathbf{Y}_1\mathbf{Z}_1}{6}l^3+\dfrac{(\mathbf{Y}_1\mathbf{Z}_1)^2}{120}l^5+\ldots\right)$
$\mathbf{C} = \mathbf{Y}_c \sinh \gamma l$	$\mathbf{D} = \mathbf{A}$
$= \mathbf{Y}\left(\dfrac{\sinh\sqrt{\mathbf{YZ}}}{\sqrt{\mathbf{YZ}}}\right)$	
$= \mathbf{Y}\left(1+\dfrac{\mathbf{YZ}}{6}+\dfrac{(\mathbf{YZ})^2}{120}+\ldots\right)$	
$= \mathbf{Y}_1\left(l+\dfrac{\mathbf{Y}_1\mathbf{Z}_1}{6}l^3+\dfrac{(\mathbf{Y}_1\mathbf{Z}_1)^2}{120}l^5+\ldots\right)$	

Equations [1.31] can be inverted to give the receiving-end conditions in terms of the sending-end conditions.

$$\mathbf{V}_R = \mathbf{D}\mathbf{V}_S - \mathbf{B}\mathbf{I}_S \quad [1.32a]$$

$$\mathbf{I}_R = -\mathbf{C}\mathbf{V}_S + \mathbf{A}\mathbf{I}_S. \quad [1.32b]$$

The proof uses the fact that for a passive line $\mathbf{AD} - \mathbf{BC} = 1$ (see section 1.4.2).

1.3.3 MEASUREMENT OF PARAMETERS

If the load impedance is $\mathbf{Z}_R = \mathbf{V}_R/\mathbf{I}_R$, the driving-point (or input) impedance on load is

$$\mathbf{Z}_d = \mathbf{V}_s/\mathbf{I}_s = \frac{(\cosh \gamma l)\mathbf{V}_R + (\sinh \gamma l)\mathbf{Z}_c\mathbf{I}_R}{(\sinh \gamma l)(\mathbf{V}_R/\mathbf{Z}_c) + (\cosh \gamma l)\mathbf{I}_R}$$

$$= \mathbf{Z}_c\left(\frac{\mathbf{Z}_R \cosh \gamma l + \mathbf{Z}_c \sinh \gamma l}{\mathbf{Z}_R \sinh \gamma l + \mathbf{Z}_c \cosh \gamma l}\right) \quad [1.33]$$

where

$$\gamma l = l\sqrt{(\mathbf{Y}_1\mathbf{Z}_1)} = \sqrt{(\mathbf{YZ})}.$$

The parameters of a uniform long line can be obtained from open-circuit and short-circuit tests carried out at one end of the line. From [1.33] the driving-point impedance on open-circuit is

$$\mathbf{Z}_{do} = \mathbf{Z}_c/\tanh \gamma l \quad [1.34]$$

and the driving-point impedance on short circuit is

$$\mathbf{Z}_{ds} = \mathbf{Z}_c \tanh \gamma l. \quad [1.35]$$

Solving these two equations gives

$$\mathbf{Z}_c = (\mathbf{Z}_{ds}\mathbf{Z}_{do})^{\frac{1}{2}} \quad [1.36]$$

$$\tanh \gamma l = (\mathbf{Z}_{ds}/\mathbf{Z}_{do})^{\frac{1}{2}} \quad [1.37]$$

from which

$$\sinh \gamma l = \{\mathbf{Z}_{ds}/(\mathbf{Z}_{do} - \mathbf{Z}_{ds})\}^{\frac{1}{2}} \quad [1.38]$$

$$\cosh \gamma l = \{\mathbf{Z}_{do}/(\mathbf{Z}_{do} - \mathbf{Z}_{ds})\}^{\frac{1}{2}}. \quad [1.39]$$

The student should now substitute $\mathbf{Z}_d = \mathbf{Z}_R$ in equation [1.33] and hence show that for this condition $\mathbf{Z}_R = \mathbf{Z}_c$. Since this result is independent of the line length l, it is true for any length x. Thus for any point x on a long line terminated by its characteristic impedance, the ratio $\mathbf{V}_x/\mathbf{I}_x$ is a constant equal to its characteristic impedance.

If the condition $\mathbf{Z}_R = \mathbf{Z}_c$ is substituted in [1.27] it will be seen that the second term on the right is zero (i.e. there are no reflected waves of voltage or current, see section 1.3.6).

1.3.4 EQUIVALENT Π AND T CIRCUITS

A uniform long line can be replaced by a circuit of lumped parameters such as the equivalent Π or T circuits shown in Fig. 1.12: the equivalence is only valid for one frequency and only for the terminal conditions. The student should be able to derive the formulae shown

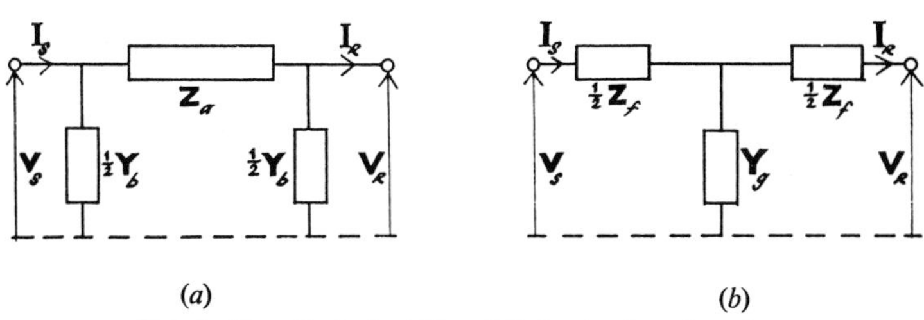

FIG. 1. 12. *Equivalent Π and T circuits for a long line.*

(*a*) *Equivalent Π*

$$\mathbf{Z}_a = \mathbf{Z}_c \sinh \gamma l = \mathbf{Z}\left(\frac{\sinh \sqrt{\mathbf{YZ}}}{\sqrt{\mathbf{YZ}}}\right)$$

$$\mathbf{Y}_b = \frac{2 \tanh (\frac{1}{2}\gamma l)}{\mathbf{Z}_c} = \mathbf{Y}\left(\frac{\tanh \frac{1}{2}\sqrt{\mathbf{YZ}}}{\frac{1}{2}\sqrt{\mathbf{YZ}}}\right)$$

(*b*) *Equivalent T*

$$\mathbf{Z}_f = 2\mathbf{Z}_c \tanh \tfrac{1}{2}\gamma l = \mathbf{Z}\left(\frac{\tanh \frac{1}{2}\sqrt{\mathbf{YZ}}}{\frac{1}{2}\sqrt{\mathbf{YZ}}}\right)$$

$$\mathbf{Y}_g = \frac{\sinh \gamma l}{\mathbf{Z}_c} = \mathbf{Y}\left(\frac{\sinh \sqrt{\mathbf{YZ}}}{\sqrt{\mathbf{YZ}}}\right)$$

in Fig. 1.12 either by equating the open-circuit and short-circuit driving-point (input) impedances for the line and for its equivalent circuit or by equating their **A** and **B** constants (see section 1.4.4, Table 1.2).

As the line becomes shorter, $\sinh \gamma l \rightarrow \gamma l$ and $\tanh (\gamma l/2) \rightarrow \gamma l/2$ and each of the factors in brackets in Fig. 1.12 tends towards unity. Thus the equivalent Π and T circuits degenerate to the nominal Π and T circuits where **Y** and **Z** are respectively the total shunt admittance and series impedance of the uniform line.

1.3.5 PROPAGATION COEFFICIENT γ: CHARACTERISTIC IMPEDANCE $\mathbf{Z}_c$

Fig. 1.13 shows a uniform transmission line of infinite length. Distance x' is now measured from the sending end. The student should be able to show that $d\mathbf{V}_{x'}/dx' = -\mathbf{Z}_1\mathbf{I}_{x'}$, $d\mathbf{I}_{x'}/dx' = -\mathbf{Y}_1\mathbf{V}_{x'}$ and $d^2\mathbf{V}_{x'}/dx'^2 = \mathbf{Y}_1\mathbf{Z}_1\mathbf{V}_{x'}$. This equation is similar to [*1.25a*] so the solution is similar to [*1.26a*] and is

$$\mathbf{V}_{x'} = \mathbf{c}\, e^{\gamma x'} + \mathbf{d}\, e^{-\gamma x'} \qquad [1.40]$$

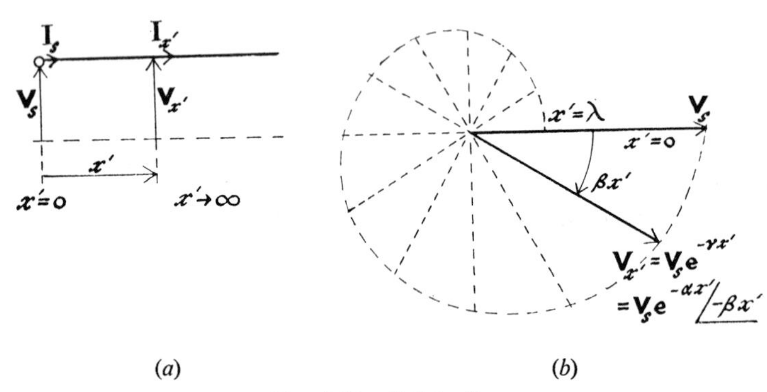

(*a*) (*b*)

FIG. 1.13. *Infinite line.*

As $x' \to \infty$, $\mathbf{c}\, e^{\gamma x'} \to \infty$ so $\mathbf{V}_{x'} \to \infty$. Since this condition is physically inadmissible, it follows that $\mathbf{c} = 0$. Thus $\mathbf{V}_{x'} = \mathbf{d}\ e^{-\gamma x'}$. When $x' = 0$, $\mathbf{V}_{x'} = \mathbf{V}_S$ so

$$\mathbf{V}_{x'} = \mathbf{V}_S\, e^{-\gamma x'} = \mathbf{V}_S\, e^{-\alpha x'} \underline{/-\beta x'}. \qquad [1.41a]$$

As x' increases, $\mathbf{V}_{x'}$ decreases in magnitude due to the attenuation coefficient α and retards in phase due to the phase-change coefficient β (see Fig. 1.13(b)).

$$\mathbf{I}_{x'} = (\mathbf{V}_S/\mathbf{Z}_c)\, e^{-\gamma x'} = (\mathbf{V}_S/\mathbf{Z}_c)\, e^{-\alpha x'} \underline{/-\beta x'}. \qquad [1.41b]$$

The driving-point (input) impedance at any distance x' from the sending end is

$$\mathbf{Z}_d = \mathbf{V}_{x'}/\mathbf{I}_{x'} = \mathbf{Z}_c. \qquad [1.42]$$

The ratio of voltage to current at any point along an infinite line is a constant equal to the characteristic impedance of the line.

Consider now two points one km apart on the line. Thus, from [*1.41a*]

$$\mathbf{V}_{x'}/\mathbf{V}_{(x'+1)} = e^{\gamma}$$

$$\gamma = \log_e (\mathbf{V}_{x'}/\mathbf{V}_{(x'+1)}). \qquad [1.43a]$$

If $\mathbf{V}_{x'}$ leads $\mathbf{V}_{(x'+1)}$ by θ radians then

$$\alpha + \mathrm{j}\,\beta = \log_e [(V_{x'}/V_{(x'+1)})\,\mathrm{e}^{\mathrm{j}\theta}]$$

$$= \log_e (V_{x'}/V_{(x'+1)}) + \mathrm{j}\theta$$

$$\alpha = \log_e (V_{x'}/V_{(x'+1)}) \text{ nepers/km} \qquad [1.43b]$$

$$\beta = \theta \text{ radians/km} \qquad [1.43c]$$

The attenuation coefficient α is the natural logarithm of the ratio of two voltages one km apart on an infinite line. It should be clear from [*1.41b*] that the voltage ratio could be replaced by the corresponding current ratio. In communication line theory, attenuation is often measured in decibels (db) where 1 neper = 8·686 db, and

$$\alpha = \log_{10}\left(\frac{V_{x'}^2/r}{V_{(x'+1)}^2/r}\right) \text{ bels/km}$$

$$= 20 \log_{10} (V_{x'}/V_{(x'+1)}) \text{ db/km} \qquad [1.44a]$$

$$= 20 \log_{10} (I_{x'}/I_{(x'+1)}) \text{ db/km} \qquad [1.44b]$$

Strictly speaking, decibels measure a power ratio and the above formulae make the usual assumption that the two voltages (or currents) are associated with equal resistances, r.

The phase-change coefficient β is the change in phase angle between two voltages (or currents) at two points one km apart on an infinite line.

Equation [*1.42*] shows that an infinite line has a driving-point (input) impedance equal to its characteristic impedance. If a finite line is terminated by its characteristic impedance that impedance could be imagined replaced by an infinite line. Thus a finite line terminated by its characteristic impedance exhibits all the properties of an infinite line.

1.3.6 TRAVELLING WAVES

The subject of travelling waves along transmission lines is dealt with in some detail in Volume 2. This section will only cover enough to explain the significance of the wavelength and velocity of propagation of electromagnetic waves along a line.

It should be clear from [*1.27a*] that the constants of integration **a** and **b** in [*1.26a*] are phasors. Assuming that **b** leads **a** by an angle ϕ

then we can write

$$\mathbf{a} = a \sin \omega t$$

$$\mathbf{b} = b \sin (\omega t+\phi)$$

$$\mathbf{V}_x = a\, e^{\alpha x} \sin (\omega t+\beta x)+b\, e^{-\alpha x} \sin (\omega t-\beta x+\phi) \qquad [1.45]$$

where x is a distance measured from the load end of the line.

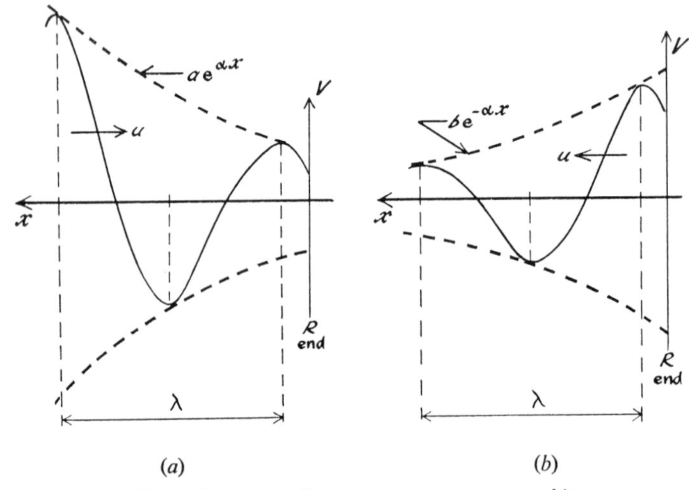

(a) (b)

FIG. 1.14. *Travelling waves (not to same scale).*
(a) Incident wave *(b) Reflected wave*

If we consider any fixed instant of time, t_1, then the first term on the right varies sinusoidally as x varies but its peak values (which occur when $(\omega t_1+\beta x)$ is an odd multiple (k) of $\pi/2$ radians) will increase exponentially as x increases, due to the factor $e^{\alpha x}$. This is shown in Fig. 1.14(a). Consider now a time t_1+dt, when a peak value occurs at a distance $x+dx$; then

$$\omega(t_1+dt)+\beta(x+dx) = k\pi/2 = \omega t_1+\beta x$$

$$dx = -\omega\, dt/\beta.$$

In a short interval of time dt the wave peak has moved a short distance dx nearer to the load. The first term therefore represents an incident or forward travelling wave moving from the sending-end (source) to the receiving-end (load). The velocity of propagation of the wave is (numerically)

$$u = dx/dt = \omega/\beta \text{ km/second} \qquad [1.46]$$

A wavelength is a distance $x = \lambda$ given by

$$\lambda\beta = 2\pi \text{ radians.} \qquad [1.47]$$

If the student examines the second term on the right of [*1.45*] he will find that it represents a reflected wave travelling back from the receiving end to the sending end and that its peak values decrease exponentially as x increases (see Fig. 1.14(b)).

For an infinite line $\mathbf{b} = 0$: there can be no reflections back from a load at an infinite distance from the sending end. Also, there can be no reflections back along a finite line terminated by the characteristic impedance of the line, $\mathbf{Z}_c$. The conclusions reached regarding voltage waves are equally valid for the corresponding current waves (see [*1.26b*] and [*1.27b*]).

1.3.7 DISTORTION-FREE LINE

A communication line transmits a signal containing several components of differing amplitude and frequency and its output should be equal to, or at least proportional to, its input. It follows that the attenuation and the wave velocity should be independent of frequency. This last condition is satisfied if the phase-change coefficient β is proportional to frequency (see [*1.46*]).

The propagation coefficient is

$$\gamma = \alpha + \mathrm{j}\,\beta = \sqrt{(\mathbf{Y}_1\mathbf{Z}_1)}$$
$$= [(G_1 + \mathrm{j}\,\omega C_1)(R_1 + \mathrm{j}\,\omega L_1)]^{\frac{1}{2}} \text{ per km}$$

It can be shown that the necessary condition is

$$R_1/G_1 = L_1/C_1. \qquad [1.48a]$$

This is the condition for distortion-free transmission and subject to this condition

$$\alpha = \sqrt{(R_1 G_1)} \text{ nepers/km} \qquad [1.48b]$$

$$\beta = \omega\sqrt{(L_1 C_1)} \text{ radians/km} \qquad [1.48c]$$

It follows from [*1.48a*] that $\omega L_1/R_1 = \omega C_1/G_1$, so $\mathbf{Z}_1$ and $\mathbf{Y}_1$ have equal phase angles. The characteristic impedance is then

$$\mathbf{Z}_c = \sqrt{(\mathbf{Z}_1/\mathbf{Y}_1)} = \sqrt{(Z/Y)} = \sqrt{(L_1/C_1)} \text{ ohms} \qquad [1.48d]$$

and is a pure resistor. Thus the voltage and current waves are in phase with each other.

1.3.8 LOSS-FREE LINE

The condition of distortion-free transmission is satisfied, for all values of L_1 and C_1, when $R = 0$ and $G = 0$. (These conditions are nearly satisfied by $4 \times 2{\cdot}6$ cm² 400-kV lines ($X/R \simeq 14$) operating under good weather conditions with negligible corona loss). Subject to the above conditions

$$\gamma = \mathrm{j}\,\omega\sqrt{(L_1C_1)} \text{ per km} \qquad [1.49a]$$

$$\alpha = 0 \qquad [1.49b]$$

$$\beta = \omega\sqrt{(L_1C_1)} \text{ radians/km} \qquad [1.49c]$$

$$Z_c = \sqrt{(L_1/C_1)} \text{ ohms} \qquad [1.49d]$$

It is shown in Chapter 2 (equation [*2.23*]) and Chapter 3 (equation [*3.20*]) that for a 3-phase overhead line (neglecting any magnetic flux inside the conductor)

$$L_1 = \frac{\mu}{2\pi}\log_e\left(\frac{D_m}{r}\right) \text{ H/m}$$

$$C_1 = \frac{2\pi\epsilon}{\log_e(D_m/r)} \text{ F/m}$$

where D_m is the geometric mean spacing between the lines and r is the radius of each conductor.

Thus

$$u = \omega/\beta = 1/(\sqrt{(L_1C_1)}) = 1/(\sqrt{(\mu\epsilon)}) \text{ m/s.} \qquad [1.50a]$$

But

$$\mu = \mu_o\mu_r \text{ where } \mu_o = 4\pi 10^{-7} \text{ H/m}$$

and

$$\epsilon = \epsilon_o\epsilon_r \text{ where } \epsilon_o = 1/(4\pi \times 9 \times 10^9) = 8{\cdot}854 \times 10^{-12} \text{ F/m}$$

so

$$u = 3 \times 10^8/\sqrt{(\mu_r\epsilon_r)} \qquad [1.50b]$$

and

$$Z_c = \frac{\sqrt{(\mu/\epsilon)}}{2\pi}\log_e(D_m/r) \text{ ohms.} \qquad [1.50c]$$

For the case of a non-magnetic power line in air, $\mu_r = \epsilon_r = 1$, and

$$u = 3 \times 10^8 \text{ m/s}$$

which is the velocity of electromagnetic propagation in air (vacuum): also

$$Z_c = 60\log_e(D_m/r) \text{ ohms.} \qquad [1.50d]$$

Typical values for Z_c for overhead lines lie between 250 and 350 ohms (and for cables, about 30 to 50 ohms—see [*1.50c*]).

It follows from [1.46] and [1.47] that

$$\lambda = 2\pi/\beta = \omega/\beta f = u/f. \qquad [1.50e]$$

Thus for a 50-Hz loss-free power line, the wavelength is 6000 km.

The effect of internal flux and the use of magnetic conductors (steel-cored aluminium—see Chapter 4) is to increase L_1 and reduce u and λ below the values given above.

The transmission equations of the loss-free line will be discussed further in section 1.6.

1.4 General 4-terminal (2-port) Network: ABCD Constants

The short, nominal Π and T, and uniform long line representations previously discussed are particular methods of solving the same problem, namely, analysing the performance of a given transmission line. The modern tendency is to find a method which is general in its application. Thus it must be clearly understood that the discussion to follow, although mainly concerned with transmission lines, can be applied to any circuit. The only restrictions are that the circuits (Fig. 1.15) must be 4-terminal (or 2-port) and must be passive (no

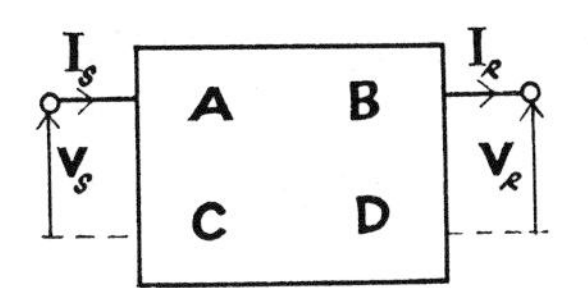

FIG. 1.15. *4-terminal network.*

internal source of e.m.f. or power) and linear (impedances constant, independent of voltage across them, current in them or any other variable) and bilateral (impedances the same for current flow in either direction). This therefore excludes circuits such as iron-cored coils and rectifiers, but transformers can be included if their no-load shunt exciting circuits are either neglected or treated as being linear.

If $\mathbf{V}_R$ and $\mathbf{I}_R$ are regarded as two independent variables to which any value could be assigned, then $\mathbf{V}_S$ and $\mathbf{I}_S$ become two dependent variables, dependent on $\mathbf{V}_R$ and $\mathbf{I}_R$ and related to them by the **Y** and **Z** of the line. The general form of the resulting simultaneous linear equations in two variables is

$$\mathbf{V}_S = \mathbf{A}\mathbf{V}_R + \mathbf{B}\mathbf{I}_R \qquad [1.52a]$$

$$\mathbf{I}_S = \mathbf{C}\mathbf{V}_R + \mathbf{D}\mathbf{I}_R. \qquad [1.52b]$$

A, **B**, **C**, and **D** are called the general 4-terminal network constants. They are merely the first four letters of the alphabet (B is not susceptance and C is not capacitance). A comparison between the **ABCD** constants of the various representations of a transmission line has already been made in section 1.3.2, Table 1.1. The units of **A**,**B**,**C** and **D** must be such as to form consistent equations. **A** and **D** must have no unit, **B** must be an impedance and **C** must be an admittance.

1.4.1 VOLTAGE DROP AND INHERENT VOLTAGE REGULATION

The voltage drop and inherent voltage regulation of a line have already been discussed in section 1.2.1. Unless otherwise qualified, both these terms are assumed to refer to rated load at the receiving-end of the line; or to an alternator delivering rated output or to a transformer supplied at rated primary voltage and delivering rated secondary current. A transformer is rated in terms of its no-load voltage ratio and its rated output which is the product of its no-load rated secondary voltage and full load secondary current (B.S. 171).

Thus for rated load, $\mathbf{V}_S$ can be calculated from [*1.52a*] and

$$\text{the voltage drop} = V_S - V_R \text{ volts.} \qquad [1.53a]$$

Holding $\mathbf{V}_S$ at its full load value, the no-load receiving-end voltage, $\mathbf{V}_{Ro}$ is given by [*1.52a*] as

$$\mathbf{V}_{Ro} = \mathbf{V}_S/\mathbf{A} \text{ volts,} \qquad [1.53b]$$

and

$$\text{the inherent voltage regulation} = V_{Ro} - V_R \text{ volts.} \qquad [1.53c]$$

Voltage drop and inherent voltage regulation are normally expressed as percentages of rated voltage. The student is reminded that when a line, having shunt capacitance, operates on no-load its no-load receiving-end voltage V_{Ro}, is usually greater than its sending-end voltage, V_S, since $A<1$.

1.4.2 SYMMETRICAL, PASSIVE NETWORK

By using the reciprocity theorem it will now be shown that for a symmetrical network

$$\mathbf{A} = \mathbf{D} \qquad [1.54]$$

and that for a passive network the determinant

$$\begin{vmatrix} \mathbf{A} & \mathbf{B} \\ \mathbf{C} & \mathbf{D} \end{vmatrix} = \mathbf{AD} - \mathbf{BC} = 1. \qquad [1.55]$$

In Fig. 1.16(a) the output is short circuited so [*1.52*] become

$$\mathbf{V} = \mathbf{B}\mathbf{I}_R$$

$$\mathbf{I}_S = \mathbf{D}\mathbf{I}_R$$

and the short-circuit driving-point (input) impedance at the *S*-end is

$$\mathbf{Z}_{ds} = \mathbf{B}/\mathbf{D}$$

In Fig. 1.16(b) the network is unchanged but the polarities of the

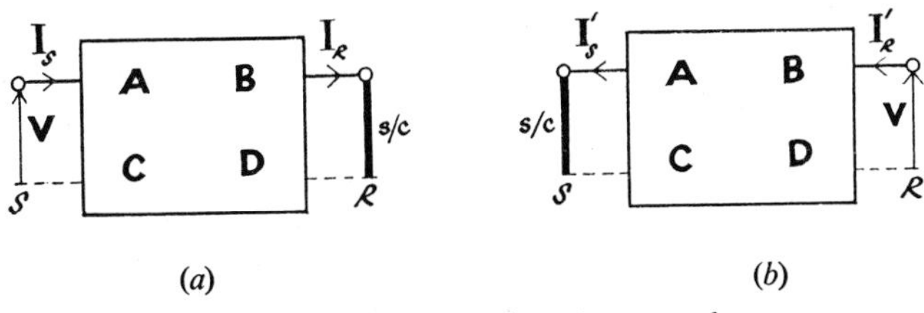

(*a*) (*b*)

FIG. 1.16. *Symmetrical passive network.*

currents have been reversed. Thus, in [*1.52*] we write $-\mathbf{I}'_S$ for $\mathbf{I}_S$ and $-\mathbf{I}'_R$ for $\mathbf{I}_R$: also $\mathbf{V}_S = 0$ and $\mathbf{V}_R = \mathbf{V}$. Thus

$$0 = \mathbf{A}\mathbf{V} - \mathbf{B}\mathbf{I}'_R$$

$$-\mathbf{I}'_S = \mathbf{C}\mathbf{V} - \mathbf{D}\mathbf{I}'_R$$

and the short-circuit driving-point (input) impedance at the *R*-end is

$$\mathbf{Z}_{ds} = \mathbf{V}/\mathbf{I}'_R = \mathbf{B}/\mathbf{A}$$

If the network is symmetrical, the two input impedances must be equal, hence $\mathbf{A} = \mathbf{D}$.

The reciprocity theorem states that if a voltage $\mathbf{V}$ in one branch of a linear passive network creates a current $\mathbf{I}_R$ in another branch, then the same voltage in the second branch will create the same current in the first branch. Therefore

$$\mathbf{I}_R = \mathbf{I}'_S$$

$$\mathbf{V}/\mathbf{B} = -\mathbf{C}\mathbf{V} + \mathbf{D}\mathbf{A}\mathbf{V}/\mathbf{B}$$

$$\mathbf{A}\mathbf{D} - \mathbf{B}\mathbf{C} = 1.$$

1.4.3 MEASUREMENT OF PARAMETERS

The driving-point (input) impedance of a 4-terminal network, to which is connected a load given by $\mathbf{Z}_R = \mathbf{V}_R/\mathbf{I}_R$, is

$$\mathbf{Z}_d = \frac{\mathbf{V}_S}{\mathbf{I}_S} = \frac{\mathbf{AV}_R+\mathbf{BI}_R}{\mathbf{CV}_R+\mathbf{DI}_R} = \frac{\mathbf{AZ}_R+\mathbf{B}}{\mathbf{CZ}_R+\mathbf{D}}. \qquad [1.56]$$

If $\mathbf{Z}_d = \mathbf{Z}_R$ then

$$\mathbf{CZ}_R^2+(\mathbf{D}-\mathbf{A})\mathbf{Z}_R-\mathbf{B} = 0.$$

If the network is symmetrical then $\mathbf{A} = \mathbf{D}$ and $\mathbf{Z}_d = \mathbf{Z}_R$ when

$$\mathbf{Z}_R = \sqrt{(\mathbf{B}/\mathbf{C})}. \qquad [1.57]$$

This particular value of $\mathbf{Z}_R$ is called the characteristic impedance of the network: for a symmetrical network the characteristic, image and iterative impedances are all identical. These impedances are of importance to a communications engineer concerned with the transfer of maximum power between two matched networks. The student should check, using [*1.28*] that $\mathbf{Z}_R$ given by [*1.57*] is equal to the characteristic impedance $\mathbf{Z}_c$ (or $\mathbf{Z}_o$ in many texts) of a uniform long line (see section 1.6 on the natural loading of power transmission lines).

When the network is open-circuited the input impedance is, from [*1.52*],

$$\mathbf{Z}_{do} = \mathbf{A}/\mathbf{C} \qquad [1.58]$$

and when it is short-circuited, the input impedance is

$$\mathbf{Z}_{ds} = \mathbf{B}/\mathbf{D}. \qquad [1.59]$$

If the network is passive and symmetrical, the student should show that

$$\mathbf{A} = \sqrt{[\mathbf{Z}_{do}/(\mathbf{Z}_{do}-\mathbf{Z}_{ds})]} = \mathbf{D} \qquad [1.60a]$$

$$\mathbf{B} = \mathbf{AZ}_{ds} \qquad [1.60b]$$

$$\mathbf{C} = \mathbf{A}/\mathbf{Z}_{do}. \qquad [1.60c]$$

Each of the parameters **A**, **B**, **C** and **D** can be given a physical interpretation in terms of data obtained from open-circuit and short-circuit tests. Thus from an open-circuit test

$$\mathbf{A} = \mathbf{V}_{So}/\mathbf{V}_{Ro} \qquad [1.61a]$$

$$= \text{open-circuit voltage transfer function.}$$

The word transfer indicates that the two voltages are simultaneously measured at the two ends of the line. For a transmission line this is hardly a practical proposition since it would involve trying to measure the phase difference between $\mathbf{V}_{So}$ and $\mathbf{V}_{Ro}$.

On short-circuit

$$\mathbf{B} = \mathbf{V}_{Ss}/\mathbf{I}_{Rs} = \mathbf{Z}_{ts} \qquad [1.61b]$$

= short-circuit transfer impedance.

On open-circuit

$$\mathbf{C} = \mathbf{I}_{So}/\mathbf{V}_{Ro} = \mathbf{Y}_{to} \qquad [1.61c]$$

= open-circuit transfer admittance.

On short-circuit

$$\mathbf{D} = \mathbf{I}_{Ss}/\mathbf{I}_{Ro} \qquad [1.61d]$$

= short-circuit current transfer function.

1.4.4 APPLICATION OF MATRIX METHODS

The equations of the 4-terminal network are well suited to manipulation by matrix methods. Thus [*1.52*] can be given as

$$\begin{bmatrix} \mathbf{V}_S \\ \mathbf{I}_S \end{bmatrix} = \begin{bmatrix} \mathbf{A} & \mathbf{B} \\ \mathbf{C} & \mathbf{D} \end{bmatrix} \begin{bmatrix} \mathbf{V}_R \\ \mathbf{I}_R \end{bmatrix}. \qquad [1.62]$$

The matrix

$$\begin{bmatrix} \mathbf{A} & \mathbf{B} \\ \mathbf{C} & \mathbf{D} \end{bmatrix}$$

is called the transfer matrix of the network.

Some special cases of 4-terminal networks will now be considered (see Table 1.2).

(*a*) Series impedance **Z**: short line or alternator.

$$\mathbf{V}_S = \mathbf{V}_R + \mathbf{Z}\mathbf{I}_R$$
$$\mathbf{I}_S = \mathbf{I}_R$$

The transfer matrix is

$$\begin{bmatrix} 1 & \mathbf{Z} \\ 0 & 1 \end{bmatrix}$$

(*b*) Shunt admittance **Y**: transformer no-load circuit.

$$\mathbf{V}_S = \mathbf{V}_R$$
$$\mathbf{I}_S = \mathbf{Y}\mathbf{V}_R + \mathbf{I}_R$$

The transfer matrix is

$$\begin{bmatrix} 1 & 0 \\ \mathbf{Y} & 1 \end{bmatrix}.$$

TABLE 1.2. *Network parameters*

	Network	A	B	C	D
(*a*) 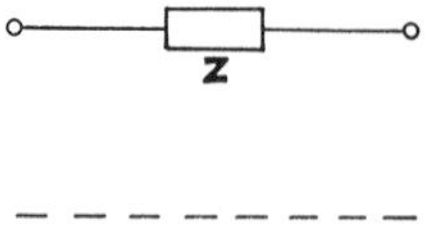	Z	1	**Z**	0	1
(*b*) 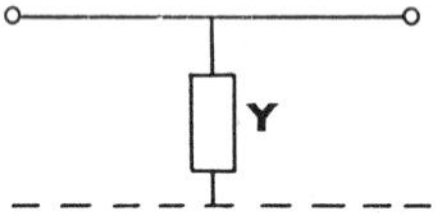	Y	1	0	**Y**	1
(*c*) 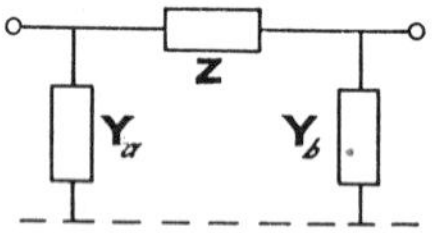	Y_a, Z, Y_b	$1+\mathbf{Y}_b\mathbf{Z}$	**Z**	$\mathbf{Y}_a+\mathbf{Y}_b+\mathbf{Y}_a\mathbf{Y}_b\mathbf{Z}$	$1+\mathbf{Y}_a\mathbf{Z}$
(*d*) 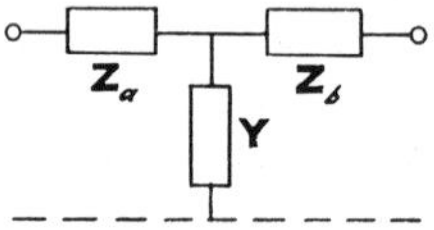	Z_a, Z_b, Y	$1+\mathbf{Y}\mathbf{Z}_a$	$\mathbf{Z}_a+\mathbf{Z}_b+\mathbf{Y}\mathbf{Z}_a\mathbf{Z}_b$	**Y**	$1+\mathbf{Y}\mathbf{Z}_b$
(*e*) 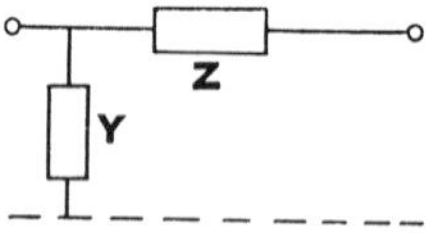	Y, Z	1	**Z**	**Y**	$1+\mathbf{YZ}$
(*f*) 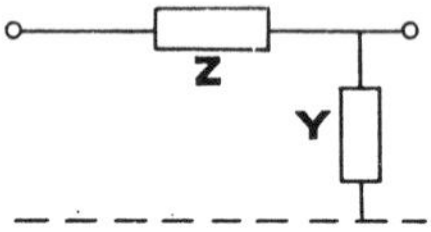	Z, Y	$1+\mathbf{YZ}$	**Z**	**Y**	1
(*g*)	Cascade Fig. 1.17.	$\mathbf{A}_1\mathbf{A}_2+\mathbf{B}_1\mathbf{C}_2$	$\mathbf{A}_1\mathbf{B}_2+\mathbf{B}_1\mathbf{D}_2$	$\mathbf{C}_1\mathbf{A}_2+\mathbf{D}_1\mathbf{C}_2$	$\mathbf{C}_1\mathbf{B}_2+\mathbf{D}_1\mathbf{D}_2$
(*h*)	Series Fig. 1.18.	$\frac{\mathbf{A}'}{\mathbf{C}'}+\frac{\mathbf{A}''}{\mathbf{C}''}$	$-\left(\frac{1}{\mathbf{C}'}+\frac{1}{\mathbf{C}''}\right)$	$\frac{1}{\mathbf{C}'}+\frac{1}{\mathbf{C}''}$	$-\left(\frac{\mathbf{D}'}{\mathbf{C}'}+\frac{\mathbf{D}''}{\mathbf{C}''}\right)$
(*i*)	Parallel Fig. 1.19.	$\frac{\mathbf{D}'}{\mathbf{B}'}+\frac{\mathbf{D}''}{\mathbf{B}''}$	$-\left(\frac{1}{\mathbf{B}'}+\frac{1}{\mathbf{B}''}\right)$	$\frac{1}{\mathbf{B}'}+\frac{1}{\mathbf{B}''}$	$-\left(\frac{\mathbf{A}'}{\mathbf{B}'}+\frac{\mathbf{A}''}{\mathbf{B}''}\right)$

Note: The elements for (*h*) are $\mathbf{Z}_{ss}$, $\mathbf{Z}_{sr}$, $\mathbf{Z}_{rs}$, $\mathbf{Z}_{rr}$ ([*1.65*]), and for (*i*) $\mathbf{Y}_{ss}$, $\mathbf{Y}_{sr}$, $\mathbf{Y}_{rs}$, and $\mathbf{Y}_{rr}$ ([*1.68*]).

(*c*) Asymmetrical Π network.

This network is a cascade connection of a shunt admittance, followed by a series impedance followed by a shunt admittance. The overall transfer matrix is obtained by multiplying together the transfer matrices of the three component parts.

$$\begin{bmatrix} 1 & 0 \\ \mathbf{Y}_a & 1 \end{bmatrix}\begin{bmatrix} 1 & \mathbf{Z} \\ 0 & 1 \end{bmatrix}\begin{bmatrix} 1 & 0 \\ \mathbf{Y}_b & 1 \end{bmatrix} = \begin{bmatrix} (1+\mathbf{Y}_b\mathbf{Z}) & \mathbf{Z} \\ (\mathbf{Y}_a+\mathbf{Y}_b+\mathbf{Y}_a\mathbf{Y}_b\mathbf{Z}) & (1+\mathbf{Y}_a\mathbf{Z}) \end{bmatrix}$$

If the network is symmetrical and $\mathbf{Y}_a = \mathbf{Y}_b = \mathbf{Y}/2$ where $\mathbf{Y}$ is the total shunt admittance, the transfer matrix reduces to that of the nominal Π transmission line (see Table 1.1).

The student should verify the formulae given for (*d*), (*e*) and (*f*) in Table 1.2. Circuits (*e*) and (*f*) give the approximate (or cantilever) representation of a transformer.

(*g*) Fig. 1.17 shows two 4-terminal networks in cascade. The student can obtain the **ABCD** constants of the total network by

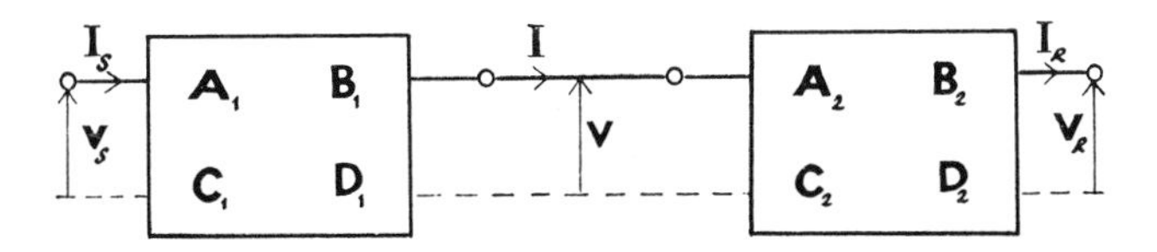

FIG. 1.17. *Two 4-terminal networks in cascade.*

expressing **V** and **I** in terms of $\mathbf{V}_R$ and $\mathbf{I}_R$, then $\mathbf{V}_S$ and $\mathbf{I}_S$ in terms of **V** and **I**, and then eliminating **V** and **I**. The transfer matrix (and hence **A**, **B**, **C** and **D**) can alternatively be obtained by multiplying the two component transfer matrices.

$$\begin{aligned} \begin{bmatrix} \mathbf{V}_S \\ \mathbf{I}_S \end{bmatrix} &= \begin{bmatrix} \mathbf{A}_1 & \mathbf{B}_1 \\ \mathbf{C}_1 & \mathbf{D}_1 \end{bmatrix}\begin{bmatrix} \mathbf{V} \\ \mathbf{I} \end{bmatrix} \\ &= \begin{bmatrix} \mathbf{A}_1 & \mathbf{B}_1 \\ \mathbf{C}_1 & \mathbf{D}_1 \end{bmatrix}\begin{bmatrix} \mathbf{A}_2 & \mathbf{B}_2 \\ \mathbf{C}_2 & \mathbf{D}_2 \end{bmatrix}\begin{bmatrix} \mathbf{V}_R \\ \mathbf{I}_R \end{bmatrix} \\ &= \begin{bmatrix} (\mathbf{A}_1\mathbf{A}_2+\mathbf{B}_1\mathbf{C}_2) & (\mathbf{A}_1\mathbf{B}_2+\mathbf{B}_1\mathbf{D}_2) \\ (\mathbf{C}_1\mathbf{A}_2+\mathbf{D}_1\mathbf{C}_2) & (\mathbf{C}_1\mathbf{B}_2+\mathbf{D}_1\mathbf{D}_2) \end{bmatrix}\begin{bmatrix} \mathbf{V}_R \\ \mathbf{I}_R \end{bmatrix}. \end{aligned} \qquad [1.63]$$

(*h*) Equations [*1.52*] can be re-arranged to express the voltages $\mathbf{V}_S$ and $\mathbf{V}_R$ in terms of the two currents $\mathbf{I}_S$ and $\mathbf{I}_R$. Since the networks are passive, $\mathbf{AD}-\mathbf{BC} = 1$ and

$$\mathbf{V}_S = (\mathbf{A}/\mathbf{C})\mathbf{I}_S-(1/\mathbf{C})\mathbf{I}_R \qquad [1.64a]$$

$$\mathbf{V}_R = (1/\mathbf{C})\mathbf{I}_S - (\mathbf{D}/\mathbf{C})\mathbf{I}_R. \qquad [1.64b]$$

These equations can be written as

$$\mathbf{V}_S = \mathbf{Z}_{SS}\mathbf{I}_S + \mathbf{Z}_{SR}\mathbf{I}_R \qquad [1.65a]$$

$$\mathbf{V}_R = \mathbf{Z}_{RS}\mathbf{I}_S + \mathbf{Z}_{RR}\mathbf{I}_R \qquad [1.65b]$$

and they can be further reduced to the matrix form

$$[\mathbf{V}] = [\mathbf{Z}][\mathbf{I}].$$

$\mathbf{Z}_{SS}$ is the open-circuit driving-point impedance at the sending-end for which the symbol $\mathbf{Z}_{do}$ was used in section 1.4.3. $\mathbf{Z}_{RS}$ and $\mathbf{Z}_{SR}$, which are equal in a bilateral network, are the open-circuit transfer impedances, so that from [*1.61c*] $\mathbf{Z}_{RS} = 1/\mathbf{Y}_{to}$. [**Z**] is called the impedance matrix of the network. Its elements, in terms of the constants **A, B, C** and **D**, can be identified from [*1.64*].

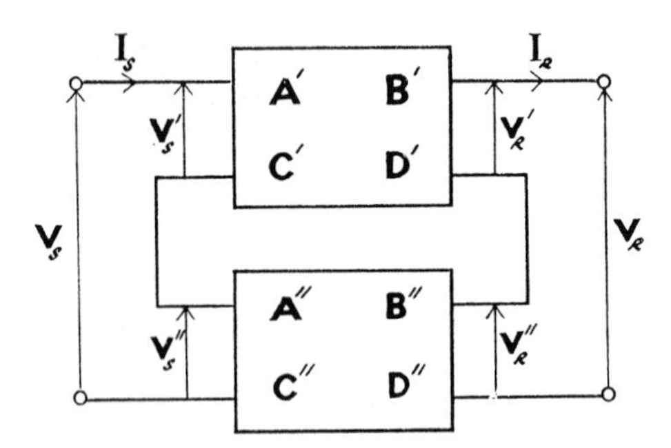

FIG. 1.18. *Two 4-terminal networks in series.*

$\mathbf{V}_S = \mathbf{V}_S' + \mathbf{V}_S''$ $\qquad$ $\mathbf{V}_R = \mathbf{V}_R' + \mathbf{V}_R''$

The impedance matrix simplifies the problem of two 4-terminal networks in series, i.e. their inputs are in series, and their outputs are in series (Fig. 1.18). If the two networks are identified by superscripts (′) and (″) then

$$\begin{bmatrix} \mathbf{V}_S' \\ \mathbf{V}_R' \end{bmatrix} = \begin{bmatrix} \mathbf{Z}_{SS}' & \mathbf{Z}_{SR}' \\ \mathbf{Z}_{RS}' & \mathbf{Z}_{RR}' \end{bmatrix} \begin{bmatrix} \mathbf{I}_S \\ \mathbf{I}_R \end{bmatrix}$$

The equation for network (″) is similar. Since both $\mathbf{V}_S$ and $\mathbf{V}_R$ are the sum of their two component parts

$$\begin{bmatrix} \mathbf{V}_S \\ \mathbf{V}_R \end{bmatrix} = \begin{bmatrix} \mathbf{V}_S' \\ \mathbf{V}_R' \end{bmatrix} + \begin{bmatrix} \mathbf{V}_S'' \\ \mathbf{V}_R'' \end{bmatrix} = \Big[[\mathbf{Z}'] + [\mathbf{Z}''] \Big] \begin{bmatrix} \mathbf{I}_S \\ \mathbf{I}_R \end{bmatrix} \qquad [1.66]$$

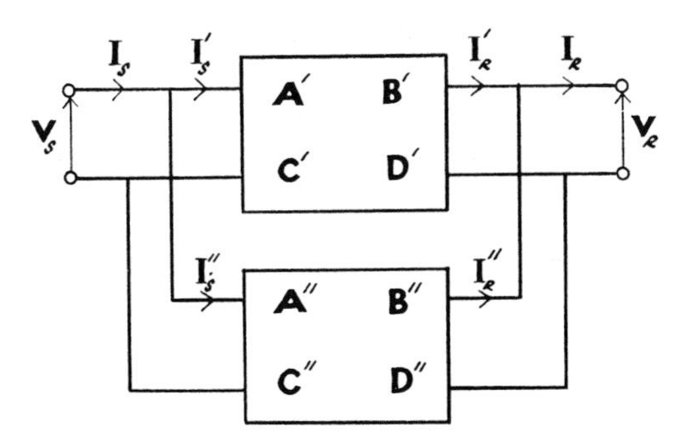

FIG. 1.19. *Two 4-terminal networks in parallel.*

$\mathbf{I}_S = \mathbf{I}'_S + \mathbf{I}''_S$ $\qquad$ $\mathbf{I}_R = \mathbf{I}'_R + \mathbf{I}''_R$

(*i*) Two networks in parallel are shown in Fig. 1.19: their inputs are in parallel and their outputs are in parallel. The admittance matrix for the overall network is given by

$$[\mathbf{I}] = [\mathbf{Y}][\mathbf{V}]$$

$$\begin{bmatrix} \mathbf{I}_S \\ \mathbf{I}_R \end{bmatrix} = \begin{bmatrix} \mathbf{Y}_{SS} & \mathbf{Y}_{SR} \\ \mathbf{Y}_{RS} & \mathbf{Y}_{RR} \end{bmatrix} \begin{bmatrix} \mathbf{V}_S \\ \mathbf{V}_R \end{bmatrix} \qquad [1.67]$$

where

$$[\mathbf{Y}] = [\mathbf{Y}'] + [\mathbf{Y}''] = \begin{bmatrix} \left(\frac{\mathbf{D}'}{\mathbf{B}'} + \frac{\mathbf{D}''}{\mathbf{B}''}\right) & -\left(\frac{1}{\mathbf{B}'} + \frac{1}{\mathbf{B}''}\right) \\ \left(\frac{1}{\mathbf{B}'} + \frac{1}{\mathbf{B}''}\right) & -\left(\frac{\mathbf{A}'}{\mathbf{B}'} + \frac{\mathbf{A}''}{\mathbf{B}''}\right) \end{bmatrix}. \qquad [1.68]$$

1.4.5 PERFORMANCE CHART FOR THE 4-TERMINAL NETWORK

The receiving-end performance chart (circle diagram), Fig. 1.20, for a long line, treated as a 4-terminal network, operating with constant receiving-end voltage, can be obtained by the method used for the short line (section 1.2.2) but using now the equation $\mathbf{V}_S = \mathbf{A}\mathbf{V}_R + \mathbf{B}\mathbf{I}_R$, where $\mathbf{A} = A\underline{/\alpha}$ and $\mathbf{B} = B\underline{/\beta}$ ohms/phase. (α and β are not the attenuation and phase change coefficients.)

For a long transmission line a typical value of **A** might be $0{\cdot}98\underline{/0{\cdot}5^\circ}$ and **B** is approximately equal to the total series impedance of the line, **Z**. To make the line charts legible it is usual to make A much smaller, and α much larger than the practical values: also BI_R is exaggerated relative to V_R.

The phasor diagram is drawn to a voltage scale, first for no-load ($\mathbf{I}_R = 0$) giving $\mathbf{V}_S = \mathbf{AV}_R$. If a unity power factor load (at the receiving end) is applied in equal increments, the locus of $\mathbf{V}_S$ will move along the power axis (P_R) in equal increments each equal (numerically) to

$$BI_p \propto V_R I_p \text{ (= power increment)}$$

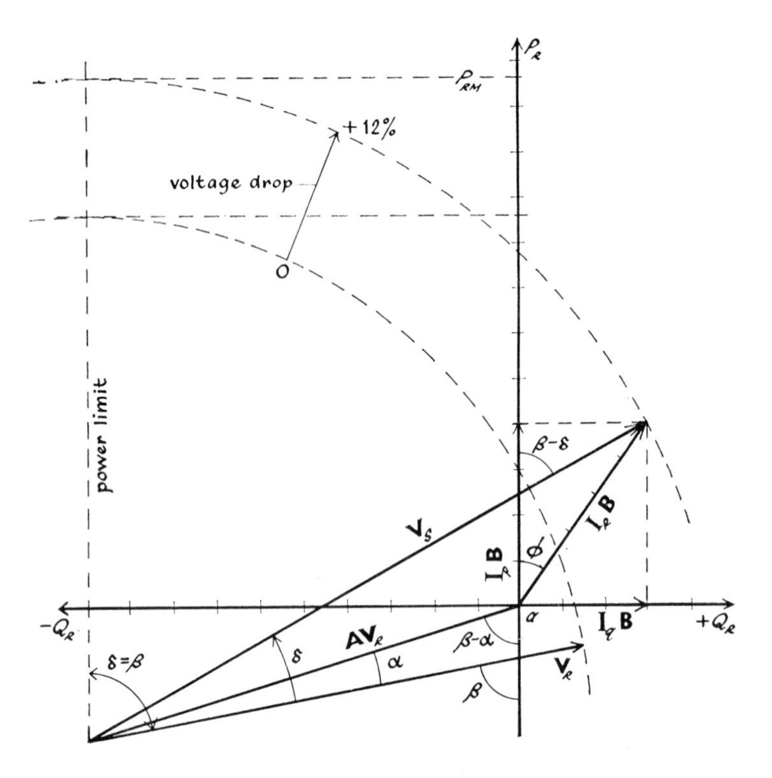

FIG. 1.20. *Performance chart for the 4-terminal network.*

for a given line operating with a constant receiving end voltage. Thus the power axis could be scaled off to represent equal increments of power. If a load at zero power factor lagging is applied, in equal increments, the locus of $\mathbf{V}_S$ will now move along the var axis ($+Q_R$) in equal increments. The var axis can therefore be scaled off to represent equal increments of vars (reactive power). The scales for power and vars will be the same since the factors B and V_R are common to both.

For a delivered load of power factor cos ϕ, the locus of $\mathbf{V}_S$ will move along the phasor $\mathbf{BI}_R$ and the scale for voltamperes will be numerically equal to the scale for power and vars.

The scale for voltamperes, power and vars is usually found by drawing the diagram, to a voltage scale, for a given load, and dividing the actual load voltamperes by the scale length of BI_R.

The conclusions to be drawn from this chart are similar to those for the short line. To deliver a constant power at a decreasing lagging power factor needs an increasing V_S and a decreasing load angle δ. To deliver a load of constant voltamperes at a varying power factor, the locus of $\mathbf{V}_S$ is a circle centred at a and of radius BI_R. If the line operates with fixed voltages at both ends, then as δ increases the delivered power will increase and the power factor of the load will change from lagging, through unity, to leading values. The student is reminded that an increase in δ involves increasing the steam supply to the alternator at the sending end (see section 1.2.2). Fig. 1.20 shows that as δ increases there is an upper limit to the power which can be delivered by a given line, and that this limit occurs when $\delta = \beta$. The power limit is a function of V_S.

By multiplying each of the phasors in Fig. 1.20 by V_R/B, the phasor $\mathbf{BI}_R$ becomes (numerically) $V_R I_R$ which is the delivered load voltamperes. In its new form the diagram can be drawn directly in a voltampere scale. The corresponding voltage scale can be found by dividing the actual load voltage by the scale length of V_R^2/B.

1.4.6 POWER FORMULAE: POWER LIMITS

Formulae for the power P_R delivered to each phase of the load, and the corresponding lagging vars Q_R, in terms of the voltages $\mathbf{V}_S$ and $\mathbf{V}_R$ and the parameters **A**, **B**, **C**, and **D**, can be obtained by the same method used for short lines (see section 1.2.3).

$$\mathbf{V}_S = \mathbf{AV}_R + \mathbf{BI}_R$$

$$\mathbf{I}_R = (\mathbf{V}_S/\mathbf{B}) - (\mathbf{AV}_R/\mathbf{B})$$

$$= (V_S/B)\underline{/\delta-\beta} - (AV_R/B)\underline{/\alpha-\beta}$$

where $\mathbf{V}_R$ is taken as the reference phasor, $\mathbf{V}_R = V_R\underline{/0^\circ}$, and δ is the load angle, the angle by which $\mathbf{V}_S$ leads $\mathbf{V}_R$: also $\mathbf{A} = A\underline{/\alpha}$ and $\mathbf{B} = B\underline{/\beta}$.

The conjugate of $\mathbf{I}_R$ is

$$\mathbf{I}_R^* = (V_S/B)\underline{/\beta-\delta} - (AV_R/B)\underline{/\beta-\alpha}$$

and the load voltamperes/phase are (see Appendix 4)

$$\mathbf{S}_R = \mathbf{V}_R\mathbf{I}_R^*$$

$$P_R = (V_S V_R/B)\cos(\beta-\delta) - (AV_R^2/B)\cos(\beta-\alpha) \qquad [1.69a]$$

$$Q_R = (V_S V_R/B)\sin(\beta-\delta) - (AV_R^2/B)\sin(\beta-\alpha). \qquad [1.69b]$$

Treating P_R as a function of δ only, P_R will be a maximum, P_{RM}, when $\delta = \beta$ and

$$P_{RM} = (V_S V_R/B) - (AV_R^2/B)\cos(\beta-\alpha). \qquad [1.70a]$$

The corresponding value of Q_R at this power limit is

$$Q_{RM} = -(AV_R^2/B)\sin(\beta-\alpha) \qquad [1.70b]$$

where the negative sign indicates lagging vars flowing from the load to the supply (leading vars from supply to load). It should be noted that Q_{RM} is independent of V_S and δ.

The formulae above can be simplified when $V_S = V_R$ and in particular

$$P_{RM} = (V_R^2/B)[1 - A\cos(\beta-\alpha)]. \qquad [1.71]$$

If the voltages at both ends of a line are equal and both are increased by 10%, the power limit is increased by about 20%: and the power limit is increased if B is decreased. **B** is the short-circuit transfer impedance (see [*1.61b*]).

If δ is eliminated between [*1.69a*] and [*1.69b*] then

$$[P_R + (AV_R^2/B)\cos(\beta-\alpha)]^2 + [Q_R + (AV_R^2/B)\sin(\beta-\alpha)]^2 = (V_S V_R/B)^2. \qquad [1.72]$$

This formula is useful when P_R, Q_R and V_S are given and it is required to find V_R.

If the components of the phasor diagram of Fig. 1.20 (which is drawn to a voltage scale with V_R = constant) are each multiplied by V_R/B, the scale is changed to a voltampere scale (the diagram is no longer a phasor diagram). All the above formulae can be interpreted in terms of this diagram; e.g. [*1.72*] is the equation of the circle V_S = constant.

In practice δ, the load angle or transmission angle, is quite small and not likely to exceed about 30° even for a very long line (500 km) transmitting rated load.

Formulae for the power and lagging vars into the sending end of a long line should be derived by the student using [*1.32a*] i.e.

$$\mathbf{V}_R = \mathbf{DV}_S - \mathbf{BI}_S$$

where $\mathbf{D} = D\underline{/\Delta}$ (for a symmetrical line $\mathbf{D} = \mathbf{A} = A\underline{/\alpha}$).

$$P_S = (DV_S^2/B)\cos(\beta-\Delta)-(V_S V_R/B)\cos(\beta+\delta) \qquad [1.73a]$$

$$Q_S = (DV_S^2/B)\sin(\beta-\Delta)-(V_S V_R/B)\sin(\beta+\delta). \qquad [1.73b]$$

P_S will be a maximum P_{SM} when $\beta+\delta = 180°$ and

$$P_{SM} = (DV_S^2/B)\cos(\beta-\Delta)+(V_S V_R/B). \qquad [1.74a]$$

The corresponding value of Q_S at this power limit is

$$Q_{SM} = (DV_S^2/B)\sin(\beta-\Delta). \qquad [1.74b]$$

It should be noted that the sending-end and receiving-end power limits do not occur at the same value of δ.

It is left as an exercise for the student to draw the sending-end performance chart (V_S = constant) of a long line, first to a voltage scale, then to a voltampere scale and hence to check the validity of the above formulae when V_S = constant.

The power loss in a long line is given by $P_S - P_R$ and the lagging vars loss by $Q_S - Q_R$, for a given value of δ.

1.5 Reactive Power Compensation

Reactive power or voltamperes reactive (vars) are defined and discussed in Appendix 4. Lagging vars flowing into an inductance are defined as positive. In power system operation, vars are by convention taken to indicate lagging vars, and the direction of flow is specified, for any piece of equipment, as an import or export of vars. Thus an inductance imports vars. A capacitance absorbs leading vars and is said to export (or generate) vars.

The performance chart of a synchronous machine is similar to that shown in Figs. 1.4, 1.5, 1.6 (for a short line) with $\mathbf{E}_o$ replacing $\mathbf{V}_S$ and $\mathbf{V}$ replacing $\mathbf{V}_R$, where the per unit synchronous impedance of an alternator is much greater than that of a short line. A synchronous machine (see Figs. 7.17, 7.18, 7.21) whether importing or exporting power, exports (or generates) vars when over-excited and imports vars when under-excited. Most loads import vars to supply the magnetising currents of transformers, induction motors, etc. At any moment in time the var demand on the grid network is fixed in magnitude and location by the voltage at the consumers' terminals (and the plant being operated at that time).

The relation between var demand from, and the voltage drop along a single short line has already been discussed in section 1.2.4 in which it was shown that the major cause of the voltage drop was the lagging reactive current flowing in the line's series inductance: and this statement is especially valid when the inductive reactance of the line is much greater than its resistance. Furthermore, reactive lagging currents produce no revenue, and moreover cause I^2R losses in the grid system. Thus vars should be generated in the power importing areas whenever this is technically possible. The purpose of the grid network is to transmit power with the var transmission reduced to the minimum possible. For a given var transmission along a line, the per-unit voltage drop is high if the line impedance is high, i.e. the fault level is low (see section 8.2.2). The reverse is true for short lengths of heavy 400-kV line transmitting large amounts of power.

The voltage, relative to nominal voltage, at all the busbars of the grid system is called the voltage profile. Since the circuit breakers, transformers and line insulation of the 400-kV system are operating close to the present limits of design, the maximum permitted voltage is 420 kV (i.e. +5%). The 400-kV grid is operated with a voltage profile of about ±5%. It follows that some 400-kV lines may have to transfer load with a voltage drop of less than 10%, and hence the load power factor must be near unity or leading. This is especially true of the longer lines (>125 km). The student is reminded of the comment in section 1.2.2 that the important factors affecting the voltage drop are the product of the load and the transmitted distance (i.e. MVA-km), and the load power factor. The 275-kV and lower voltage systems normally operate within about ±10%.

The performance chart of a long line treated as a 4-terminal network is shown in Fig. 1.20. At times of maximum load the vars absorbed ($2\pi fLI^2$) by the series inductance of the line are greater than the vars generated ($2\pi fCV^2$) by the shunt capacitance of the line: and $V_S > V_R$ especially for a lagging power factor load. In an extreme case it may be necessary for the line to import vars at both ends (i.e. the delivered load is at a leading power factor in order to reduce the voltage drop: in addition the consumers' demand for vars must be met.

At times of light load the line generates vars and if these exceed the consumers' demand the excess vars must be absorbed by operating the generators under-excited. There is also a tendency for the system voltage to rise ($V_R > V_S$). Switching some lines out of service and reducing the grid voltage might ease the problem.

When the load at the receiving end of a line is equivalent to that taken by the characteristic impedance of the line, $\mathbf{Z}_c$, the line is said to be operating at its natural load. It is shown in section 1.6 that at natural load the vars absorbed by the series inductance equal those generated by the shunt capacitance, and $V_S \simeq V_R$.

Reactive power is normally obtained from the generators connected to the system. These have rated power factors varying from about 0·8 to 0·9 lagging. During maximum load on the grid with the generators at or near rated power, there is a limit to the vars which they can generate. During minimum load on the grid, with the generators under-excited but operating near rated power (since there are only a few operating), there is a limit to the vars they can absorb since their load angle (rotor angle) should not exceed about 75°.

Thus it is sometimes necessary to provide additional equipment, called reactive power compensation equipment, to generate or absorb vars. A shunt-connected inductance absorbs vars while a shunt-connected capacitor generates vars. Specially designed synchronous machines, called synchronous compensators (capacitors or phase modifiers) can generate, or absorb, vars if over-, or under-excited. It is desirable that compensation equipment should be transportable since its optimum location is liable to change as the system load pattern changes.

In the discussion to follow, reference will be made to compensation equipment being slow or fast. For system voltage control during normal load changes it is sufficient to be able to operate the compensation equipment within a few minutes, e.g. the time required to operate an on-load tap-changer (see Chapter 6). During system faults the operation of the compensation equipment should be almost instantaneous and automatic to ensure system stability (see Volume 2), and to maintain the system voltage which tends to fall after lines have been switched out, when the grid is heavily loaded.

1.5.1 STATIC COMPENSATION

Fig. 1.21 shows static compensation equipment using a shunt connected inductor and capacitors. This equipment is often connected, at the receiving end of the line, to the tertiary delta winding of the transformer. The capacitors are used when the load demand is heavy. If they are manually switched in groups the control is slow. During maximum system load, when the system voltage is liable to be low, the var output of these capacitors is a minimum. This aggravates the problem. The inductor, which is used when the system load is low, can be a coreless type, a gapped-core type, a saturable reactor or a transductor. The coreless type is linear and is used for the larger ratings and voltages (e.g. 100 MVA at 400 kV). The gapped-core

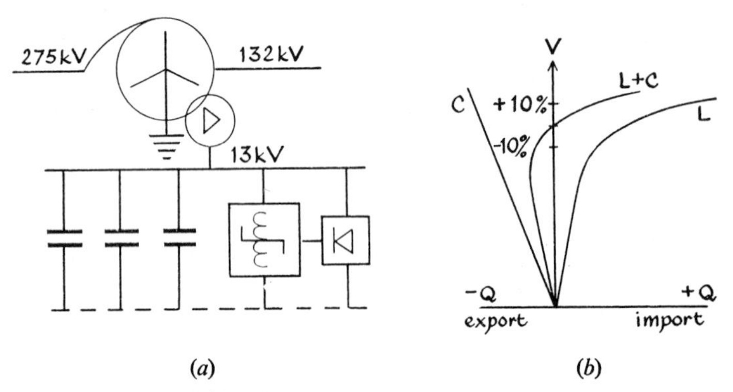

(*a*) (*b*)

FIG. 1.21. *Static, reactive power compensation.*

type is designed to be linear up to a voltage of about 1·3 per unit. Both these types will absorb vars proportional to the square of the terminal voltage (for a given reactance). For slow control an on-load tap-changer can be fitted to the reactor or to an auxiliary transformer (the latter would increase the capital and running costs of the scheme). An alternative scheme uses an auxiliary tap-changing, star-connected transformer, which can be tapped down to zero volts, to supply via a two-way, off-load selector isolator, either a fixed capacitor or a fixed inductor. The above schemes are relatively cheap, simple and need little maintenance.

It is possible to use a tapped saturable reactor which saturates at selected voltages within the range $\pm 10\%$ of rated voltage. When saturated, the vars absorbed by the reactor change rapidly for small changes in voltage. They are used in parallel with capacitors. The voltage/exported vars characteristic of a capacitor is a straight line through the origin. The voltage/imported vars characteristic of a saturable reactor is similar to the magnetisation curve of a transformer. When saturable reactors and capacitors operate in parallel, the characteristic, as the voltage rises, is initially similar to that of the capacitor then swings over to that of the saturable reactor at a voltage between 90 and 110% of nominal. Thus at low voltage the equipment exports vars while at high voltage it imports vars (Fig. 1.21(b)).

Another possibility is to use a transductor, the d.c. supply for the control winding being obtained from static rectifying equipment (as shown in Fig. 1.21). When used in parallel with capacitors the characteristic is similar to that of the saturable reactor but the control of the characteristic is finer and faster. Transductor equipment is expensive and would only be used where fast control is essential to ensure system stability.

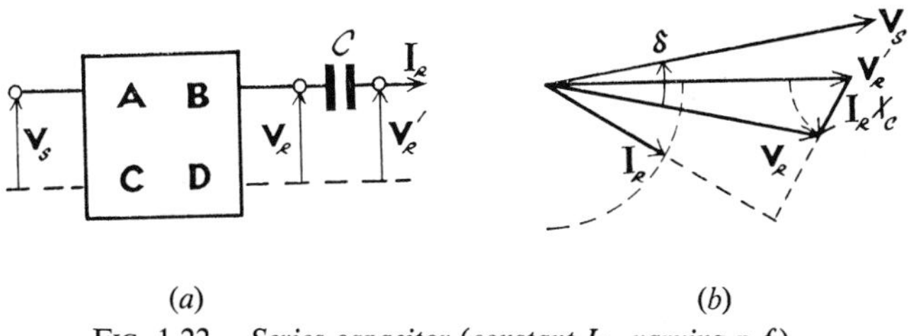

(*a*) (*b*)

FIG. 1.22. *Series capacitor (constant I_R, varying p.f.).*

The bulk supply into large urban areas is often carried by underground cables (up to about 20 km long) operating at 132 kV and above. These cables have a large capacitance and are a source of vars even when carrying rated current (see section 1.6). These vars are absorbed by connecting shunt inductors at intervals of a few km along the cable.

Although they are not yet used in the British grid system, it is appropriate to mention series-capacitors which are connected in series with the line. They generate vars proportional to the square of the current in them. Their primary use is to reduce the series impedance of the line, by cancelling about 50% of the line's series inductance, in order to raise the power limit (section 1.2.3) and to improve the system stability (see Volume 2). No British grid lines are long enough to justify their use. The effect of a series capacitor on the system voltage is shown in Fig. 1.22. For a load of lagging power factor $V_R' > V_R$ and the voltage drop between V_S and V_R' is less than the voltage drop between V_S and V_R. A problem with series capacitors is that when they carry fault current the voltage across them generally exceeds the voltage rating of the capacitor. Series capacitors are therefore shunted by arc-gaps, set to flash-over when the fault current is about three times the rated current of the capacitor, in series with a relay which closes a normally-open circuit breaker shunted across the capacitor: alternatively the arc-gap can be made self-extinguishing when the voltage across the capacitor has fallen to a safe value.

1.5.2 SYNCHRONOUS COMPENSATORS

Synchronous compensators are basically synchronous motors with no mechanical output. They are usually of salient pole design (6 or 8 poles) with ratings up to about 60 MVA, 13 kV (they are connected to higher voltage systems by transformers). Several methods are available for starting a compensator: reduced voltage using a tap-changer on the transformer and a cage winding in the pole faces, or a pony motor which can be a cage-, or wound-rotor induction motor or a synchronous-induction motor. The starting sequence is usually under automatic control. When running, the control of the compensator is also automatic. The excitation can be set to give automatically, either constant terminal voltage or constant vars, subject to an over-excitation limit and an under-excitation limit (to prevent unstable operation). The methods of obtaining and controlling the d.c. excitation supply of the synchronous compensator are similar to those for alternators (see section 7.9). The control of a synchronous compensator is fast and continuous from rated vars export to about half rated vars import. But, unlike static plant, they contribute current to a system fault (raise the fault level), are liable to become unstable and need relatively high maintenance. Loads such as steel mills and arc furnaces, which fluctuate widely frequently and rapidly, require fast and automatic compensation equipment placed as close as possible to them. In the case of very long overhead lines (not applicable to the U.K.) synchronous compensators are shunt-connected across the line at intervals of about 150 km.

1.6 Natural Load: Surge Impedance Load

A line is said to be operating at its natural load when the effective impedance of the delivered load equals the characteristic or surge impedance of the line, i.e. $\mathbf{V}_R/\mathbf{I}_R = \mathbf{Z}_R = \mathbf{Z}_c$.

The analysis to follow is simplified by assuming that the line is loss-free (see section 1.3.8). For a loss-free line, the characteristic impedance is a pure resistance and this will be emphasised by using Z_c rather than $\mathbf{Z}_c$. When $\alpha = 0$, $\gamma = \mathrm{j}\,\beta$ and [*1.28c*] and [*1.28d*] become

$$\sinh(\mathrm{j}\,\beta l) = \mathrm{j}\sin\beta l \qquad [1.75a]$$

$$\cosh(\mathrm{j}\,\beta l) = \cos\beta l. \qquad [1.75b]$$

Substituting these values in [*1.28a*] and [*1.28b*] gives

$$\mathbf{V}_S = (\cos\beta l)\mathbf{V}_R + (\mathrm{j}\,Z_c\sin\beta l)\mathbf{I}_R \qquad [1.76a]$$

$$\mathbf{I}_S = \left(\frac{\mathrm{j}\sin\beta l}{Z_c}\right)\mathbf{V}_R + (\cos\beta l)\mathbf{I}_R. \qquad [1.76b]$$

The quantities in the brackets are the **A, B, C** and **D** constants for a loss-free line: in this context β is the phase-change coefficient and not the phase angle of **B**. By regarding l as variable, [*1.76*] give the voltage and current at any point distant l from the receiving end.

If the loss-free line is operating at its natural load then $\mathbf{V}_R = Z_c\mathbf{I}_R$ and [*1.76*] reduce to

$$\mathbf{V}_S = \mathbf{V}_R(\cos\beta l + \mathrm{j}\sin\beta l) = \mathbf{V}_R\,\mathrm{e}^{\mathrm{j}\beta l} = \mathbf{V}_R\underline{/\beta l} \qquad [1.77a]$$

$$\mathbf{I}_S = \mathbf{I}_R(\cos\beta l + \mathrm{j}\sin\beta l) = \mathbf{I}_R\,\mathrm{e}^{\mathrm{j}\beta l} = \mathbf{I}_R\underline{/\beta l}. \qquad [1.77b]$$

Thus the voltage is constant in magnitude at all points along the line but $\mathbf{V}_S$ advances in phase relative to $\mathbf{V}_R$ by β radians per km: the corresponding comment is true for current. The driving-point (or input) impedance at any point on the line equals $\mathbf{V}_R/\mathbf{I}_R = Z_c$. For a transmission distance equal to a wave length, $l = \lambda$, (for 50-Hz line $\lambda = 6000$ km) and $\beta\lambda = 2\pi$ radians. The phase shift of voltage, and current, is 360° for $l = \lambda$ and in proportion for other distances: e.g. 6° per 100 km. The locus of $\mathbf{V}_S$ is the arc of a circle, of radius V_R = constant, as shown in Fig. 1.23. The phasor diagram for current is identical but with $\mathbf{I}_S$ and $\mathbf{I}_R$ replacing $\mathbf{V}_S$ and V_R.

It is convenient to discuss natural loading on a per-unit basis with rated phase voltage as base voltage and natural load power at rated voltage, $P_N = V_R^2/Z_c$, as base power. Since the line current is numerically constant along the whole length of the line, the per-unit total vars absorbed by the line's series inductance are $I_R^2\omega L/P_N$ (where L is the total inductance of the line) which expression can be reduced to βl radians. Similarly the per-unit total vars generated by the numerically constant voltage across the total shunt capacitance of the line are also equal to βl radians. Thus for a loss-free line operating at natural load, the sending-end voltage, current and power are delivered unchanged at the receiving end (apart from the phase displacement βl).

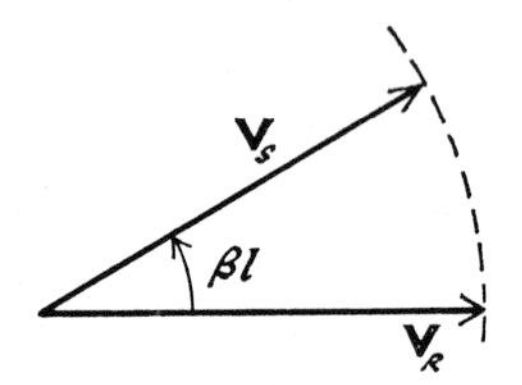

FIG. 1.23. *Loss-free line at natural load.*

The performance of a loss-free line at any load will now be discussed with the delivered load expressed in per-unit of natural load. The student should check that [*1.69*] reduces to

$$P_R^2+(Q_R+\cot \beta l)^2 = V_S^2/\sin^2 \beta l \text{ per unit} \qquad [1.78]$$

providing the receiving-end voltage is rated voltage: where P_R and Q_R are respectively the power and vars delivered at the load end of the line.

Assuming now a unity-power-factor delivered load, [*1.78*] can be reduced to

$$V_S^2 = P_R^2 \sin^2 \beta l+\cos^2 \beta l \text{ per unit.} \qquad [1.79]$$

The student should use [*1.79*] for a line of length one-twelfth of a wavelength, i.e. 500 km where $\beta l = 360°/12 = 30°$. By taking values of P_R from zero in steps of 0·5 per unit, he should show that when $P_R = 0$, V_S is about 14% below rated voltage and that when $P_R = 1{\cdot}5$ per unit, V_S is about 14% above rated voltage.

Clearly, when $V_S = V_R = 1$ per unit, [*1.78*] reduces to

$$(Q_R+\cot \beta l)^2 = (1/\sin^2 \beta l)-P_R^2 \text{ per unit.} \qquad [1.80]$$

Since the right hand side of [*1.80*] must be positive, it follows that the maximum value of P_R, for $V_S = V_R$, is given by

$$P_{RM} = \sqrt{(1/\sin^2 \beta l)} \text{ per unit.} \qquad [1.81]$$

For the 500-km line, $P_{RM} = 2$ per unit. Taking values of P_R from 0 to 2 per unit, the student should find the corresponding values of Q_R. Since a square root has two possible values, there are two possible sets of answers to this problem, but by considering the case of $P_R = 1$ per unit, the student should be able to choose the correct set of results and conclude that when $P_R = 0$, $Q_R = 0{\cdot}27$ per unit and that when $P_R = 2$ per unit, $Q_R = -1{\cdot}73$ per unit, and that in this last case the power factor of the delivered load is 0·758 leading.

The student should go through the same analysis for a line of length 50 km and deduce that very long lines can only be operated (within reasonable technical and economic limitations, e.g. voltage drop and the cost of reactive power compensation equipment) up to loads slightly in excess of natural load while short lines can be operated up to several times natural load. The longest line in the British grid system is under 320 km route length: the majority of drop and the cost of reactive power compensation equipment
been erected: a major problem with such lines is the stability of the two interconnected a.c. systems (see Volume 2).

Clearly the natural load P_N is a constant for a given line operating at a given rated voltage. A long line intended to operate at its natural load would not need any reactive power compensation equipment, but any deviation from natural load would cause large changes in voltage.

For underground cables, the characteristic impedance is so low that the natural load greatly exceeds the rated load of the cable. Thus a cable is always a source of lagging vars.

REFERENCES

BRITISH INSULATED CALLENDER'S CABLES LTD. 1965. *The application of power capacitors*. London.

COX, E. H. 'Overhead-line practice', 1976, *Proc. I.E.E.*, **122**, IOR, pp 1009–1017.

INSTITUTION OF ELECTRICAL ENGINEERS. 1964. *Abnormal loads on power systems*. I.E.E. Conference Report Series No. 8, London.

INSTITUTION OF ELECTRICAL ENGINEERS. 1966. *Design criteria and equipment for transmission at 400 kV and higher voltages*. I.E.E. Conference publication No. 15, London.

LEWIS, W. E. & PRYCE, D. G. 1965. *The application of matrix theory to electrical engineering*. Spon, London.

MIDDLETON, A. G., SEALY, T. & HOLDER, F. E., 'Control communications and protection for the El Chocon–Buenos Aires 500 kV transmission system', 1975, *Proc. I.E.E.*, **122**, pp 1405–1415.

MORTON, A. H. 1966. *Advanced electrical engineering*. Pitman, London.

OAKESHOTT, D. F. & WHITTINGTON, H. W. 'Impulse performance of a tower for transmission at 1300 kV', 1974, *Proc. I.E.E.*, **121**, pp 191–196.

STEVENSON, W. D. 1962. *Elements of power system analysis*. McGraw-Hill, New York.

TROPPER, A. M. 1962. *Matrix theory for electrical engineering students*. Harrap, London.

WADDICOR, H. 1964. *The principles of electric power transmission*. Chapman and Hall, London.

WESTINGHOUSE (Ed). 1964. *Electrical transmission and distribution reference book*. Westinghouse Electric Corporation, Pennsylvania, U.S.A.

WOOD, A. B., TAYLOR, J. V. & LIPTROT, F. J. 'Design, testing and construction of the overhead lines for the El Chocon to Buenos Aires transmission system', 1976, *Proc. I.E.E.*, **123**, pp 51–59.

I.E.E. Conference Publication No. 107, 1974, 'High-voltage d.c. and/or a.c. power transmission.'

Examples

1. A 3·3-kV, 3-phase ring distributor *ABC* is fed at *A* at rated voltage. The balanced loads at *B* and *C* are 200 A at power factor 0·707 lagging and 200 A at power factor 0·8 lagging respectively, both power factors being expressed relative to the voltage at *A*. The impedances of the three sections *AB*, *BC* and *CA* are 1+j 1, 0·867+j 1·5 and 1·5+j 0·867 ohms/phase respectively. Calculate the currents $\mathbf{I}_{AB}$, $\mathbf{I}_{BC}$, and $\mathbf{I}_{AC}$: also the voltage at *C*.

(L.C.T.) ($213\underline{/-47{\cdot}25^\circ}$, $15{\cdot}5\underline{/-79{\cdot}17^\circ}$, $189\underline{/-33{\cdot}7^\circ}$A, 2·74 kV)

2. Two 3-phase generators *A* and *B* can be considered to have negligible internal impedances and terminal voltages of 11 kV with *A* leading *B* by 14°. Both generators supply a substation *C* via a ring *ABC* whose section impedances *AB*, *BC* and *CA* are 2+j 4, 1+j 2 and 1+j 3 ohms/phase respectively. If a symmetrical 3-phase overload current of 2−j 1·5 kA flows out of *C* find the power supplied by the generator at *B* and the power loss in the section *AB*.

(I.E.E.) (13·1, 0·72 MW)

3. A star-connected, 3-phase, 4-wire supply has zero internal impedance, a symmetrical voltage of 230 V per phase and phase sequence *ABC*. It is connected, via a 4-core cable whose line cores each have a resistance of 0·5 ohm and whose neutral core has a resistance of 1 ohm, to a 4-wire star load as follows: (3·5+j 3) ohm to *A*; (4·5+j 5) ohm to *B*; and 12·5 ohm to *C*. The supply neutral *N* is connected to the star-point *S*. Calculate the current in the *A* line.

(L.C.T.) (46·4 A)

4. The following data is taken from tables relating to a 40-km length of 33-kV line: series impedance 0·125 + j 0·295 ohms/phase-km; rated current 400 A; receiving-end load, for a voltage drop of 10%, 18·0 MW at unity power factor and 7·75 MW at power factor 0·8 lagging; power limit for $V_S = V_R$ is 52 MW. Draw the receiving-end performance chart (V_R = 33 kV) for this line, (*a*) to a voltage scale and (*b*) to a voltampere scale, and check the data given. If the line delivers rated current, at power factor 0·8 lagging, what is the percentage voltage drop?

(L.C.T.) (7·9%)

5. A 275-kV, twin 2·6 cm^2 line (treated as a short line) has a series impedance of 0·03 + j 0·265 ohms/phase-km, a cold weather rating of 1600 A (760 MVA at 275 kV) and a route length of 140 km. Draw the receiving-end performance chart for V_R= 275 kV and hence estimate (*a*) the loading of reactance compensation plant at the receiving-end which will limit the line's voltage drop to 10% when the consumer's delivered load is 600 MW at power factor 0·9 lagging, (*b*) the delivered load MVA at power factor 0·8 lagging for a voltage regulation of 10% (without any compensation plant) and (*c*) the MVAr loading of the reactance compensation plant to hold the voltage drop at 10% when the delivered load is zero.

(L.C.T.) (221 MVAr export, 286 MVA, 207 MVAr import)

6. Two 132-kV busbars *A* and *B* are interconnected by two identical lines in parallel, each of reactance 20 ohms and negligible resistance. The total input at *A* is 200 MVA, 132 kV, power factor 0·8 lagging. If into one circuit at end *A* is injected an inphase boost e.m.f. of 5%, boosting from *A* towards *B*, calculate power and reactive power fed into the boosted and unboosted circuit, assuming the input remains unchanged.

(L.C.T.) (80·0+j 82·0, 80·1+j 38·3 MVA)

7. Two 3-phase lines *A* and *B* operate in parallel. Their reactances are 0·7 and 1·3 ohms/phase respectively, and their resistances are negligible. The total input is 30 MVA, power factor 0·8 lagging, 11 kV, and this input can be assumed constant. Find the load, that each line carries, in the form (P+j Q) MVA. A boost e.m.f. is now injected into line *B*, at the sending end, boosting towards the receiving end, and will be expressed in volts to neutral and with reference to

the sending-end voltage. Calculate the quadrature boost e.m.f. which will cause the lines to carry equal power. Calculate the boost e.m.f. which will cause the lines to carry identical currents. For all three cases, calculate the total vars absorbed in the system.

(L.C.T.) (15·62+j 11·72, 8·38+j 6·28 MVA; j 380 V; 286+j 382 = 477 V; 1·131, 1·202, 1·25 MVAr)

8. An 11-kV generator *A* supplies two substations *B* and *C* via a ring *ABC*. The impedances of the sections *AB*, *BC* and *CA* are 0·5+j 1, 0·75+j 1·25 and 0·5+j 0·75 ohms/phase respectively. The load taken from *B* is 600 A at power factor 0·8 lagging, and that from *C* is 480 A at power factor 0·9 lagging, both power factors being expressed relative to the voltage at *A*. Calculate the current $\mathbf{I}_{AB}$. A booster is now installed at *A* in the line *AB* boosting from *A* to *B*. If the booster gives an in-phase boost of 5% of rated voltage recalculate $\mathbf{I}_{AB}$. What boost voltage, expressed as a percentage of rated voltage, will reduce the current $\mathbf{I}_{BC}$ to zero?

(L.C.T.) (426−j 314, 380−j 235 A, 3·55+j 1·28%)

9. The 11-kV 3-phase busbars in each of two separated substations *S* and *R* are joined by two parallel circuits 1 and 2. Each circuit consists of an 11/132-kV transformer, a 132-kV transmission line and a 132/11-kV transformer. The total impedances of the circuits, including both transformers and the transmission line, and referred to 11 kV are

$$\mathbf{Z}_1 = (0{\cdot}2+\mathrm{j}\,0{\cdot}4) \quad \text{and} \quad \mathbf{Z}_2 = (0{\cdot}2+\mathrm{j}\,0{\cdot}6) \text{ ohms/phase}$$

(*a*) If busbar *S* is at 11 kV and is sending 30 MW at 0·8 power factor leading, find the individual currents into each transformer at *S*.

(*b*) If the transformer at *S* in circuit 2 is fitted with tappings on the 11 kV side, what percentage tapping would be required to make each circuit carry equal reactive power and what power would circuit 2 then be sending?

(I.E.E.) (1152, 814 A; +2·89% (increased turns), 11·92 MW)

10. A balanced 3-phase load of 100 MVA at 0·8 lagging power factor is transmitted from a substation by two circuits in parallel; each circuit consists of a transformer/transmission-line/transformer. The total impedance per phase of each circuit is $\mathbf{Z}_a = (5+\mathrm{j}\,20)\%$ and $\mathbf{Z}_b = (5+\mathrm{j}\,10)\%$ referred to 100 MVA.

(*a*) If all the transformers have equal turns ratios, find the individual loads transmitted by each circuit, in form $(P \pm j\,Q)$ MVA.

(*b*) What percentage tapping change would be needed on the secondary of one of the substation transformers so that both circuits will transmit numerically equal MVA? What are then the individual loads in each circuit in the form $(P \pm j\,Q)$ MVA?

(I.E.E.) (25+j 25; 55+j 35 MVA; 5·77%; 30·8+j 42·3; 49·2+j 17·7 MVA)

11. A 132-kV line has a series impedance of 0·125 + j 0·335 ohms/phase-km and a route length of 100 km. For a voltage drop of 10% of rated voltage, calculate the receiving-end power if its power factor is (*a*) unity, (*b*) 0·8 lagging. Calculate also the power limit of the line if the voltages at the two ends are both equal to rated voltage.

(L.C.T.) (110, 46, 318 MW)

12. A short line has negligible series resistance and a series reactance of 16 ohms/phase. The input to the line is 20 MW/phase and 16 MVAr lagging/phase. If $\mathbf{V}_R = (100+j\,0)$ kV/phase calculate $\mathbf{V}_S$ and $\mathbf{I}_S$.

(I.E.E.) (102·4+j 3·2 kV/phase, 200−j 150 A)

13. A 132-kV line has a series impedance of 10+j 40 ohms/phase. The delivered load is 200 MW at power factor 0·8 lagging. If the voltages at the ends of the line are both 132 kV, calculate the loading of the required reactance compensation plant, and the load angle.

(I.E.E.) (265 MVAr export, 31·77°)

14. A 3-phase transmission circuit supplies 600 MW to a load point at 0·98 power factor lagging. The reactance of the circuit at the supply frequency is 21·83 Ω/phase and its resistance may be neglected. Determine the reactive power supply required at the receiving end if the voltage at both sending and receiving ends is to be maintained at 275 kV.

(I.E.E.) (178 MVAr export)

15. A transmission circuit has a series reactance per phase of 24 Ω at the supply frequency, and its series resistance and shunt capacitance may be neglected. If the voltage at the receiving end is 132 kV and that at the sending end is 140 kV when the magnitude of the

transmitted current is 500 A, calculate the corresponding active and reactive power delivered at the receiving end.

(I.E.E)) (49·7 MW, 1·54 MVAr leading)

16. A 66-kV short line has a series impedance of 0·125 + j 0·295 ohm/phase-km and a route length of 60 km. It is operating with 66 kV at both ends and a load angle of 10°. Calculate the receiving-end power and power factor.

(L.C.T.) (34·9 MW, 0·877 leading)

17. A short line has a series impedance of 10+j 30 ohms/phase, sending and receiving voltages of 140 and 120 kV respectively, and a load angle of 5°. Calculate the receiving-end power and power factor.

(L.C.T.) (67·8 MW, 0·783 lagging)

18. A 3-phase line has a total series impedance of $300\underline{/78^\circ}$ Ω/phase and a total shunt admittance of $24\times10^{-4}\underline{/90^\circ}$ siemens/phase. The voltage at the receiving end is 220 kV but there is no load at that end. A load of 100 MW at unity power factor is connected at the mid-point of the line. Using a nominal-Π representation, calculate the sending-end voltage.

(I.E.E.) (316·5 kV)

19. A 3-phase, 50-Hz interconnector transmission line, 48 km long, operates at 66 kV (line) at each end. The resistance of each conductor is 0·181 Ω/km and the inductance is 1·16 mH/km; the capacitance is negligible. In-phase and quadrature boost transformers are used at the sending end to control the power flow. Ignoring transformer losses, determine (graphically or otherwise) the transmitted power, the receiving-end power, and the line current for the following conditions: in-phase boost 0·06 p.u. and leading quadrature boost 0·08 p.u. of the sending-end voltage; the sending-end busbar voltage leads the corresponding receiving-end voltage by 10°.

(University of London) (62·7 MW, 55·6 MW, 520 A)

20. Two parallel feeders have the following constants:

Constants	*Feeder I*	*Feeder II*
A = **D**	$0{\cdot}816\underline{/4{\cdot}35^\circ}$	$0{\cdot}871\underline{/2{\cdot}5^\circ}$
B	$227{\cdot}2\underline{/72{\cdot}3^\circ}$ Ω :	$196\underline{/73^\circ}$ Ω
C	$15{\cdot}7\times10^{-4}\underline{/91{\cdot}4^\circ}$ siemen	$12{\cdot}94\times10^{-4}\underline{/92{\cdot}35^\circ}$ siemen

Calculate the sending-end impedance of the system when the receiving-end is on open-circuit, and determine the sending-end voltage, if the receiving-end voltage is 132 kV.

(University of Leeds) ($295\underline{/-86^\circ\ 26'}\ \Omega$, 111·8 kV)

21. A short 3-phase transmission line delivers 1000 kVA at 10 kV, 0·8 power factor lagging to a balanced load. The resistance per conductor is 4 Ω and the reactance is 6 Ω. Calculate the voltage at the middle of the line.

A balanced, star-connected load of resistance 100 Ω/phase and inductive reactance 200 Ω/phase in series, is now connected at the middle of the line, the voltage at the generator end being changed to keep the voltage unchanged at the middle. Calculate the new current and its power factor at the generator end of the line.

(University of London) (10 320 V, 82·6 A, 0·69 lagging)

22. A 3-phase transmission line has a resistance/phase of 10 Ω and inductive reactance of 30 Ω. Calculate the maximum power which can be transmitted when the receiving-end and sending-end voltages are 132 kV and 135 kV respectively.

(University of Leeds) (390 MW)

23. A 3-phase transmission line delivers 1000 kW at 11 kV and 0·8 power factor lagging. The line has resistance of 4 Ω and reactance of 6 Ω/conductor. Calculate the voltage at the sending end of the line, if the load voltage is maintained constant when connection of an additional load increases the line current by 50 A and raises the power factor to 0·9 lagging. Calculate also the equivalent series resistance and reactance of the added load.

(University of London) (12 200 V, 116·5+j 25·7 Ω)

24. Three 3-phase transmission lines in parallel deliver 1000 kVA at 6·6 kV and 0·8 power factor lagging.

Line A, resistance/conductor = 1 Ω: reactance/conductor = 0·4 Ω
Line B, resistance/conductor = 0·5 Ω: reactance/conductor = 0·5 Ω
Line C, resistance/conductor = 0·4 Ω: reactance/conductor = 0·6Ω

Calculate the current delivered by line B and its phase relationship to the total load current.

(University of London) (34·1 A, lagging 1° 31′)

25. A 3-phase transmission line has a resistance of 4 Ω and induc-

tive reactance of 12 Ω/conductor. It delivers 5000 kVA at 0·8 p.f. lagging. The line voltage halfway along the line is 30 kV. Calculate the voltage at the load end and the power factor halfway along the line.

(29 200 V, 0·79 lagging)

26. A 3-phase transmission line has a resistance of 4 Ω and inductive reactance of 6 Ω per conductor. It is connected to a balanced load which would take 2000 kVA at 0·8 p.f. (lagging), if the line voltage at the load end of the line were 30 kV. Calculate the equivalent resistance and reactance of the load 'to neutral'.

A new balanced load is added, having a series resistance of 250 Ω and an inductive reactance of 500 Ω 'to neutral'. If the voltage at the load end of the line is 30 kV, calculate for these conditions (*a*) the line current and (*b*) the voltage, at the generator end of the line.

(360 Ω, 270 Ω, 68 A, 30·8 kV)

27. The constants of a transmission line are **A** = $0{\cdot}90\underline{/2^\circ}$ and **B** = $140\underline{/70^\circ}$ Ω/phase. At the receiving end the line voltage is 132 kV and the load is 60 MVA at 0·8 lagging power factor. Calculate the sending-end line voltage and the load angle.

(I.E.E.) (176·5 kV, 12·8°)

28. A long 132-kV line has constants of **A** = $0{\cdot}98\underline{/3^\circ}$ and **B** = $110\underline{/75^\circ}$ Ω/phase. Draw the receiving-end performance chart for V_R = 132 kV and hence estimate the power limit of the line, and the loading of reactance compensation plant at the load end, both for zero voltage drop. The delivered load is 50 MVA, power factor 0·8 lagging, 132 kV.

(L.C.T.) (112 MW, 45·6 MVAr export)

29. A long, 3-phase 132-kV line has constants **A** = $0{\cdot}98\underline{/3^\circ}$ and **B** = $110\underline{/75^\circ}$ Ω/phase. The receiving end load is 50 MVA, power factor 0·8 lagging, 132 kV. Calculate the sending-end voltage, current and power factor and the nett vars absorbed by the line; also the loading of reactance compensation plant (across the load) required to reduce the sending-end voltage to 140 kV.

(L.C.T.) (166 kV, 254 A, 0·894 lagging, 2·8 MVAr, 35·5 MVAr export)

30. A 275-kV 3-phase transmission line has parameters **A** =

$0{\cdot}95\underline{/3^\circ}$, **B** = $80\underline{/75^\circ}$ ohm/phase. For a receiving-end voltage of 275 kV: (*a*) calculate the sending-end voltage required if the receiving-end load is 250 MW at 0·85 lagging power factor; (*b*) calculate the additional reactive power required at the receiving end if the sending-end voltage is 300 kV and the load is 400 MVA at power factor 0·9 lagging.

(L.C.T.) (331 kV, 222 MVAr export)

31. A 3-phase line has general constants **A** = $0{\cdot}856\underline{/6{\cdot}72^\circ}$, and **B** = $101\underline{/81{\cdot}46^\circ}$ ohms/phase. Between the receiving end of the line and the load is connected a series capacitor of reactance 50 ohms (one in series with each conductor). If the load delivered is 150 MW, 220 kV, power factor 0·8 lagging, calculate the sending-end voltage.

(I.E.E.) (236 kV)

32. A long, symmetrical, 3-phase overhead line, open-circuited at the load end, gave the following test data (all per phase star, for the same phase): load voltage $200\underline{/0^\circ}$ kV, sending voltage $180\underline{/5^\circ}$ kV, sending current $200\underline{/92^\circ}$ A. Find the constants **A, B, C,** and **D** and hence the input voltage, power and power factor for a load of 200 kV per phase, 40 MW/phase, unity power factor.

(L.C.T.) ($0{\cdot}9\underline{/5^\circ}$, $248\underline{/53{\cdot}2^\circ}$ Ω, $1\underline{/92^\circ}$ millisiemen, $0{\cdot}9\underline{/5^\circ}$, 216 kV, 48·5 MW, 0·808 leading)

Chapter 2

TRANSMISSION LINE INDUCTANCE

2.1 Introduction

When the magnetic flux linked with any electrical circuit changes, an e.m.f. is induced in the circuit, which is proportional to the rate of change of flux linkages. In general, the flux linked with a circuit is given, at any instant, by integrating the normal flux density over any surface bounded by the conductors composing the circuit. In cases where the circuit has turns which can be clearly defined, the flux linkages can be represented simply by the sum of the fluxes linked with each turn of the circuit, i.e. by the total number of turns multiplied by the average flux linked with each turn. There are cases, however, where it is not possible to define the effective flux linkages in this simple way, such as when an appreciable amount of linkage arises from flux which links with part of the circuit current *inside* the conductors. Here it is convenient to divide the current flow into elements, each of which links with some flux, and then to sum the current-flux linkages as shown in section 2.2.1.

If an instantaneous flux of ϕ Wb links completely with N turns the flux linkages are ϕN weber-turns, and the instantaneous e.m.f. induced in the circuit comprising these turns is $e = -\mathrm{d}(\phi N)/\mathrm{d}t$ volts, and if, as is usual, the number of turns is constant, then $e = -N(\mathrm{d}\phi/\mathrm{d}t)$. The instantaneous e.m.f. is 1 volt if the flux changes at that instant at the rate of 1 weber-turn/second. The negative sign indicates that the e.m.f. is in such a direction as to tend, as stated by Lenz's Law, to circulate a current which would oppose the change of flux causing it.

If ϕN is directly proportional to the instantaneous exciting current i in the circuit of N turns—i.e. if there is no saturation of ferromagnetic material, so that the magnetic circuit has constant reluctance, then the constant of proportionality may be written as the inductance L henry. Thus $\phi N = Li$. A circuit thus has a self-inductance of 1 H, if a change in current in it of 1 A causes a change of flux linkage of 1 Wb turn. The e.m.f. induced by self-induction is therefore $e = -\mathrm{d}(Li)/\mathrm{d}t$ volts and if L is constant $e = -L(\mathrm{d}i/\mathrm{d}t)$ volts.

When a magnetic circuit contains ferromagnetic material (as in steel-cored aluminium (or s.c.a.) line conductors), so that the flux is no longer proportional to the magnetising current, the inductance is not defined as the incremental value $\mathrm{d}(\phi N)/\mathrm{d}i$, but as the change in flux linkages upon current reversal divided by the current change. This definition in terms of current reversal rather than reduction of current to zero, relates the effective inductance to the peak flux density but not to the residual magnetism left when the current is reduced to zero.

When the magnetising current varies sinusoidally with time at a frequency of f Hz and has an r.m.s. value $\mathbf{I}$ amperes, the flux linkages also are sinusoidal and of r.m.s. value $= \mathbf{\Phi} N$. The r.m.s. induced e.m.f. $\mathbf{E} = \mathrm{j}\,\omega L\mathbf{I} = \mathrm{j}\,\omega\mathbf{\Phi} N$ volts, where $\omega = 2\pi f$ rad/s, and

$$L = (\Phi N/I). \qquad [2.1]$$

Considering next the form of the field distribution, the magnetic circuit law may be used, which states that the line integral of magnetising force $\mathbf{H}$ round any closed path, is equal to the total current linked with the path, i.e.

$$\oint \mathbf{H}_{tx} \mathrm{d}x = \mathbf{I}N.$$

If a circular flux path of radius x metres is taken, around a single long conductor very far removed from the return conductor, with each conductor carrying $\mathbf{I}$ r.m.s. amperes, then $2\pi x \mathbf{H}_{tx} = \mathbf{I}$. The suffix in $\mathbf{H}_{tx}$ denotes that the r.m.s. magnetising force at radius x is tangential at all points around the circular path. Thus

$$\mathbf{H}_{tx} = (\mathbf{I}/2\pi x). \qquad [2.2]$$

2.2 Two-wire Line

We shall consider the magnetic field set up by each conductor separately in its contribution to the circuit inductance; in fact as the conductors of a single-phase line will have equal radii, then each contributes half the circuit inductance. Fig. 2.1 shows typical circular paths which flux lines would follow due to the alternating r.m.s. current **I** in conductor 1, and external to it, temporarily neglecting the current $-\mathbf{I}$ in conductor 2. It is not possible to assume that the conductor radius r is negligibly small, so that the current flows in a filament along its axis. (On this assumption the flux linked

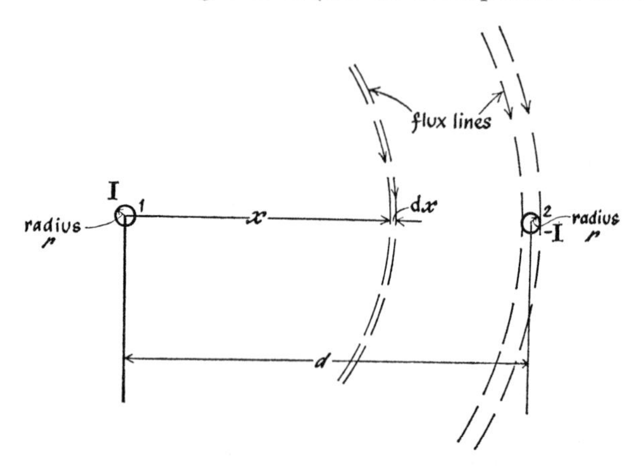

FIG. 2.1. *Two-wire line showing magnetic flux lines due to current* **I** *in conductor* 1 *alone.*

with the circuit composed of conductors 1 and 2 due to **I** in conductor 1, would from equation [*2.2*] be

$$\frac{\mu_o \mathbf{I}}{2\pi}\int_o^d \frac{dx}{x}$$

which is infinite.) We must therefore examine the flux linkages within a conductor of finite radius r.

2.2.1 FLUX LINKAGES WITHIN THE CONDUCTOR PRODUCING THE FLUX

It is assumed that the current density over the cross-section is constant. This would only be so if (*a*) skin effect at 50Hz, though it may become noticeable in the cross-section needed to carry large line currents, is neglected; (*b*) the non-uniformity due to stranding or due to the magnetisation of a steel core is neglected; (*c*) the proximity effect of current distribution in conductor 1 being disturbed by the magnetic field of $-\mathbf{I}$ in conductor 2 is neglected.

The magnetic lines of flux inside conductor 1 are assumed to be circles concentric with the conductor axis. Each of these lines of flux will link only with a part of the total conductor current **I**, e.g. the line at radius $x(x<r)$ links only with

$$\mathbf{I}(x^2/r^2) = \mathbf{I}_x, \qquad [2.3]$$

and the definition of inductance as $L = N\Phi/I$ for N turns, and $L = \Phi/I$ for the single-turn of a single-phase line, is now found to be too limited to deal with the internal flux linkages.

If tubes of current and tubes of flux, each with equal elemental amounts of current and flux of δI amperes and $\delta\Phi$ Wb/m respectively, make up the distribution within the conductor, then from a consideration of the energy stored in the magnetic field of $\frac{1}{2}LI^2$ joules, it is possible to write the internal self-inductance in two alternative ways:

$$L_i = \frac{1}{I^2}\Sigma(\Phi\delta I) \text{ H/m.} \qquad [2.4a]$$

$$L_i = \frac{1}{I^2}\Sigma(I_x\delta\Phi) \text{ H/m.} \qquad [2.4b]$$

We shall use [2.4b], where the flux is divided into elements $\delta\Phi$, each of which is multiplied by the amount of current I_x which is linked with it. When these have been summed over the whole conductor cross-section, the result is divided by the square of the current. Effectively, this process involves thinking of the total conductor current I flowing in a complete single turn, so that when the partial flux linkages $I_x\delta\Phi_x$ each of which links part of the turn only, have been integrated to give the total partial flux-current linkages, we must divide by I to obtain the flux deemed to link with the whole turn carrying the current I. These flux linkages must then be divided by the current I in the usual way to give the inductance, so that the expression for inductance has become

$$L = \frac{\int \text{flux-current linkages}}{I^2}.$$

If conductor 1, the cross-section of which is shown in Fig. 2.2, and which is one of the pair shown in Fig. 2.1, with its return conductor d metres away, where $d \gg r$, carries a total current I uniformly distributed throughout the section, then the flux density at a radius x where $x<r$ is

$$B_x = \frac{\mu_o I_x}{2\pi x} \text{ tesla} \qquad [2.5]$$

where I_x is the current flowing within radius x and is given by [2.3] therefore

$$B_x = \frac{\mu_o I x}{2\pi r^2} \text{ tesla}$$

If the flux is divided into annular rings of radius x, radial width dx and axial length 1 metre, then each contains flux $\delta\Phi$, where

$$\delta\Phi = \frac{\mu_o I x \, dx}{2\pi r^2} \text{ Wb.}$$

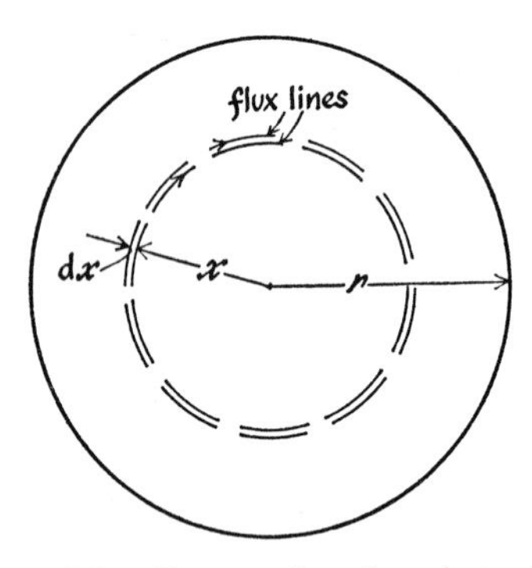

FIG. 2.2. *Cross-section of conductor* 1.

This element of flux $\delta\Phi$/metre length of conductor links with the current I_x flowing within its radius, therefore

$$I_x \delta\Phi = \frac{\mu_o I^2 x^3 \, dx}{2\pi r^4} \text{ current linkages/m.}$$

The inductance/m length of conductor 1 due to internal flux linkages caused by its own current, is therefore given by [2.4b] as

$$L_i = \frac{\mu_o}{2\pi r^4} \int_o^r x^3 \, dx$$

$$= \frac{\mu_o}{8\pi} = \frac{10^{-7}}{2} \text{ H/m.} \qquad [2.6]$$

It must be noted that the above analysis assumes a non-magnetic homogeneous conductor. The effect of stranding and of a steel core is discussed briefly in section 2.7.

2.2.2 FLUX LINKAGES OUTSIDE THE CONDUCTOR PRODUCING THE FLUX

As Fig. 2.1 indicates, flux produced by current in conductor 1 links with the whole conductor current I for values of x between r and $(d-r)$. For $(d-r)<x<(d+r)$ the fraction of current linked falls from 1 to 0. If it is assumed that the flux links the whole current I from r to d, and no current at all from d to $(d+r)$, then for $d \gg r$, little error occurs in calculating the effective external inductance L_e, and this assumption is generally made.

The flux in an annular element of radius x with radial width dx and axial length 1 metre, where $x>r$, due to the current I in conductor 1, can be written from equation [*2.2*] as $\mu_o I\,\mathrm{d}x/2\pi x$. This flux is assumed to link with a current I in a single-turn circuit for $r<x<d$, so from equation [*2.1*]

$$L_e = \frac{\mu_o}{2\pi}\int_r^d \frac{\mathrm{d}x}{x}$$

$$= \frac{\mu_o}{2\pi}\log_e \frac{d}{r} \text{ H/m.} \qquad [2.7]$$

L_i+L_e gives the total circuit inductance due to the current **I** in conductor 1. As we have assumed equal conductor radii, the current $-$**I** in conductor 2 will contribute an exactly similar m.m.f. in phase with the other, and as constant permeability has also been assumed, the flux linkages and hence the inductances may be added directly.

The single-phase circuit or loop therefore possesses a total inductance of $2(L_i+L_e)$ for each metre length of its route, and half this may be called the inductance of each conductor. The expression

$$(L_i+L_e)\text{ H/m} = L = \frac{\mu_o}{2\pi}\left[\log_e\left(\frac{d}{r}\right)+\frac{1}{4}\right]\text{H/m} \qquad [2.8]$$

may therefore be used either

(*a*) to give the inductance of one conductor/metre of route length: here the actual length of one conductor or the route length is used. (The justification for applying the concept of inductance which is a property of a circuit, to a single conductor of a 3-phase line, is given in section 2.4),

or (*b*) to give the inductance of the whole loop comprising both go and return conductors/metre of length measured round the loop: here the loop length which is twice the route length is used.

In order that [2.8] may be re-written as

$$L = \frac{\mu_o}{2\pi} \log_e \frac{d}{r_m} \text{ H/m} \qquad [2.9]$$

a modified radius r_m is introduced, where

$$r_m = r\,e^{-\frac{1}{4}},$$

and this is called the geometric mean radius of the conductor (G.M.R.), or the self geometric mean distance from itself (self G.M.D.). Comparison with [2.7] shows that r_m may then be regarded as the effective radius of a conductor for which the internal flux linkages may be neglected.

The student should prove for himself that in general

$$r_m = r\,e^{-\mu_c/4},$$

for a single-strand circular section conductor in air, with an effective conductor relative permeability of μ_c. The effects of multiple stranding and of a steel core are given in detail by Butterworth (1954). The student should also verify as an exercise, that if the two conductors have differing radii of r_1 and r_2, the average inductance due to external flux will be

$$\frac{\mu_o}{2\pi} \log_e \frac{d}{\sqrt{(r_1 r_2)}} \text{ H/m}.$$

2.3 General Multi-conductor System

Most of the multi-conductor systems which will be referred to here will be single-circuit 3-phase transmission lines so that interest is mainly in 3- or 4-conductor systems. It is beyond the scope of this chapter to consider in great detail, mutual effects between double-circuit 3-phase lines, between power transmission lines and other circuits, e.g. interference with telephone lines, and bundle-conductor transmission lines, i.e. where each phase has more than one conductor. The latter and double-circuit lines are however mentioned briefly in sections 2.8 and 2.9.

Interference, where for example 50 Hz signals may be induced in such circuits as telephone lines running near overhead power lines, is not discussed. It may however be noted that [2.11] developed below, may be used to determine the power frequency voltage induced into a neighbouring circuit, and some examples are included at the end of the chapter on this kind of calculation.

Fig. 2.3 shows a 3-phase line in air with a neutral conductor n, where (if the earth itself can be neglected), $\mathbf{I}_a+\mathbf{I}_b+\mathbf{I}_c+\mathbf{I}_n = 0$. If R is a dummy return conductor, then each of the four currents

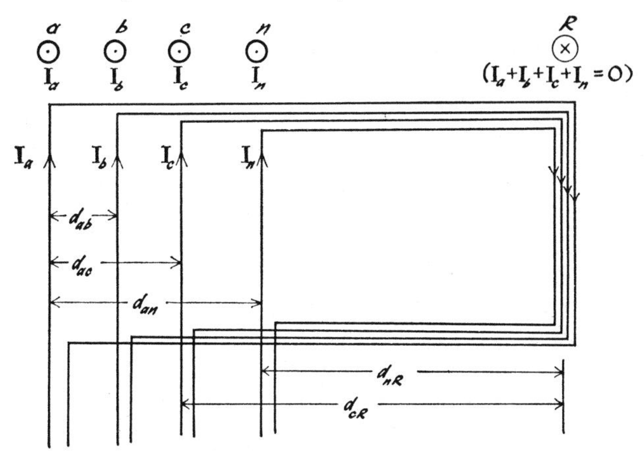

FIG. 2.3. *Single-circuit three-phase line with earth wire and a dummy return conductor.*

can be imagined to circulate in a single-turn circuit, e.g. $\mathbf{I}_a$ flowing in a and R, $\mathbf{I}_b$ in b and R, etc. Since the current in R will at all times be zero it is an imaginary conductor with an arbitrary position relative to the real conductors. Since all circuits have a single turn, the flux linkages with any circuit are equal to the flux linked with that circuit, e.g. a flux $\mathbf{\Phi}_a$ links the circuit composed of a and R.

From [*2.1*] and [*2.9*] the flux linked with the circuit composed of a and R due to the current $\mathbf{I}_a$ in conductor a is

$$\mathbf{\Phi}_a' = \frac{\mu_o \mathbf{I}_a}{2\pi} \log_e \frac{d_{aR}}{r_m} \text{ Wb/m} \qquad [2.10]$$

where r_m is the G.M.R. of all three phase conductors, and the neutral conductor has a G.M.R. of r_m'.

The flux $\mathbf{\Phi}_a''$ linked with the same circuit due to $\mathbf{I}_b$ flowing in conductor b occurs as circular flux lines with the axis of conductor b as centre and between the limiting radii of d_{ab} and d_{bR}, so that

$$\mathbf{\Phi}_a'' = \frac{\mu_o}{2\pi}\mathbf{I}_b \log_e \frac{d_{bR}}{d_{ab}} \text{ Wb/m.} \qquad [2.11]$$

Summing [2.10] and [2.11] together with the two contributions due to $\mathbf{I}_c$ and $\mathbf{I}_n$, the total r.m.s. value of the flux linked with the circuit of a and R is given by

$$\mathbf{\Phi}_a = \frac{\mu_o}{2\pi}\left(\mathbf{I}_a \log_e \frac{d_{aR}}{r_m} + \mathbf{I}_b \log_e \frac{d_{bR}}{d_{ab}} + \mathbf{I}_c \log_e \frac{d_{cR}}{d_{ac}} + \mathbf{I}_n \log_e \frac{d_{nR}}{d_{an}}\right) \qquad [2.12]$$

$$\mathbf{\Phi}_a = \frac{\mu_o}{2\pi}\left(\mathbf{I}_a \log_e \frac{1}{r_m} + \mathbf{I}_b \log_e \frac{1}{d_{ab}} + \mathbf{I}_c \log_e \frac{1}{d_{ac}} + \mathbf{I}_n \log_e \frac{1}{d_{an}} + \right.$$
$$\left. \mathbf{I}_a \log_e d_{aR} + \mathbf{I}_b \log_e d_{bR} + \mathbf{I}_c \log_e d_{cR} + \mathbf{I}_n \log_e d_{nR}\right) \qquad [2.13]$$

and substituting $\mathbf{I}_n = -(\mathbf{I}_a + \mathbf{I}_b + \mathbf{I}_c)$ into the last term of [2.13]

$$\mathbf{\Phi}_a = \frac{\mu_o}{2\pi}\left(\mathbf{I}_a \log_e \frac{1}{r_m} + \mathbf{I}_b \log_e \frac{1}{d_{ab}} + \mathbf{I}_c \log_e \frac{1}{d_{ac}} + \mathbf{I}_n \log_e \frac{1}{d_{an}} + \right.$$
$$\left. \mathbf{I}_a \log_e \frac{d_{aR}}{d_{nR}} + \mathbf{I}_b \log_e \frac{d_{bR}}{d_{nR}} + \mathbf{I}_c \log_e \frac{d_{cR}}{d_{nR}}\right). \qquad [2.14]$$

As the dummy conductor R has no physical reality it has an arbitrary position. When it recedes further from the real conductors, the last three terms in [2.14] each approach zero, since when R is at infinity all four conductors are equidistant from it. Thus $\mathbf{\Phi}_a$ which is the flux linked with conductor a becomes

$$\mathbf{\Phi}_a = \frac{\mu_o}{2\pi}\left(\mathbf{I}_a \log_e \frac{1}{r_m} + \mathbf{I}_b \log_e \frac{1}{d_{ab}} + \mathbf{I}_c \log_e \frac{1}{d_{ac}} + \mathbf{I}_n \log_e \frac{1}{d_{an}}\right). \qquad [2.15]$$

It should be noted that if r_m, the self geometric mean distance of the conductor a from itself, is written as d_{aa}, a consistent double subscript notation results.

The fluxes linked with the other conductors may then be obtained as follows: to obtain $\mathbf{\Phi}_b$ interchange a and b in [2.15]; to obtain $\mathbf{\Phi}_c$ interchange a and c in [2.15], and so on. The four equations for $\mathbf{\Phi}_a$, $\mathbf{\Phi}_b$, $\mathbf{\Phi}_c$ and $\mathbf{\Phi}_n$ which result, and of which [2.15] is the first, can be written more conveniently and compactly in matrix form as in [2.16] rather than as separate equations.

For the benefit, however, of students unfamiliar with matrix notation at this stage, the last three equations of the set of four represented by [2.16] are now written separately. By comparing [2.15] and these three equations with the matrix in [2.16], the student should observe how to write individual equations from the matrices used. The rule of matrix multiplication which is applied here, is that the first element in the column matrix on the left-hand side of [2.16], is the sum of each of the elements in the first row of the first matrix on the right-hand side multiplied by the corresponding element of the final column matrix of currents, i.e. (element of column 1) × (element of row 1) + (element of column 2) × (element of row 2) + ... etc. This is repeated for each row to build up the other elements of the left-hand side column matrix. This is all that is required at this stage; further application of matrix algebra to power system analysis is not dealt with in this Volume but appears in Volume 2. The last three equations of [2.16] are:

$$\mathbf{\Phi}_b = \frac{\mu_o}{2\pi}\left(\mathbf{I}_a \log_e \frac{1}{d_{ba}} + \mathbf{I}_b \log_e \frac{1}{r_m} + \mathbf{I}_c \log_e \frac{1}{d_{bc}} + \mathbf{I}_n \log_e \frac{1}{d_{bn}}\right)$$

$$\mathbf{\Phi}_c = \frac{\mu_o}{2\pi}\left(\mathbf{I}_a \log_e \frac{1}{d_{ca}} + \mathbf{I}_b \log_e \frac{1}{d_{cb}} + \mathbf{I}_c \log_e \frac{1}{r_m} + \mathbf{I}_n \log_e \frac{1}{d_{cn}}\right)$$

$$\mathbf{\Phi}_n = \frac{\mu_o}{2\pi}\left(\mathbf{I}_a \log_e \frac{1}{d_{na}} + \mathbf{I}_b \log_e \frac{1}{d_{nb}} + \mathbf{I}_c \log_e \frac{1}{d_{nc}} + \mathbf{I}_n \log_e \frac{1}{r'_m}\right)$$

$$\begin{bmatrix} \mathbf{\Phi}_a \\ \mathbf{\Phi}_b \\ \mathbf{\Phi}_c \\ \mathbf{\Phi}_n \end{bmatrix} = \frac{\mu_o}{2\pi} \log_e \begin{bmatrix} \frac{1}{r_m} & \frac{1}{d_{ab}} & \frac{1}{d_{ac}} & \frac{1}{d_{an}} \\ \frac{1}{d_{ba}} & \frac{1}{r_m} & \frac{1}{d_{bc}} & \frac{1}{d_{bn}} \\ \frac{1}{d_{ca}} & \frac{1}{d_{cb}} & \frac{1}{r_m} & \frac{1}{d_{cn}} \\ \frac{1}{d_{na}} & \frac{1}{d_{nb}} & \frac{1}{d_{nc}} & \frac{1}{r'_m} \end{bmatrix} \cdot \begin{bmatrix} \mathbf{I}_a \\ \mathbf{I}_b \\ \mathbf{I}_c \\ \mathbf{I}_n \end{bmatrix} \qquad [2.16]$$

It should be noted that the flux linking each conductor is due in part to the current in that conductor (a self-inductance effect), and in part to the current in all the other conductors (a mutual inductance effect). The diagonal elements (i.e.

$$\frac{1}{r_m}, \frac{1}{r_m}, \frac{1}{r_m} \text{ and } \frac{1}{r'_m}; \text{ or } \frac{1}{d_{aa}}, \frac{1}{d_{bb}}, \frac{1}{d_{cc}} \text{ and } \frac{1}{d_{nn}}$$

as they could be written) in [2.16] represent the self-inductance effects, and the remaining elements represent the mutual inductance effects.

The flux $\mathbf{\Phi}_a$ is that linked with the circuit composed of conductor a and conductor R, but since the latter conductor has no physical reality, the self-inductance of the circuit aR may be called the self-inductance of conductor a and this will be denoted by L_{aa}. Similarly the mutual inductance between the circuits aR and bR may be called the mutual inductance L_{ab} between conductors a and b.

The matrix of coefficients in [2.16] is symmetrical about the diagonal, because, provided the permeance of the magnetic circuit is constant, as it must be for non-magnetic conductors in air, the mutual inductance between two circuits is reciprocal, i.e. $L_{ab} = L_{ba}$, $L_{ac} = L_{ca}$, etc.

Since two conductors of a transmission line constitute effectively a single-turn circuit, [2.16] may then be written from [2.1], using L_{aa}, L_{bb}, etc. as symbols for self-inductance and L_{ab}, L_{bc} etc. as symbols for mutual inductance, as:

$$\begin{bmatrix} \mathbf{\Phi}_a \\ \mathbf{\Phi}_b \\ \mathbf{\Phi}_c \\ \mathbf{\Phi}_n \end{bmatrix} = \begin{bmatrix} L_{aa} & L_{ab} & L_{ac} & L_{an} \\ L_{ab} & L_{bb} & L_{bc} & L_{bn} \\ L_{ac} & L_{bc} & L_{cc} & L_{cn} \\ L_{an} & L_{bn} & L_{cn} & L_{nn} \end{bmatrix} \cdot \begin{bmatrix} \mathbf{I}_a \\ \mathbf{I}_b \\ \mathbf{I}_c \\ \mathbf{I}_n \end{bmatrix} \qquad [2.17]$$

The symmetry about the diagonal self-inductance terms should be noted.

In most cases, the neutral wire n will be an aerial earth wire running above the three phase conductors in order to afford some shielding from direct lightning strokes (see Volume 2), and need not be taken into account as far as inductance is concerned. This is because for both positive- and negative-sequence currents (see Chapter 8) in the line conductors, the sum of the three line currents is zero, and the earth wire carries no neutral current, except that caused to flow by induction due to the combined magnetic field of the three line conductors. This current will be very small because, as [2.2] shows, the magnetic field produced by a single line current is inversely proportional to distance from the conductor axis, and at the earth wire the m.m.f.s. due to three positive- or negative-sequence line currents will have 120° phase displacement from one another and will be nearly equal. Thus the resultant m.m.f. will be relatively small. The earth plane can also be neglected when calculating positive- or negative-sequence inductance, because even if it were a perfectly conducting plane, which it is not, it would only introduce a term $-\mu_o/2\pi \log_e(d'/2h)$ into [2.9] for a single-phase line. h is the height of the two conductors above earth, and d' is the distance from conductor 1 to the image of conductor 2. This extra term, which may be obtained using the method of images (Carter, 1967), is very small compared with the expression in [*2.9*].

When the three phase currents contain a zero-sequence current (see Chapter 8) due to an earth fault, the earth itself may offer a path of much lower impedance than the earth wire and this is dealt with in section 2.6.

A three conductor system will now be considered so that conductor n in Fig. 2.3 will be ignored.

2.4 Effective Conductor Self-inductance

Consider a 3-phase 3-wire system shown in Fig. 2.4, and assume that each pair of conductors *ab*, *bc* and *ca* taken in turn, acts as a go

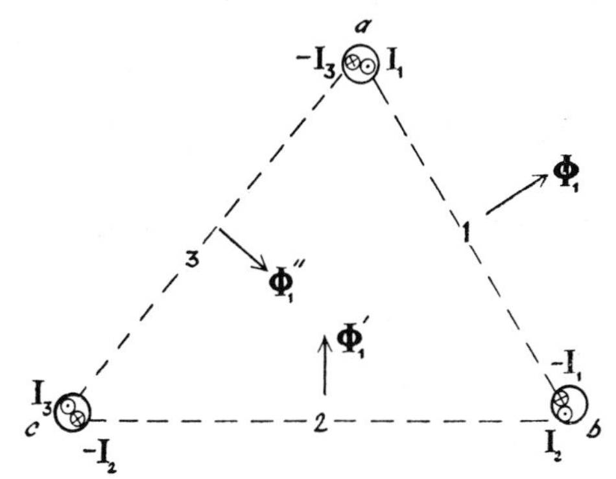

FIG. 2.4. *Three conductors shown as constituting three effective circuits.*

and return circuit, denoted by 1, 2 and 3 respectively, where each of these circuits supplies a current $\mathbf{I}_1$, $\mathbf{I}_2$ and $\mathbf{I}_3$ respectively to an equivalent delta-connected load. The actual currents in the three line conductors *a, b* and *c*, which flow into a load which may or may not be balanced, are $\mathbf{I}_a$, $\mathbf{I}_b$ and $\mathbf{I}_c$, where $\mathbf{I}_a = \mathbf{I}_1 - \mathbf{I}_3$, $\mathbf{I}_b = \mathbf{I}_2 - \mathbf{I}_1$ and $\mathbf{I}_c = \mathbf{I}_3 - \mathbf{I}_2$.

The self inductance L_{11} of circuit 1 is equal to the flux linkage set up in that circuit by 1 A flowing in a and -1 A flowing in b. The flux linkage set up in circuit 1 by 1 A flowing in a is $L_{aa} - L_{ab}$, and that set up by -1 A in b is $L_{bb} - L_{ab}$.

Thus $L_{11} = L_{aa} + L_{bb} - 2L_{ab}$ and from [*2.16*] and [*2.17*]

$$L_{11} = \frac{\mu_o}{\pi} \log_e \frac{d_{ab}}{r_m} \text{ H/m.}$$

The mutual inductance between circuits *ab* and *ac* = L_{13}. The flux linkage set up in *ac* by 1 A flowing in conductor *a* is $L_{aa}-L_{ac}$. The flux linkage set up in *ac* by -1 A flowing in conductor *b* is $L_{bc}-L_{ab}$. Thus

$$L_{13} = L_{aa}+L_{bc}-L_{ab}-L_{ac}. \qquad [2.18]$$

When $\mathbf{I}_1$ flows in *a* and $-\mathbf{I}_1$ in *b* and $\mathbf{I}_1 = 1$ A, then fluxes of L_{11}, L_{12} and L_{13} are set up in circuits 1, 2 and 3 in the directions shown by the arrows in Fig. 2.4, i.e. $\mathbf{\Phi}_1 = L_{11}$, $\mathbf{\Phi}_1' = L_{12}$ and $\mathbf{\Phi}_1'' = L_{13}$, and $L_{11} = L_{12}+L_{13}$ since there is no other flux path.

When currents $\mathbf{I}_1$, $\mathbf{I}_2$ and $\mathbf{I}_3$ flow as shown in the three circuits the flux linked with circuit 1 is $L_{11}\,\mathbf{I}_1-L_{12}\mathbf{I}_2-L_{13}\mathbf{I}_3$, since positive current in circuit 2 or 3 sets up negative flux (i.e. with a component in the direction opposite to that of the arrow on flux $\mathbf{\Phi}_1$ set up by the current flow shown in circuit 1 and taken as positive), linked with circuit 1 as shown in Fig. 2.4. Replacing L_{11} by $L_{12}+L_{13}$, the flux linked with circuit 1 is

$$L_{12}(\mathbf{I}_1-\mathbf{I}_2)+L_{13}(\mathbf{I}_1-\mathbf{I}_3) = -L_{12}\mathbf{I}_b+L_{13}\mathbf{I}_a.$$

Thus e.m.f. is induced in circuit *ab* as though conductor *a* were a single-turn circuit of inductance L_{13} and conductor *b* were a single-turn circuit of inductance L_{12}, without any mutual inductance between *a* and *b*.

A *3-phase transmission line*, whether the currents are balanced or not provided that $\mathbf{I}_a+\mathbf{I}_b+\mathbf{I}_c = 0$, can therefore be treated as though *each phase conductor has an effective self-inductance.*

$$\begin{aligned} &\text{the effective self-inductance of conductor } a = L'_{aa} = L_{13} \\ &\text{the effective self-inductance of conductor } b = L'_{bb} = L_{12} \qquad [2.19] \\ &\text{the effective self-inductance of conductor } c = L'_{cc} = L_{23} \end{aligned}$$

From [2.18]

$$L'_{aa} = L_{13} = L_{aa}+L_{bc}-L_{ab}-L_{ac}$$

and from [2.16] and [2.17] values may be substituted for the terms on the right-hand side, giving

$$L'_{aa} = \frac{\mu_o}{2\pi}\log_e\frac{d_{ab}d_{ac}}{r_m d_{bc}}\ \text{H/m}. \qquad [2.20]$$

Similarly

$$L'_{bb} = \frac{\mu_o}{2\pi}\log_e\frac{d_{ab}d_{bc}}{r_m d_{ac}}\ \text{H/m}. \qquad [2.21]$$

$$L'_{cc} = \frac{\mu_o}{2\pi} \log_e \frac{d_{bc} d_{ac}}{r_m d_{ab}} \text{ H/m.} \qquad [2.22]$$

When the phase conductors are not at the corners of an equilateral triangle, and in practice they will not be so placed, the effective self-inductance of each phase differs from that of the others, as shown by equations [*2.20*], [*2.21*] and [*2.22*].

2.5 Positive- or Negative-sequence Inductance of a Transposed Single-circuit 3-phase Line

If the line is sufficiently long and the unbalance in the inductance (and capacitance) between phases is so large as to justify the cost, the line may be transposed. In this case the phase conductors have their positions interchanged at special transposition towers, and the conductor of each phase occupies each of the three positions 1, 2 and 3 for about one-third of its length, as illustrated in Fig. 2.5.

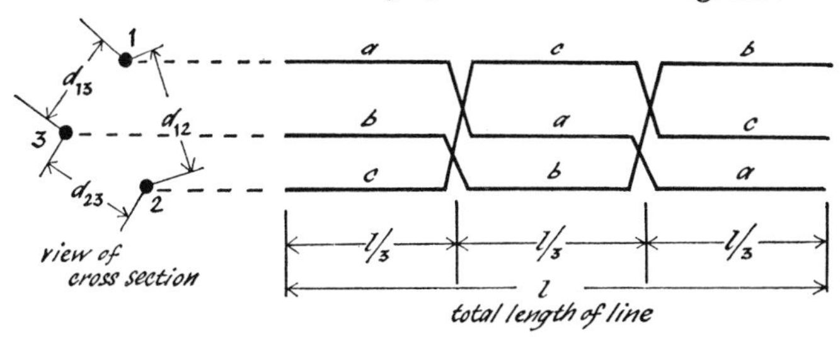

FIG. 2.5. *Three transposed conductors.*

The flux linkages Φ_a/metre of route length of phase conductor a, averaged over its whole length l, can be written from [*2.16*], where the spacings must now be written as d_{12}, d_{23} and d_{13} in place of d_{ab}, d_{bc} and d_{ac} since the latter now vary.

$$\Phi_a = \frac{\mu_o}{2\pi}\frac{1}{3}\left(\mathbf{I}_a \log_e \frac{1}{r_m} + \mathbf{I}_b \log_e \frac{1}{d_{13}} + \mathbf{I}_c \log_e \frac{1}{d_{12}} + \right.$$

$$\mathbf{I}_a \log_e \frac{1}{r_m} + \mathbf{I}_b \log_e \frac{1}{d_{23}} + \mathbf{I}_c \log_e \frac{1}{d_{13}} +$$

$$\left. \mathbf{I}_a \log_e \frac{1}{r_m} + \mathbf{I}_b \log_e \frac{1}{d_{12}} + \mathbf{I}_c \log_e \frac{1}{d_{23}} \right)$$

$$= \frac{\mu_o}{2\pi}\left(\mathbf{I}_a \log_e \frac{1}{r_m} + (\mathbf{I}_b + \mathbf{I}_c) \log_e \frac{1}{\sqrt[3]{(d_{12} d_{23} d_{13})}}\right).$$

Providing there is no earth fault on the system and only positive- and negative-sequence currents flow, $(\mathbf{I}_b+\mathbf{I}_c) = -\mathbf{I}_a$ and

$$\mathbf{\Phi}_a = \frac{\mu_o}{2\pi}\mathbf{I}_a \log_e \frac{D_m}{r_m}$$

where

$$D_m = \sqrt[3]{(d_{12}d_{23}d_{13})}$$

and is called the geometric mean spacing or distance between the conductors.

Thus the effective self-inductance of each phase conductor of a 3-phase line, which is independent of the direction of phase rotation and is therefore the same for both positive- and negative-sequence currents (see Chapter 8), is given by:

$$L = \frac{\mu_o}{2\pi} \log_e \frac{D_m}{r_m} \text{ H/m.} \qquad [2.23]$$

This expression may generally be used, even if the line is not transposed, as it commonly is not, since then the unbalance between L'_{aa}, L'_{bb} and L'_{cc} is small, and the approximation is justified of taking [*2.23*] which is the average of the values of L'_{aa}, L'_{bb} and L'_{cc} as may be seen from [*2.20*], [*2.21*] and [*2.22*].

Since D_m/r_m is of the order of 300 to 400 for very high voltage lines, the inductance is about 1·2 mH/km for each of the three conductors of a 3-phase line.

It may be noted that [*2.23*] could be written as

$$L = \frac{\mu_o}{2\pi} \log_e \frac{1}{r_m} + \frac{\mu_o}{2\pi} \log_e D_m.$$

The first term could then be termed the 'inductance at 1 metre spacing', which may be encountered in some tables, and depends only on the conductor and frequency. The second term in the reactance is then independent of the type of conductor and involves spacing and frequency.

2.6 Zero-sequence Inductance of a Transposed Single-circuit 3-phase Line

It is suggested that students may wish to study some parts of Chapter 8 before reading this section.

For zero-sequence current

$$\mathbf{I}_a = \mathbf{I}_b = \mathbf{I}_c = \mathbf{I}_o.$$

If the return current $3\mathbf{I}_o$ is assumed to flow entirely in the earth, and not in an aerial earth wire strung above the phase conductors, then this current $3\mathbf{I}_o$ can be assumed to flow in an equivalent earth return conductor at a depth D_e below the phase conductors. Since $D_e \gg d_{ab}$, d_{bc} or d_{ac} the distance from all three phase conductors to this equivalent earth conductor may be considered to be D_e.

It has been found that

$$D_e \simeq k\sqrt{(\rho/f)}$$

where ρ is the earth resistivity which is very variable, e.g.

Nature of ground	ρ (*order of magnitude*)
swampy ground	10–100 ohm metres
dry earth	1000 ohm metres
slate	10^7 ohm metres
sandstone	10^9 ohm metres

f is the supply frequency
for 50-Hz supply

$$D_e \simeq 93\rho^{\frac{1}{2}} \text{ metres.}$$

For the first of the transposed positions shown in Fig. 2.5, i.e. when $d_{ab} = d_{13}$, $d_{ac} = d_{12}$, $d_{bc} = d_{23}$, the flux linkages with the three conductors may be written from [*2.16*] as

$$\begin{bmatrix} \mathbf{\Phi}'_a \\ \\ \mathbf{\Phi}'_b \\ \\ \mathbf{\Phi}'_c \end{bmatrix} = \frac{\mu_o}{2\pi} \log_e \begin{bmatrix} \frac{1}{r_m} & \frac{1}{d_{13}} & \frac{1}{d_{12}} & \frac{1}{D_e} \\ \frac{1}{d_{13}} & \frac{1}{r_m} & \frac{1}{d_{23}} & \frac{1}{D_e} \\ \frac{1}{d_{12}} & \frac{1}{d_{23}} & \frac{1}{r_m} & \frac{1}{D_e} \end{bmatrix} \cdot \begin{bmatrix} \mathbf{I}_o \\ \\ \mathbf{I}_o \\ \\ \mathbf{I}_o \\ \\ =-3\mathbf{I}_o \end{bmatrix}$$

In the second transposed position, i.e. when $d_{ab} = d_{23}$ etc.

$$\mathbf{\Phi}''_a = \frac{\mu_o \mathbf{I}_o}{2\pi}\left(\log_e \frac{1}{r_m} + \log_e \frac{1}{d_{23}} + \log_e \frac{1}{d_{13}} - 3 \log_e \frac{1}{D_e}\right)$$

and in the third position

$$\mathbf{\Phi}'''_a = \frac{\mu_o \mathbf{I}_o}{2\pi}\left(\log_e \frac{1}{r_m} + \log_e \frac{1}{d_{12}} + \log_e \frac{1}{d_{23}} - 3 \log_e \frac{1}{D_e}\right).$$

The average flux linkages with phase a conductor are

$$\frac{1}{3}(\Phi'_a+\Phi''_a+\Phi'''_a)=\frac{1}{3}\frac{\mu_o\mathbf{I}_o}{2\pi}\left(3\log_e\frac{D_e}{r_m}+2\log_e\frac{D_e^3}{d_{12}d_{23}d_{13}}\right)$$

$$=\frac{1}{3}\frac{\mu_o}{2\pi}\mathbf{I}_o\log_e\frac{D_e^9}{r_m^3D_m^6}.$$

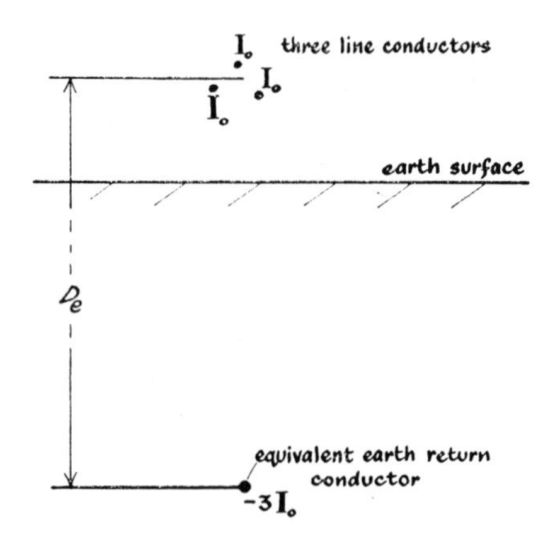

FIG. 2.6. *Three line conductors carrying zero-sequence current with the equivalent earth return conductor.*

Thus the zero-sequence inductive reactance of each phase conductor is

$$X_o=2\pi f\left(3\frac{\mu_o}{2\pi}\log_e\frac{D_e}{\sqrt[3]{(r_mD_m^2)}}\right)\Omega/\text{m}. \qquad [2.24]$$

It can be seen from [*2.24*] that the zero-sequence series reactance of a transmission line varies appreciably with the earth resistivity. The zero-sequence reactance is usually between 2 and 3·5 times the positive-sequence series reactance X_1, for single-circuit lines, since

$$\log_e\frac{D_e}{\sqrt[3]{(r_mD_m^2)}}$$

does not generally differ greatly from $\log_e D_m/r_m$ (cf equation [*2.23*]). It will be lower when there is an earth wire than when there is not, as this effectively reduces D_e and tends to reduce the magnetic field set up by the current in the phase conductors. If, for example, there is a non-magnetic earth wire, X_o will be about $2X_1$ and if there is no earth wire it will be about $3{\cdot}5X_1$. For double-circuit lines X_o will be about $3X_1$ and $5{\cdot}5X_1$ respectively for these two cases.

2.7 Stranded and Steel-cored Line Conductors

The line conductors have been assumed in the preceding sections to be a single strand of homogeneous non-magnetic material. They will in fact be non-homogeneous and frequently contain steel. Multi-stranded copper or very much more commonly steel-cored aluminium conductors may be used. The latter consists of steel strands wound with alternate layers spiralled in opposite directions, forming a core on which aluminium strands are similarly wound.

When each phase conductor is composed of a number n of identical non-magnetic filaments or strands, each of which is assumed to carry the same current, then writing down the flux linkages for each filament, taking the average inductance of one filament, and connecting them in parallel gives an average inductance for each phase conductor which is that given by [*2.23*].

However, the geometric mean distance D_m is now written as the n^2 root of the product of the distances from all n filaments of one phase conductor to all n filaments of the conductor of another phase. If the three phase conductors are not at the corners of an equilateral triangle, then taking the n^2 root for each pair would give three different distances and D_m would be the cube root of their products as before.

The G.M.R. or self geometric mean distance between the filaments making up the conductor, is no longer $r\,e^{-\frac{1}{4}}$, i.e. $0{\cdot}779\,r$, as it is for a single non-magnetic strand. Instead it is the n^2 root of the product for all n filaments in one conductor, of the geometric mean radius $r_m = 0{\cdot}779\,r$ for each strand, multiplied by the distances from that strand to all others in the conductor.

This is illustrated for the 7-stranded copper or aluminium conductor shown in Fig. 2.7, where the geometric mean radius

$$r'_m = \sqrt[49]{([r_m(2r)^3(2\sqrt{3}r)^2 4r]^6[r_m(2r)^6])}$$

$$r'_m = r\sqrt[49]{(2^{48}3^6 0{\cdot}779^7)}$$

$$r'_m = 2{\cdot}177r$$

i.e. the geometric mean radius for this conductor is 0·7255 × overall radius.

As the number of strands increases, the ratio approaches the value 0·779 which is that for a solid conductor, e.g. for 3 strands the ratio is 0·678 and for 169 strands it is 0·777.

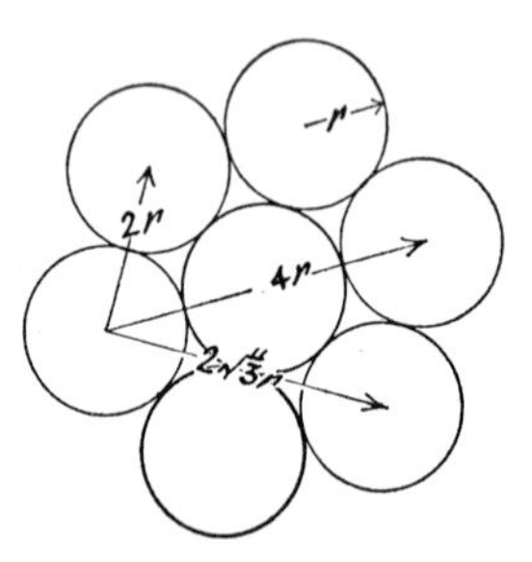

FIG. 2.7. *Cross-section of conductor with seven strands.*

The inductance of steel-cored aluminium cables is sometimes calculated neglecting the steel strands. If they are considered, they are assumed to carry no current at first and corrections are then made for current in the steel and for spiralling, with an assumed permeability. The results of such calculations are available in standard tables (e.g. those given by Butterworth (1954)).

2.8 Bundle or Multiple Conductors

The extension of the grid system at higher voltage levels has led to the introduction of bundle conductors, where each phase has two or more conductors: e.g. the 275-kV system has two conductors 30 cm apart, and 400-kV lines may have four conductors per phase at the corners of a 30-cm square.

These bundle conductors not only reduce corona loss and interference with other systems such as communications circuits, but because the effective G.M.R., or self-geometric mean distance for each phase, is increased the inductance is reduced. The G.M.R. for each of the steel-cored aluminium conductors may be calculated as indicated in the previous section and the effective G.M.R. for the phase may now be found in the same way. If, for example, each phase has two conductors each having a G.M.R. of r_m' metres and these two conductors are S metres apart, the G.M.R. is $\sqrt{(r_m' S)}$. Since S is of the order of 15 to 20 times the conductor radius, the G.M.R. is increased by a factor of about 4. If four conductors are used per phase at the corners of a square of side S, the G.M.R.

$$= \sqrt[4]{(r'_m \times S \times S \times \sqrt{2}S)} \simeq 8r'_m.$$

The positive-sequence inductance/phase of a line with two conductors/phase obtained from [2.23] is

$$L = \frac{\mu_o}{2\pi} \log_e \frac{D_m}{\sqrt{(r'_m S)}} \text{ H/m} \qquad [2.25]$$

and if it is assumed that D_m is virtually the same for a corresponding line with single conductor/phase (i.e. with same total cross-section) then the latter would have an inductance of

$$\frac{\mu_o}{2\pi} \log_e \frac{D_m}{\sqrt{2}r'_m} \text{ H/m.} \qquad [2.26]$$

Subtracting [2.25] from [2.26] gives

$$\frac{\mu_o}{2\pi} \log_e \frac{\sqrt{(r'_m S)}}{\sqrt{2}\, r'_m}$$

and the reduction as a percentage of the inductance of the single conductor value is

$$\frac{\log_e \dfrac{\sqrt{(r'_m S)}}{\sqrt{2}\, r'_m}}{\log_e \dfrac{D_m}{\sqrt{2}\, r'_m}} \times 100$$

Evaluation of this reduction for a 275 kV line with two 2 cm diameter conductors/phase spaced 30 cm apart, with a mean interphase spacing of 7 m, shows the reduction to be about 20%.

In obtaining [2.26], the assumption was made that D_m was the same for a line with two conductors/phase, as it was for a line with only one conductor/phase. This is in fact very nearly true, but D_m is given by

$$D_m = \sqrt[3]{(d_{ab} d_{ac} d_{bc})}$$

where

$$d_{ab} = \sqrt[4]{(d_{a_1b_1} d_{a_1b_2} d_{a_2b_1} d_{a_2b_2})}$$

$$d_{ac} = \sqrt[4]{(d_{a_1c_1} d_{a_1c_2} d_{a_2c_1} d_{a_2c_2})}$$

and

$$d_{bc} = \sqrt[4]{(d_{b_1c_1} d_{b_1c_2} d_{b_2c_1} d_{b_2c_2})}$$

$d_{a_1b_1}$, $d_{a_1b_2}$ are the spacings between conductors of different phases, as illustrated in Fig. 2.8, where for simplicity the conductors are all shown in one horizontal plane.

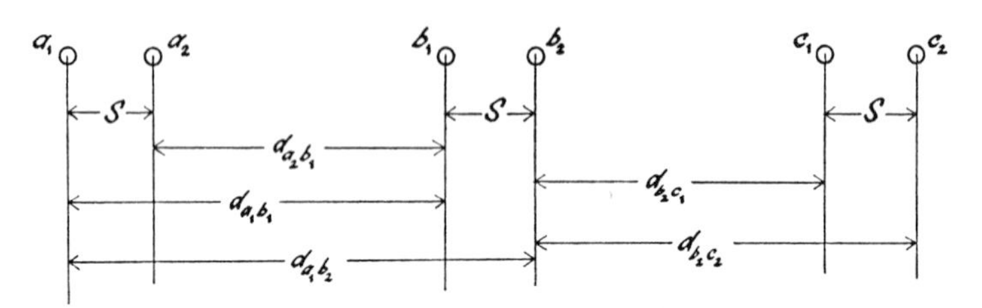

FIG. 2.8. *Single-circuit line with two conductor/phase in one horizontal plane.*

2.9 Double-circuit 3-phase Lines

Transmission towers sometimes carry two 3-phase circuits operating in parallel. If the lines had been so far apart that mutual effects could be neglected, e.g. on separate towers some distance from each other, or were completely transposed (i.e. No. 1 circuit is completely transposed to its three positions with No. 2 circuit in the original position, then No. 2 is transposed once followed by a cycle of transposition of No. 1 and following another transposition of No. 2, the first circuit is again transposed three times), then the inductance of the two circuits would be exactly half that of one of them alone. Similarly, the capacitance would be twice that value, and no power would be transferred between the two circuits.

Equal frequency transposition is however more likely to be used because it is cheaper and easier than complete transposition. In this method both circuits are transposed together, i.e. the current originally in the conductor in position 1 (see Fig. 2.5) flows for equal successive lengths in the conductors in positions 2, 3, 1, etc. Similarly the currents in the other five conductors are advanced or retarded by 120° at each transposition. Each phase of circuit No. 1 has in turn and for equal lengths, the reactances and resistances appropriate to positions *a*, *b* and *c*, and so do the phases of circuit No. 2. Thus over any multiple of three transpositions, the reactances and resistances of all phases in any one circuit are the same.

If within a single un-transposed 3-phase circuit the series impedance of each of the conductors is considered, it is found to contain resistive terms of the form $k \log_e d_{12}/d_{13}$ which represent power transfer from one phase to another. These terms clearly sum to zero over the three phases. In a similar way, power transfer resistances arise in double-circuit lines unless they are completely transposed. A small amount of power is transferred from one circuit to the other, and the inductance may be increased and the capacitance decreased or vice versa, depending upon which of a number of methods of spacing and transposition is adopted.

Conductors a_1, b_1 and c_1 are the three conductors of circuit No. 1 and a_2, b_2 and c_2 are those of circuit No. 2. In the arrangement shown

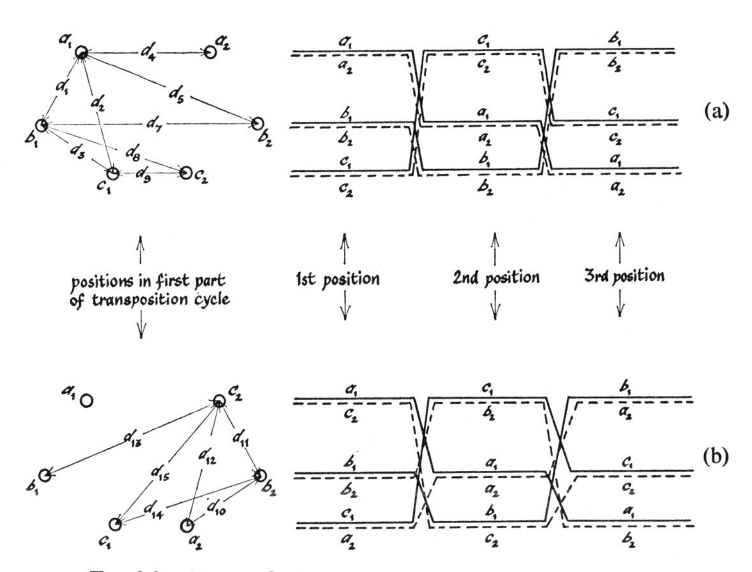

FIG. 2.9. *Two methods of transposing a double-circuit line.*

in Fig. 2.9(a), the conductors at the same level are always those of the same phase. Symmetry about a vertical plane through the tower centre, shows that conductors of the same phase will have similar currents and charges.

If, on the other hand, the two conductors of each phase are separated as widely as possible from each other, the self geometric mean distance r_m' for that phase of the two lines operating in parallel is increased. If the distances between the different phases are kept as small as possible, then the geometric mean spacing D_m is reduced.

Both of these effects which reduce line inductance can be better achieved by the arrangement shown in Fig. 2.9(b).

For either of the arrangements shown in Fig. 2.9, the positive-sequence inductance can be calculated using [*2.23*], if D_m is replaced by an equivalent spacing

$$D_{eq} = \sqrt[3]{(D_{ab}D_{bc}D_{ca})} \qquad [2.27]$$

and if r_m is replaced by a self G.M.D.

$$r_{eq} = \sqrt[3]{(r_1 r_2 r_3)} \qquad [2.28]$$

where D_{ab} = mutual geometric mean spacing between phases a and b in position 1, i.e. for Fig. 2.9(a),

$$D_{ab} = \sqrt[4]{(d_1 d_5 d_{11} d_{13})}$$

or for Fig. 2.9(b)

$$D_{ab} = \sqrt[4]{(d_1 d_5 d_8 d_{10})}.$$

Similarly for phases b and c in position 1, for Fig. 2.9(a),

$$D_{bc} = \sqrt[4]{(d_3 d_8 d_{10} d_{14})}$$

or for Fig. 2.9(b),

$$D_{bc} = \sqrt[4]{(d_3 d_{11} d_{13} d_{14})}$$

and for phases c and a in position 1, for Fig. 2.9(a),

$$D_{ca} = \sqrt[4]{(d_2 d_6 d_{12} d_{15})}$$

or for Fig. 2.9(b),

$$D_{ca} = \sqrt[4]{(d_2 d_4 d_9 d_{12})}.$$

Thus for Fig. 2.9(a),

$$D_{eq} = \sqrt[12]{(d_1 d_2 d_3 d_5 d_6 d_8 d_{10} d_{11} d_{12} d_{13} d_{14} d_{15})} \qquad [2.29]$$

or for Fig. 2.9(b),

$$D_{eq} = \sqrt[12]{(d_1 d_2 d_3 d_4 d_5 d_8 d_9 d_{10} d_{11} d_{12} d_{13} d_{14})}. \qquad [2.30]$$

If the two sets of conductor positions are symmetrical about a vertical centre-line of the tower as is usual, then the above expressions reduce to, for Fig. 2.9(a),

$$D_{eq} = \sqrt[6]{(d_1 d_2 d_3 d_5 d_6 d_8)} \qquad [2.31]$$

or for Fig. 2.9(b),

$$D_{eq} = \sqrt[12]{(d_1^2 d_2^2 d_3^2 d_4 d_5^2 d_8^2 d_9)}. \qquad [2.32]$$

It should be noted that D_{eq} has the same value for each case in the other positions 2 and 3 of the transposition cycles shown in Fig. 2.9.

If the self G.M.D. or G.M.R. of a single conductor is r_m', then for the phase a consisting of conductors a_1 and a_2, and in the first position of the transposition cycle, the self G.M.D. is, for Fig. 2.9(a),

$$r_1 = \sqrt{(r'_m d_4)}$$

or for Fig. 2.9(b),

$$r_1 = \sqrt{(r'_m d_6)}$$

Similarly in position 2 of the transposition cycle, the self G.M.D. of phase *a* is, for both Figs. 2.9(a) and (b),

$$r_2 = \sqrt{(r'_m d_7)}$$

and for position 3, the self G.M.D. of phase *a* is, for Fig. 2.9(a),

$$r_3 = \sqrt{(r'_m d_9)}$$

or for Fig. 2.9(b),

$$r_3 = \sqrt{(r'_m d_{15})}.$$

Therefore,

$$r_{eq} = \sqrt[3]{(r_1 r_2 r_3)}$$

for Fig. 2.9(a),

$$r_{eq} = (r'_m)^{\frac{1}{2}} \sqrt[3]{(d_4^{\frac{1}{2}} d_7^{\frac{1}{2}} d_9^{\frac{1}{2}})} \qquad [2.33]$$

or for Fig. 2.9(b),

$$r_{eq} = (r'_m)^{\frac{1}{2}} \sqrt[3]{(d_6^{\frac{1}{2}} d_7^{\frac{1}{2}} d_{15}^{\frac{1}{2}})} \qquad [2.34]$$

and these values apply equally to positions 2 and 3 of the transposition cycle.

Consider a 132-kV line with symmetry about a vertical plane and $d_1 = 4{\cdot}25$ m, $d_2 = 7{\cdot}75$ m, $d_3 = 3{\cdot}75$ m, $d_4 = 7{\cdot}5$ m, $d_5 = 9{\cdot}3$ m, $d_6 = 11{\cdot}4$ m, $d_7 = 9{\cdot}3$ m, $d_8 = 9{\cdot}15$ m, $d_9 = 7{\cdot}5$ m. The arrangement of Fig. 2.9(a) gives an inductance which is about 3 to 4% greater than if the circuits were on separate towers some way apart, and that of Fig. 2.9(b) gives a reduction of about 3 to 4% in the inductance. The arrangement of Fig. 2.9(b) is more often used than that of Fig. 2.9(a), so that the inductive reactance of two lines in parallel on the same towers is frequently about 4% less than half the reactance of one line alone.

REFERENCES

BUTTERWORTH, S. 1954. *Electrical characteristics of overhead lines.* E.R.A. Report O/T4.

CARTER, G. W. 1967. *The electromagnetic field in its engineering aspects.* Longmans Green, London, 2nd edition.

STEVENSON, W. D. 1962. *Elements of power system analysis.* McGraw-Hill (International students edition), New York, 2nd edition.
WADDICOR, H. 1964. *The principles of electric power transmission.* Chapman and Hall, London, 5th edition.

Examples

1. A 3-phase 132-kV line with balanced load has solid conductors of 2 cm diameter, spaced at distances between centres of 3·65, 5·5 and 8·2 m, and is transposed along its length. Calculate the positive-sequence inductance of each conductor per km.

(1·3 mH)

2. A 50-Hz, 3-phase transmission line is 16 km long and has solid conductors 2·5 cm diameter with a geometric mean spacing of 4·2 cm. The line may be considered to be transposed. Calculate the sending-end voltage when 100,000 kVA is delivered at zero power factor lagging and at a voltage of 132 kV. The line resistance and capacitance may be neglected.

(University of Leeds) (136·4 kV)

3. Show that the G.M.R. of a hollow conductor of inner radius r and outer radius R is given by $R\,e^{-K\mu_c}$, where μ_c is the relative permeability of the conductor material and

$$K = \frac{(R^4/4) - R^2r^2 + (3r^4/4) + r^4 \log_e R/r}{(R^2 - r^2)^2}$$

4. Calculate the zero-sequence inductive reactance/km for each conductor of a transposed 3-phase 50-Hz overhead line with solid conductors of 1 cm radius. The distances between the centres of the conductors are 7·25, 3·9 and 6·4 m and the depth of the equivalent earth conductor is 935 m.

(University of Leeds) (1·38Ω)

5. A 3-phase, 50-Hz transmission line has conductors composed of 12 strands of aluminium wound round a core of 7 strands of steel. Each aluminium and steel strand has a diameter of 0·28 cm. The geometric mean spacing between phase conductors is 6·1 m and the line may be treated as though it is transposed.

Allowing for the steel having an electrical conductivity which is 16% that of aluminium and that standard lays are used, tables show that the inductive reactance of each conductor of this line is 0·525Ω/km.

Calculate the percentage error which would arise if this inductance were calculated instead on the assumption that the steel strands carry no current.

(−17%)

6. A single-phase 50-Hz power transmission line has its conductors 3·05 m apart and at the same height above earth. The two conductors of a telephone circuit are 0·61 m apart and they are both at the same height above earth but 3·05 m below the power line. The horizontal distance between the central plane of the power line and the central plane of the telephone line is 6·1 m. If the lines run parallel to each other for one km, calculate the 50-Hz voltage induced in the telephone circuit, when the power line carries 500 A.

(5·1 volts)

7. A 3-phase, 50-Hz power transmission line has its conductors at the corners of an equilateral triangle 3·05 m apart. The two lower conductors are at the same height above earth and 3·65 m above the two conductors of a telephone line. These latter conductors are 0·61 m apart. If the lines run parallel to each other for one km, calculate the 50-Hz voltage induced in the telephone circuit, when the power line carries 500 A.

(3·15 volts)

8. A 3-phase transmission line with two conductors per phase is arranged as shown in Fig. 2·8, with $S = 30{\cdot}5$ cm and with the centre line of the b phase 6·1 m from the centre lines of the a and c phases. Each conductor may be regarded as being the equivalent of solid copper of 2·54 cm diameter. Calculate the average inductance/metre of each phase.

($9{\cdot}86 \times 10^{-7}$ H/m)

9. A solenoid is wound on a long former, square in section and containing no magnetic material. It is bent round into a toroid of internal and external radii r_1 and r_2. A straight, thin cable of infinite length passes along the axis of the toroid, at right angles to its plane.

Show that the mutual inductance between the cable and the solenoid is:

$$M = \mu_o n \frac{(r_2^2 - r_1^2)}{2} \log_e \frac{r_2}{r_1} \text{H}$$

where n = mean number of turns/metre.

Calculate the peak e.m.f. induced in the coil if $r_1 = 0{\cdot}09$ metre, $r_2 = 0{\cdot}11$ metre, $n = 1000$, and the sinusoidal current in the cable has a peak amplitude of 1000 A at 50-Hz.

(University of Leeds) (0·1575 volts)

(This arrangement of a toroidal coil around a conductor, known as a Rogowski coil, has sometimes been used for current measurement.)

10. A 3-phase transmission line carrying balanced 3-phase currents, consists of three equilaterally-spaced, parallel conductors. Show that at a point equidistant from all three conductors, there is a pure rotating magnetic field, and find an expression for the field strength and for its angular velocity of rotation.

Determine the value of the magnetic field strength when the spacing between conductors is 1 m and the line is carrying 18 MVA at a line voltage of 33 kV.

(L.C.T.) ($\phi = 1{\cdot}5\Phi_{max}\, e^{-j\omega t}$, where Φ_{max} is peak flux due to one conductor and $\omega = 2\pi \times$ supply frequency. 184·5 A/m)

Chapter 3

TRANSMISSION LINE CAPACITANCE

3.1 Introduction

Two overhead line conductors with air insulation between them, constitute a capacitance which, when connected to an alternating voltage supply, will take a charging current which will flow even under no-load conditions. This charging current will be greatest at the sending end and will diminish to zero at the receiving end. The capacitance is in effect a leading power-factor load on the line, and the effect of this on the steady-state performance of the line has been discussed in Chapter 1.

The line construction may for example consist of a double-circuit line with two conductors/phase, so that there are twelve conductors and one or two aerial earth wires and earth itself, all constituting a complicated capacitance system, comprising capacitances between pairs of line conductors and between line conductors and earth. However complicated it is though, an effective capacitance to earth may be obtained, and the chief principles involved in this reduction are discussed in this chapter. Apart from its role in the steady-state operation of the line as discussed in Chapter 1, this effective capacitance to neutral is of prime importance under transient conditions, such as over-voltages arising when opening or closing a circuit-breaker connected to a long unloaded line or when travelling waves of voltage occur in lines due for example to lightning (see Volume 2).

As an aid to visualising the distribution of electric field between two charged line conductors, lines of electric flux are drawn in Fig. 3.1 for two long parallel conductors with equal and opposite charges. Each pair of flux lines contain a definite amount of electric flux, so that for any point in the field there may be assigned a vector **D** which is the electric flux density.

At this point it is necessary to comment upon the notation used for vectors (i.e. quantities which vary in magnitude and direction according to the position in space being considered) and phasors (i.e. quantities which vary with time, and here the variation considered unless expressly stated otherwise, is a simple sinusoid). In this chapter and indeed throughout this book, alternating and not direct voltages and currents are dealt with, since no d.c. components of a power system are discussed (with the exception of cables, section 5.9). The bold-face type denotes an r.m.s. value of a quantity varying sinusoidally in time, e.g. **I**, and shows that it must in that context be treated as a phasor. When shown italicised, *I*, only the magnitude of the r.m.s. quantity is important (this must not be confused with the amplitude or peak value of the alternating quantity, viz. $\sqrt{2}\, I$). Since the charges in the system arise from alternating voltages they too are sinusoids, and vectors representing electric flux density and electric force are also varying sinusoidally in time. In this case therefore the symbols for these space vectors are also shown bold-face, **D** and **E**, and it must be remembered that they vary in time as well as in space. (The equations which follow in this section are more commonly written, e.g. in texts on electromagnetism, in terms of instantaneous and scalar quantities. Here, however, boldface type is used because it is necessary for developing the equations in following sections to work in alternating quantities.)

Gauss' theorem states that at any instant of time, the total electric flux through any closed surface A, is equal to the total charge enclosed by that surface, i.e. if $\mathbf{D}_n$ is the flux density normal to an element of surface of area da then

$$\int_A \mathbf{D}_n \, da = \Sigma \mathbf{Q}$$

where $\Sigma\mathbf{Q}$ is the net sum of all charges enclosed by the surface. Electric flux density **D** may therefore be measured in coulombs/m^2.

At any point in the electric field there is a scalar (i.e. a value which does not vary with direction, so that it has a unique magnitude at any given point in the field) potential V with respect to some datum zero potential. Since however, the potential varies sinusoidally in time its r.m.s. value is shown **V**. The potential difference between two points is defined as the work done in joules in moving a positive charge of one coulomb, from one point towards the other more positively-charged point, against the force of the electric field. The zero datum potential may be chosen arbitrarily, e.g. as the potential at infinity. In power transmission line work, it is taken to be that of earth which is assumed for this purpose to be perfectly conducting. The work is done in moving a charge against a force **E** which is a vector measured in volts/m, and may be termed electric force, electric field intensity, or potential gradient, since $\mathbf{E}_x = -\partial \mathbf{V}/\partial x$ where ∂x is an element of distance. The negative sign is a consequence of the fact that if x increases away from a positively-charged conductor, the potential decreases so that $\partial \mathbf{V}/\partial x$ is negative while **E**

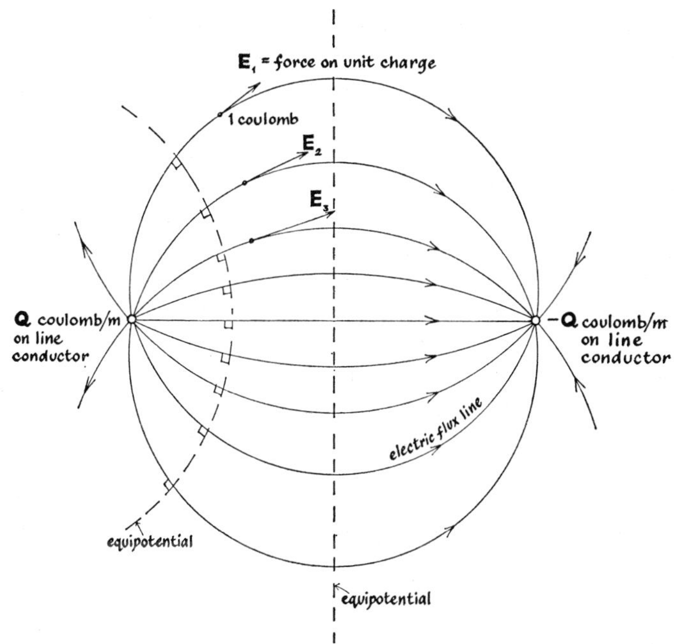

FIG. 3.1. *Electric field between two line conductors.*

is positive, since a positive charge experiences a force away from a positively-charged conductor. As illustrated in Fig. 3.1, the magnitude and direction of **E** varies from point to point in the field. Integrating the above equation gives $\mathbf{V} = \int \mathbf{E}\, \partial x$ which defines potential in terms of work done. There will be lines in the field on which the potential is the same at any instant and two of these lines, called equipotentials, are shown in Fig. 3.1. They cross the flux lines at 90°.

At any point in the field $\mathbf{D}/\mathbf{E} = \epsilon$ where $\epsilon = \epsilon_o \epsilon_r$ and

$$\epsilon_o = 1/(36\pi 10^9) = 8{\cdot}854 \times 10^{-12} \text{ F/m}$$

and is known as the permittivity of free space, or primary electric constant. ϵ_r is a dimensionless property of the medium in which the electric field is being examined, and is known as the relative permittivity or dielectric constant. For vacuum $\epsilon_r = 1{\cdot}00$, and for air at normal temperature and pressure it is sufficiently accurate for most purposes to take ϵ_r as unity.

If a single long straight conductor of radius r metres in air has a charge $\mathbf{Q}$ coulombs/metre of length, uniformly distributed over its surface, then at a position remote from its ends, a cylindrical surface of x metres radius ($x > r$) may be imagined to be co-axial with the conductor. Since the flux density vector at x, $\mathbf{D}_x$, is radial and normal to the surface and constant over it, then

$$\mathbf{D}_x = \frac{\mathbf{Q}}{2\pi x} \text{ C/m}^2$$

and

$$\mathbf{E}_x = \frac{\mathbf{Q}}{2\pi \epsilon_o x} \text{ V/m} \qquad [3.1]$$

$$= -\frac{\partial \mathbf{V}}{\partial x}.$$

Integration gives

$$\mathbf{V} = \frac{-\mathbf{Q}}{2\pi \epsilon_o} \log_e x + K \qquad [3.2]$$

where K is an arbitrary constant which depends upon the choice of the value of x at which the potential is zero.

If there are a number of long straight parallel conductors in air, then provided that their charges can be assumed to be distributed uniformly over their surfaces (i.e. that their distances from each other and from earth are large compared with their radii), the potential at a point which is distant x_1, x_2, x_3, etc., from the axes of the conductors carrying line charges $\mathbf{Q}_1$, $\mathbf{Q}_2$, $\mathbf{Q}_3$, etc./unit length can be written from [3.2] as

$$\mathbf{V}_x = -\frac{1}{2\pi \epsilon_o}[\mathbf{Q}_1 \log_e x_1 + \mathbf{Q}_2 \log_e x_2 + \mathbf{Q}_3 \log_e x_3 + \ldots] + K. \qquad [3.3]$$

Similarly the potential at another point distant y_1, y_2, y_3, etc. from the line charges is

$$\mathbf{V}_y = -\frac{1}{2\pi\epsilon_o}[\mathbf{Q}_1 \log_e y_1 + \mathbf{Q}_2 \log_e y_2 + \mathbf{Q}_3 \log_e y_3 + \ldots] + K,$$

i.e. the potential of x relative to y is

$$\mathbf{V}_{xy} = \frac{1}{2\pi\epsilon_o}\left[\mathbf{Q}_1 \log_e \frac{y_1}{x_1} + \mathbf{Q}_2 \log_e \frac{y_2}{x_2} + \mathbf{Q}_3 \log_e \frac{y_3}{x_3} + \ldots\right]. \qquad [3.4]$$

3.2 Two-wire Line

If there are only two conductors carrying charges $+\mathbf{Q}$ and $-\mathbf{Q}$ respectively then equation [*3.3*] gives the potential at any point in the field distant x_1 and x_2 from the two charges as

$$\mathbf{V} = -\frac{1}{2\pi\epsilon_o}(\mathbf{Q} \log_e x_1 - \mathbf{Q} \log_e x_2) = \frac{\mathbf{Q}}{2\pi\epsilon_o} \log_e \frac{x_2}{x_1}. \qquad [3.5]$$

At infinity $x_1 = x_2$ and $V = 0$, so that $K = 0$.

In [*3.5*] x_1 and x_2 must be larger than the actual radii of the conductors, since when x_1 or x_2 is less than the radius r of the conductors the potential is constant, whereas [*3.5*] gives it as a variable and if $x_1 = 0$ the potential is made to appear to be infinite. Equation [*3.5*] enables Fig. 3.1 to be drawn. The equipotential lines are such that x_2/x_1 = constant, and the flux lines are perpendicular to the equipotential lines (orthogonal system). The conductors cannot be taken as line charges of infinitely small radius because [*3.5*] only applies for $x > r$. In Fig. 3.2, the two parallel long conductors a and b, each of radius r metres are shown with their axes a distance d metres apart.

If there were only a charge $\mathbf{Q}$ on a and no charge on b, and b were infinitely thin, then the equipotentials would be circles, concentric with the axis of a. The metal conductor b of finite radius r distorts the field because it introduces a region $2r$ wide between equipotentials, all of which is at one potential. This potential may be taken to be that of a circular equipotential which would have passed through the axis of the conductor b if it were removed from the field or made infinitely thin. At a distance of several radii from the finite conductor this equipotential will become a circle concentric with the axis of a and distant d from it.

The potential difference between two points, since it is equal to the work done in moving a unit charge from one of the points to the other, is independent of the path taken in the field between the points. Thus the p.d. of a relative to b, $\mathbf{V}_{ab}$, may be found by integrating [3.1] between the surface of conductor a where $x = r$, to $x = d$ for the charge $\mathbf{Q}$ and adding the similar contribution from the charge $-\mathbf{Q}$ on conductor b. Alternatively, from [3.5] putting $x_1 = r$ and

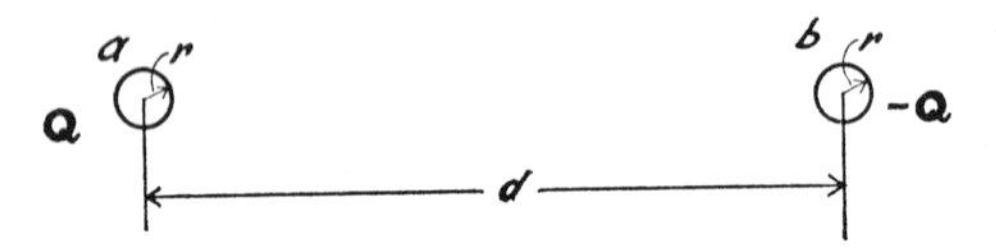

FIG. 3.2. *Two-wire line.*

$x_2 = d$, the potential of a, $\mathbf{V}_a$ due to its own charge $+\mathbf{Q}$/unit length and $-\mathbf{Q}$ on b is given by

$$\mathbf{V}_a = \frac{\mathbf{Q}}{2\pi\epsilon_o} \log_e \frac{d}{r} \qquad [3.6]$$

the potential of b, $\mathbf{V}_b$ due to both charges is

$$\mathbf{V}_b = \frac{\mathbf{Q}}{2\pi\epsilon_o} \log_e \frac{r}{d}$$

thus

$$\mathbf{V}_{ab} = \mathbf{V}_a - \mathbf{V}_b = \frac{\mathbf{Q}}{2\pi\epsilon_o} \log_e \frac{d^2}{r^2}$$

$$= \frac{\mathbf{Q}}{\pi\epsilon_o} \log_e \frac{d}{r}. \qquad [3.7]$$

This is the value given by [3.4] if x_1, the distance of conductor a from its charge is its radius, and y_2, the distance of conductor b from its charge is its radius.

The capacitance C/metre length between the two conductors is

$$C = \frac{Q}{V_{ab}} = \frac{\pi\epsilon_o}{\log_e d/r} \text{ F/m} \qquad [3.8]$$

There are several approximations involved in applying [3.8] to an actual transmission line:

(*a*) The charge distribution over the line conductor surface has been assumed to be uniform, and this is only strictly true if all other charged bodies are removed to infinity, i.e. the error becomes smaller as d/r is increased. For high-voltage transmission lines the ratio of the distance of the nearest conductor to its radius is often of the order of 300 to 400, and the error in capacitance is then negligibly small, e.g. for a 2-wire line with no other conductors or earth, equation [3.8] should strictly be replaced by

$$C = \frac{\pi\epsilon_o}{\log_e [d/2r + \sqrt{(d^2/4r^2 - 1)}]},$$

but Stevenson (1962) quotes data showing how the error increases as d/r decreases. Taking the 'worst' case of minimum d and maximum r, e.g. $d = 0{\cdot}61$ m and $r = 0{\cdot}63$ cm, then $d/r = 100$ and the error if [3.8] is used is only 0·002%.

(*b*) The distance d between the centres of the two conductors, has been used as one limit of integration rather than $(d-r)$, which is again justifiable when $d \gg r$.

(*c*) The actual overall radius of a stranded conductor may be used to replace r which has been used as the radius of a smooth cylindrical conductor. Since the disturbance in the electric field due to stranding occurs only very close to the conductor, this approximation is again justified when calculating the line capacitance, though stranding affects the onset of corona (section 3.8).

It is suggested that the student should derive a formula for the capacitance between the conductors of Fig. 3.2, assuming they are of different radii r_1 and r_2, which will show that in [3.8], r is replaced by $\sqrt{(r_1 r_2)}$.

The capacitance is proportional to the route length of the line so that the capacitive reactance is inversely proportional to length and so cannot be expressed in ohms/km. Instead the capacitive susceptance may be used and quoted in siemens/km.

3.3 General Multi-conductor System

Consider a number of (single) conductors which run parallel to themselves and to a perfectly conducting earth plane as shown in Fig. 3.3. (Bundle conductor systems, where each phase has more than one conductor are discussed in section 3.9.) The potential of the latter may be taken as zero, and if the conductors are initially uncharged they are also at zero potential.

If a charge $\mathbf{Q}_1$ is taken from earth to conductor 1 then the potential of all conductors will be proportional to $\mathbf{Q}_1$. Writing p as the constant of proportionality, the potential of conductor 1 relative to earth will be given by $V'_{10} = p_{11}Q_1$, whilst the potentials of all other conductors relative to earth will be given by $V'_{20} = p_{21}Q_1$, $V'_{30} = p_{31}Q_1$ and so on. The dimensions of p will be volts/coulomb, i.e. the reciprocal of capacitance.

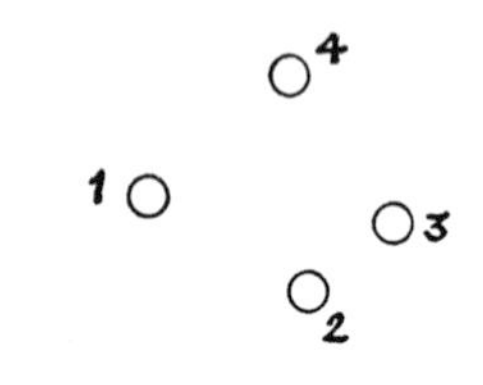

FIG. 3.3. *General multi-conductor system illustrated with four conductors only.*

Similarly if Q_2 is transferred from earth to conductor 2, the potential of 1 with respect to earth rises by $V''_{10} = p_{12}Q_2$, the potential of 2 with respect to earth rises by $V''_{20} = p_{22}Q_2$ and so on. Thus if charges Q_1, Q_2, Q_3 and Q_4 are transferred simultaneously from earth to conductors 1, 2, 3 and 4, then by superposition the potentials of the conductors relative to earth rise to

$$
\begin{aligned}
V_{10} &= p_{11}Q_1+p_{12}Q_2+p_{13}Q_3+p_{14}Q_4 \\
V_{20} &= p_{21}Q_1+p_{22}Q_2+p_{23}Q_3+p_{24}Q_4 \\
V_{30} &= p_{31}Q_1+p_{32}Q_2+p_{33}Q_3+p_{34}Q_4 \\
V_{40} &= p_{41}Q_1+p_{42}Q_2+p_{43}Q_3+p_{44}Q_4
\end{aligned}
$$

It will be seen that it is only necessary to write the first equation, as the others may be obtained by changing the first subscript in each double subscript.

Matrix notation enables the equations to be written in a more compact form as:

$$
\begin{bmatrix} V_{10} \\ V_{20} \\ V_{30} \\ V_{40} \end{bmatrix} = \begin{bmatrix} p_{11} & p_{12} & p_{13} & p_{14} \\ p_{21} & p_{22} & p_{23} & p_{24} \\ p_{31} & p_{32} & p_{33} & p_{34} \\ p_{41} & p_{42} & p_{43} & p_{44} \end{bmatrix} \cdot \begin{bmatrix} Q_1 \\ Q_2 \\ Q_3 \\ Q_4 \end{bmatrix} \qquad [3.9]
$$

The p coefficients are constant for a given configuration as they depend only on the geometry of the conductors and earth plane. This may be seen by comparing [3.9] with [3.4].

If the matrix of p coefficients in [3.9] is inverted (i.e. the equations are solved to give $\mathbf{Q}$ in terms of $\mathbf{V}$), another set of coefficients will be obtained which again must be dependent only on the size and position of the conductors: these new coefficients will have the dimensions of capacitance.

$$\begin{bmatrix} \mathbf{Q}_1 \\ \mathbf{Q}_2 \\ \mathbf{Q}_3 \\ \mathbf{Q}_4 \end{bmatrix} = \begin{bmatrix} c_{11} & c_{12} & c_{13} & c_{14} \\ c_{21} & c_{22} & c_{23} & c_{24} \\ c_{31} & c_{32} & c_{33} & c_{34} \\ c_{41} & c_{42} & c_{43} & c_{44} \end{bmatrix} \cdot \begin{bmatrix} \mathbf{V}_{10} \\ \mathbf{V}_{20} \\ \mathbf{V}_{30} \\ \mathbf{V}_{40} \end{bmatrix} \qquad [3.10]$$

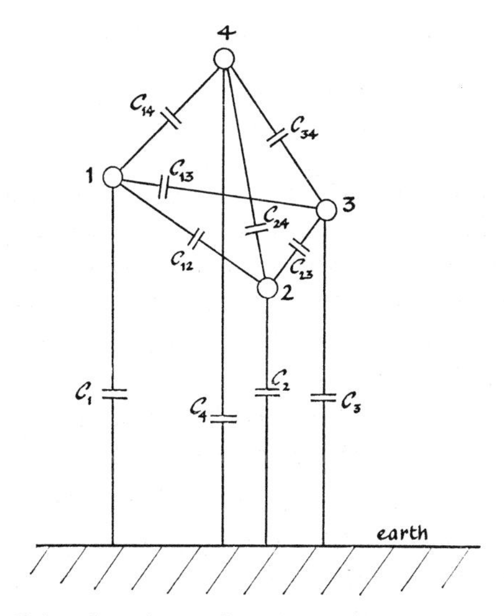

FIG. 3.4. *Capacitances between conductors and to earth.*

Since the energy stored in the electric field when these charges are transferred is independent of the order in which work is done to move the charges, then $p_{12} = p_{21}$ and $c_{12} = c_{21}$, i.e. the matrices in [3.9] and [3.10] are symmetrical about the leading diagonal (top left to bottom right). For three line conductors, i.e. for a single-circuit 3-phase line with earth conductor, the first equation in [3.10] may be re-written

$$\mathbf{Q}_1 = (c_{11}+c_{12}+c_{13}+c_{14})\mathbf{V}_{10}-c_{12}(\mathbf{V}_{10}-\mathbf{V}_{20})-$$
$$c_{13}(\mathbf{V}_{10}-\mathbf{V}_{30})-c_{14}(\mathbf{V}_{10}-\mathbf{V}_{40})$$

$$\mathbf{Q}_1 = C_1\mathbf{V}_{10}+C_{12}\mathbf{V}_{12}+C_{13}\mathbf{V}_{13}+C_{14}\mathbf{V}_{14} \qquad [3.11]$$

where

$$C_1 = (c_{11}+c_{12}+c_{13}+c_{14})$$

and

$$C_{12} = -c_{12}.$$

Similarly

$$\mathbf{Q}_2 = C_{12}\mathbf{V}_{21}+C_2\mathbf{V}_{20}+C_{23}\mathbf{V}_{23}+C_{24}\mathbf{V}_{24}$$

where

$$C_2 = (c_{21}+c_{22}+c_{23}+c_{24})$$

and $\mathbf{Q}_3$ and $\mathbf{Q}_4$ may similarly be written in this form. These four equations are those of a system of capacitors shown in Fig. 3.4, e.g. for conductor 1, its charge is made up of the four components in [*3.11*], viz. the p.d. between itself and earth multiplied by a capacitance C_1 to earth, and three components each the product of the p.d. between itself and another conductor multiplied by a capacitance between these two conductors.

3.4 Effect of Earth

If a thin conducting sheet is introduced mid-way between two similar cylindrical conductors in air carrying charges $\pm\mathbf{Q}$, and normal to the line joining their centres as shown in Fig. 3.5, then since it occupies a position which is an equipotential before placing it there, the electric field is not disturbed by its presence. From [*3.4*] the p.d. $\mathbf{V}_{10}$ between conductor 1 and the conducting sheet 0 is given by

$$\mathbf{V}_{10} = \frac{1}{2\pi\epsilon_0}\left(\mathbf{Q}_1 \log_e \frac{y_1}{x_1} + \mathbf{Q}_2 \log_e \frac{y_2}{x_2}\right).$$

y_1 and y_2 are the distances from any point on the conducting plane to the axes of the conductors carrying the charges $\mathbf{Q}_1$ and $\mathbf{Q}_2$ and these are equal for all points in the plane.

$$\mathbf{Q}_1 = +\mathbf{Q} \quad \text{and} \quad \mathbf{Q}_2 = -\mathbf{Q}$$

x_1 is the distance from the axis of conductor 1 to its own charge, which was shown (see equation [*3.7*]) to be equal to its radius, and

x_2 is the distance from the axis of conductor 1 to the axis of the conductor 2, i.e. $x_2 = d$.

Thus

$$V_{10} = \frac{Q}{2\pi\epsilon_o} \log_e \frac{d}{r} \qquad [3.12]$$

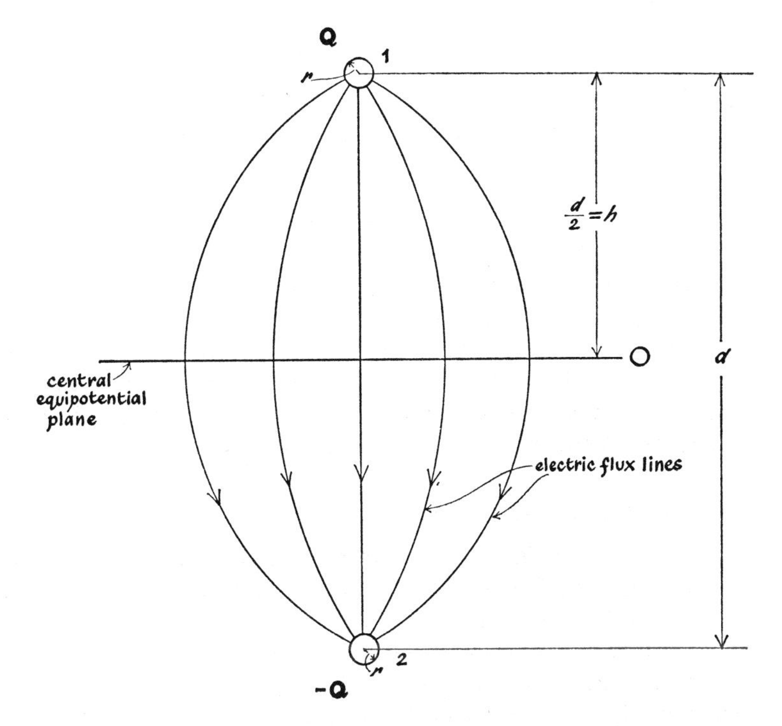

FIG. 3.5. *Central equipotential plane between the two conductors of a single-phase line, or earth plane with image charge of one line conductor.*

The capacitance from an air-insulated line conductor to a perfectly conducting plane h metres below it is therefore

$$C = \frac{2\pi\epsilon_o}{\log_e \dfrac{2h}{r}} \text{ F/m} \qquad [3.13]$$

This could have been deduced directly from [3.8], since the capacitance between the two conductors in Fig. 3.5 is that of two equal capacitors in series, one the capacitance of line 1 to a central conducting sheet and the other that of line 2 to the sheet, where each capacitor has a value of

$$\frac{2\pi\epsilon_o}{\log_e d/r} \text{ F/m.}$$

It may be noted that this value is then exactly similar to the capacitance to neutral of a 3-phase line given later in [*3.20*]. If the effect of earth on the electric field of transmission lines is required, this may be represented by an imaginary line conductor (image) carrying an equal and opposite charge to that of each conductor, and positioned as far below the earth surface as the real conductor is above it.

$p_{11}\mathbf{Q}_1$ was defined as the change in potential of conductor 1 due to its acquisition of a charge $\mathbf{Q}_1$/metre length. It can now be seen that if the conductor is h_1 metres above the earth, this potential change is due to the charge $\mathbf{Q}_1$ on the conductor and the image charge $-\mathbf{Q}_1$ vertically below the conductor and h_1 metres below earth surface, i.e. $2h_1$ metres below conductor 1. Thus from [*3.13*]

$$p_{11}\mathbf{Q}_1 = \frac{\mathbf{Q}_1}{\dfrac{2\pi\epsilon_o}{\log_e 2h_1/r}}$$

and

$$p_{11} = \frac{\log_e (2h_1/r)}{2\pi\epsilon_o} \qquad [3.14]$$

Similarly $p_{21}\mathbf{Q}_1$ is the potential of conductor 2 due to the charge $\mathbf{Q}_1$ on conductor 1/metre length, at a distance d_{12}, and the image charge $-\mathbf{Q}_1$ distant $d_{1'2}$. From [*3.3*]

$$p_{21}\mathbf{Q}_1 = -\frac{\mathbf{Q}_1}{2\pi\epsilon_o}(\log_e d_{12} - \log_e d_{1'2}) = \frac{\mathbf{Q}_1}{2\pi\epsilon_o}\log_e \frac{d_{1'2}}{d_{12}}$$

and therefore

$$p_{21} = \frac{\log_e (d_{1'2}/d_{12})}{2\pi\epsilon_o}. \qquad [3.15]$$

3.5 General Single-circuit 3-phase Line

Fig. 3.6 shows a single-circuit three-phase line without earth conductor. Each line conductor *a*, *b* and *c* has its image conductor carrying equal and opposite charge. The distance between a phase conductor and its image is twice the height of the conductor above earth, e.g. $d_{aa'} = 2h_a$, etc. All phase conductors have the same radius r. It can be seen by symmetry that $d_{ab'} = d_{ba'}$, etc.

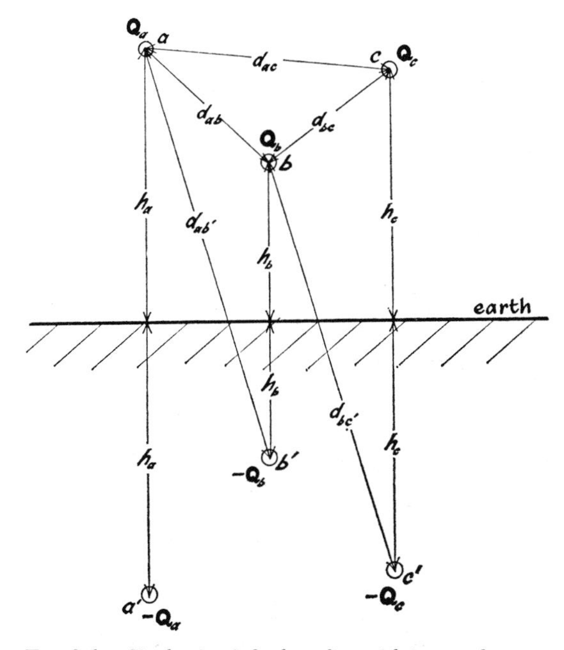

FIG. 3.6. *Single-circuit 3-phase line with image charges.*

Taking earth potential as zero, the potential of a conductor, say the a phase conductor, is due to:

(1) charges $\mathbf{Q}_a$ and $-\mathbf{Q}_a$ on a and a'
(2) charges $\mathbf{Q}_b$ and $-\mathbf{Q}_b$ on b and b'
(3) charges $\mathbf{Q}_c$ and $-\mathbf{Q}_c$ on c and c'

The contribution of (1) may be seen from [*3.14*] to be

$$\frac{\mathbf{Q}_a}{2\pi\epsilon_o} \log_e \frac{2h_a}{r}.$$

(2) contributes

$$\frac{\mathbf{Q}_b}{2\pi\epsilon_o} \log_e \frac{d_{ab'}}{d_{ab}}$$

(see equation [3.15]), and (3) contributes

$$\frac{\mathbf{Q}_c}{2\pi\epsilon_o}\log_e\frac{d_{ac'}}{d_{ac}}.$$

Thus the potential of the a phase conductor relative to earth is given by:

$$\left.\begin{aligned}
\mathbf{V}_a &= \frac{1}{2\pi\epsilon_o}\left[\mathbf{Q}_a\log_e\frac{2h_a}{r}+\mathbf{Q}_b\log_e\frac{d_{ab'}}{d_{ab}}+\mathbf{Q}_c\log_e\frac{d_{ac'}}{d_{ac}}\right]\\
&\text{Similarly}\\
\mathbf{V}_b &= \frac{1}{2\pi\epsilon_o}\left[\mathbf{Q}_a\log_e\frac{d_{ab'}}{d_{ab}}+\mathbf{Q}_b\log_e\frac{2h_b}{r}+\mathbf{Q}_c\log_e\frac{d_{bc'}}{d_{bc}}\right]\\
&\text{and}\\
\mathbf{V}_c &= \frac{1}{2\pi\epsilon_o}\left[\mathbf{Q}_a\log_e\frac{d_{ac'}}{d_{ac}}+\mathbf{Q}_b\log_e\frac{d_{bc'}}{d_{bc}}+\mathbf{Q}_c\log_e\frac{2h_c}{r}\right]
\end{aligned}\right\} \quad [3.16]$$

Re-writing these equations in matrix form, the symmetry enables the effect of an earth wire of radius r_e and height h_e above earth to be included by inspection as follows:

$$\begin{bmatrix}\mathbf{V}_a\\ \mathbf{V}_b\\ \mathbf{V}_c\\ \mathbf{V}_e=0\end{bmatrix} = \frac{1}{2\pi\epsilon_o}\log_e\begin{bmatrix}\frac{2h_a}{r} & \frac{d_{ab'}}{d_{ab}} & \frac{d_{ac'}}{d_{ac}} & \frac{d_{ae'}}{d_{ae}}\\ \frac{d_{ab'}}{d_{ab}} & \frac{2h_b}{r} & \frac{d_{bc'}}{d_{bc}} & \frac{d_{be'}}{d_{be}}\\ \frac{d_{ac'}}{d_{ac}} & \frac{d_{bc'}}{d_{bc}} & \frac{2h_c}{r} & \frac{d_{ce'}}{d_{ce}}\\ \frac{d_{ae'}}{d_{ae}} & \frac{d_{be'}}{d_{be}} & \frac{d_{ce'}}{d_{ce}} & \frac{2h_e}{r_e}\end{bmatrix}\cdot\begin{bmatrix}\mathbf{Q}_a\\ \mathbf{Q}_b\\ \mathbf{Q}_c\\ \mathbf{Q}_e\end{bmatrix} \quad [3.17]$$

i.e.

$$[\mathbf{V}] = [p]\,.\,[\mathbf{Q}]$$

and

$$[\mathbf{Q}] = [p]^{-1}[\mathbf{V}] = [c]\,.\,[\mathbf{V}].$$

Thus inverting the matrix in [3.17] gives equations similar to [3.11] with values of capacitance between phase conductors and between the conductors and earth. Each of these two sets of capacitances will, in general, be unbalanced due to the unequal spacings between conductors.

The capacitances between the pairs of conductors could be represented by delta-star transformation, as three capacitances to neutral, and where the neutral is earthed or is at earth potential, these capacitances could then be added to those existing between each conductor and earth. The system of capacitances found by inverting the matrix of p coefficients in [*3.17*] can thus be reduced to three capacitances to neutral, one from each line conductor.

From this point on, only these capacitances to neutral will be considered and they will be taken as balanced, i.e. equal for all three phases. This will be nearly true if the lines are too short for the conductors to be transposed and will be true if they are transposed.

3.6 Positive- or Negative-sequence Capacitance to Neutral of a Transposed Single-circuit 3-phase Line

When calculating the positive- (or negative-) sequence capacitance the line is usually assumed to be transposed, and also the effect of any over-running aerial earth wire can generally be neglected (e.g. the presence of the earth wire of a typical 132-kV single-circuit line

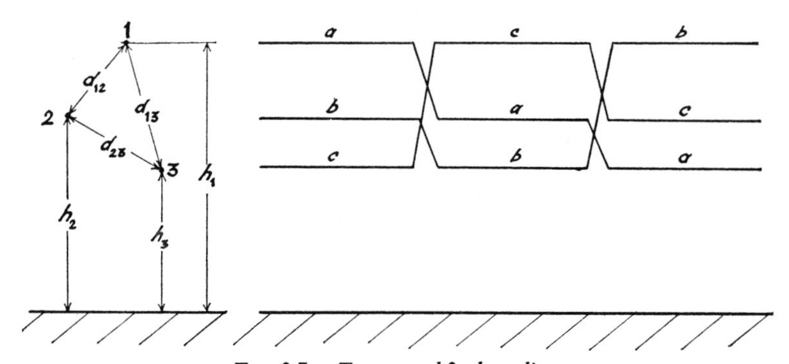

FIG. 3.7. *Transposed 3-phase line.*

reduces the capacitance to neutral by about 1%). If the line is transposed as shown in Fig. 3.7, then for the first one third of the length, the conductors of phases *a*, *b* and *c* occupy positions 1, 2 and 3 as shown, and so on for the other two positions of the cycle.

It should be noted that it is assumed here for simplicity that the charge/unit length on conductor *a* is $\mathbf{Q}_a$ for all three transposition positions. This is not strictly true since if the voltage drop due to line series impedance is negligibly small, the phase voltage to neutral $\mathbf{V}_a$

is the same for all three positions. Since the capacitance to neutral of the phase conductor is different for these positions, then the charge/unit length must differ also. It is much simpler however, to take $\mathbf{Q}_a$ as constant and $\mathbf{V}_a$ as variable, and to find an average value of $\mathbf{V}_a$ for the three positions. The difference between the capacitance calculated here and that calculated more rigorously by taking $\mathbf{V}_a$ as constant, is negligibly small.

For the first position, [*3.16*] gives the potential of conductor a with respect to neutral as,

$$\mathbf{V}'_a = \frac{1}{2\pi\epsilon_o}\left[\mathbf{Q}_a \log_e \frac{2h_1}{r} + \mathbf{Q}_b \log_e \frac{d_{12'}}{d_{12}} + \mathbf{Q}_c \log_e \frac{d_{13'}}{d_{13}}\right]$$

For the other two positions

$$\mathbf{V}''_a = \frac{1}{2\pi\epsilon_o}\left[\mathbf{Q}_a \log_e \frac{2h_2}{r} + \mathbf{Q}_b \log_e \frac{d_{23'}}{d_{23}} + \mathbf{Q}_c \log_e \frac{d_{12'}}{d_{12}}\right] \qquad [3.18]$$

(again using the fact that $d_{21'} = d_{12'}$, etc.) and

$$\mathbf{V}'''_a = \frac{1}{2\pi\epsilon_o}\left[\mathbf{Q}_a \log_e \frac{2h_3}{r} + \mathbf{Q}_b \log_e \frac{d_{13'}}{d_{13}} + \mathbf{Q}_c \log_e \frac{d_{23'}}{d_{23}}\right]$$

the average value is

$$\mathbf{V}_a = \tfrac{1}{3}(\mathbf{V}'_a + \mathbf{V}''_a + \mathbf{V}'''_a).$$

For positive- or negative-sequence voltages applied to the lines

$$\mathbf{Q}_a + \mathbf{Q}_b + \mathbf{Q}_c = 0$$

so that

$$\mathbf{V}_a = \frac{1}{3}\frac{\mathbf{Q}_a}{2\pi\epsilon_o}\left[\log_e \frac{8h_1h_2h_3}{r^3} - \log_e \frac{d_{12'}d_{13'}d_{23'}}{d_{12}d_{13}d_{23}}\right]$$

$$= \frac{\mathbf{Q}_a}{2\pi\epsilon_o}\log_e\left[\frac{\sqrt[3]{(d_{12}d_{13}d_{23})}}{r} \cdot \frac{2\sqrt[3]{(h_1h_2h_3)}}{\sqrt[3]{(d_{12'}d_{13'}d_{23'})}}\right]$$

The capacitance to neutral of each phase conductor which may be assumed to be balanced for a short line, or may be made equal if the line is transposed, is therefore given by

$$C = \frac{2\pi\epsilon_o}{\log_e\left[\dfrac{D_m}{r} \cdot \dfrac{2\sqrt[3]{(h_1h_2h_3)}}{\sqrt[3]{(d_{12'}d_{13'}d_{23'})}}\right]} \text{ F/m} \qquad [3.19]$$

where D_m is the geometric mean spacing between conductors and is given by

$$D_m = \sqrt[3]{(d_{12}d_{13}d_{23})}.$$

The greater the height of the line above earth compared with the spacing between conductors, the more nearly

$$\frac{2\sqrt[3]{(h_1 h_2 h_3)}}{\sqrt[3]{(d_{12'} d_{13'} d_{23'})}}$$

approaches unity.

If the effect of earth is neglected then the capacitance to neutral is

$$C = \frac{2\pi\epsilon_o}{\log_e (D_m/r)} \text{ F/m.} \qquad [3.20]$$

Since $\log_{10} D_m/r$ is usually between 2 and 3, a kilometre of overhead line has a capacitance to neutral of the order of 0·01 μF.

For a typical single-circuit 3-phase 132-kV line, the capacitance given by [*3.19*] exceeds that given by [*3.20*] by only about 1 %, so that for positive- and negative-sequence voltages, it is generally sufficient to use [*3.20*]. Both earth wire and the earth itself are therefore frequently neglected.

It can be seen from [*3.20*] and [*3.8*], that if a 3-phase line had its conductors at the corners of an equilateral triangle of side d metres, i.e. $D_m = d$, and if a single-phase line operated with a neutral mid-way in potential between the two conductors, then in both cases the capacitance to neutral would be given by

$$\frac{2\pi\epsilon_o}{\log_e d/r} \text{ F/m.}$$

The positive-sequence capacitance given in [*3.20*] for a single-circuit 3-phase line without the effect of earth or earth wire, can be calculated more directly than above. Instead of using [*3.5*] and [*3.6*], which give the potential of a line with respect to earth due to pairs of charges and their images due to the earth, charges equal and opposite to those on the three line conductors may be placed on a dummy conductor which is so far from the real conductors that it may be said to be distant S metres from each of them.

Equation [*3.18*] is then replaced by

$$\mathbf{V}'_a = \frac{1}{2\pi\epsilon_o}\left[\mathbf{Q}_a \log_e \frac{S}{r} + \mathbf{Q}_b \log_e \frac{S}{d_{12}} + \mathbf{Q}_c \log_e \frac{S}{d_{13}}\right]$$

$$\mathbf{V}_a'' = \frac{1}{2\pi\epsilon_o}\left[\mathbf{Q}_a \log_e \frac{S}{r} + \mathbf{Q}_b \log_e \frac{S}{d_{23}} + \mathbf{Q}_c \log_e \frac{S}{d_{12}}\right]$$

$$\mathbf{V}_a''' = \frac{1}{2\pi\epsilon_o}\left[\mathbf{Q}_a \log_e \frac{S}{r} + \mathbf{Q}_b \log_e \frac{S}{d_{13}} + \mathbf{Q}_c \log_e \frac{S}{d_{23}}\right]$$

If the terms are separated, those containing S vanish since

$$\mathbf{Q}_a + \mathbf{Q}_b + \mathbf{Q}_c = 0$$

and

$$\mathbf{V}_a = \tfrac{1}{3}(\mathbf{V}_a' + \mathbf{V}_a'' + \mathbf{V}_a''') = \frac{\mathbf{Q}_a}{2\pi\epsilon_o} \log_e \frac{\sqrt[3]{(d_{12}d_{13}d_{23})}}{r}$$

which gives the result of [*3.20*].

If the line is operating at a line voltage $\mathbf{V}_L$ and the voltage drop on its series impedance is so small compared with $\mathbf{V}_L$ that the latter may be considered to be the voltage between a pair of lines at all points along the route, then the charging current flowing in each of the three line conductors at the sending end of the line is

$$\mathbf{I}_c = \mathrm{j}\,\omega C l \frac{\mathbf{V}_L}{\sqrt{3}} \text{ amperes,}$$

where C is given by [*3.20*] and l is the route length of the line in metres.

3.7 Zero-sequence Capacitance to Neutral of a Transposed Single-circuit 3-phase Line

3.7.1 WITHOUT EARTH WIRE

If the line shown in Fig. 3.7 has zero-sequence voltage only applied to it then $\mathbf{V}_{ao} = \mathbf{V}_{bo} = \mathbf{V}_{co}$ so that

$$\mathbf{Q}_a = \mathbf{Q}_b = \mathbf{Q}_c = \mathbf{Q}$$

then

$$\mathbf{V}_{ao} = \frac{1}{3}\frac{\mathbf{Q}}{2\pi\epsilon_o}\left[\log_e \frac{2h_1}{r} + \log_e \frac{d_{12'}}{d_{12}} + \log_e \frac{d_{13'}}{d_{13}} + \log_e \frac{2h_2}{r} + \log_e \frac{d_{23'}}{d_{23}} + \log_e \frac{d_{12'}}{d_{12}} + \log_e \frac{2h_3}{r} + \log_e \frac{d_{13'}}{d_{13}} + \log_e \frac{d_{23'}}{d_{23}}\right]$$

$$= \frac{\mathbf{Q}}{2\pi\epsilon_o} \log_e \left(\frac{2\sqrt[3]{(h_1h_2h_3)}}{r} \cdot \sqrt[3]{\left[\left(\frac{d_{12'}d_{13'}d_{23'}}{d_{12}d_{13}d_{23}}\right)^2\right]}\right)$$

The zero-sequence capacitance to neutral is

$$C_o = 2\pi\epsilon_o \frac{1}{\log_e \left(\frac{2\sqrt[3]{(h_1 h_2 h_3)}}{r} \cdot \sqrt[3]{\left[\left(\frac{d_{12'} d_{13'} d_{23'}}{d_{12} d_{13} d_{23}} \right)^2 \right]} \right)} \text{ F/m.} \qquad [3.21]$$

When h_1, h_2 and $h_3 \gg d_{12}$, d_{13} and d_{23}

$$\sqrt[3]{(8h_1h_2h_3(d_{12'}d_{13'}d_{23'})^2)} \simeq 8h_1h_2h_3$$

so that

$$C_o \simeq 2\pi\epsilon_o \frac{1}{\log_e (8h_1h_2h_3/rD_m^2)} \text{ F/m.} \qquad [3.22]$$

3.7.2 WITH EARTH WIRE

In the case of the zero-sequence capacitance, the earth wire strung above the phase conductor to reduce direct lightning strokes, may increase the capacitance by an amount of the order of 10%, so that its presence cannot be neglected, as may be done for the positive-sequence. The earth wire has a radius r_e and is h_e above earth. d_{1e} is the distance between the conductor in position 1 (see Fig. 3.7) and earth wire, and $d_{1e'}$ is the distance from 1 to the image of the earth wire, etc.

In order to allow for its presence the charge $\mathbf{Q}_e$/unit length of earth wire must be found in terms of the charge $\mathbf{Q}$ C/m existing on each of the phase conductors. This may be done from [*3.17*]. The equations now written, relate to the first position in the transposition cycle, i.e. $d_{ab'} = d_{12'}$, $d_{ab} = d_{12}$, $h_a = h_1$ etc. (see Fig. 3.7).

Equation [*3.17*] with $\mathbf{Q}_a = \mathbf{Q}_b = \mathbf{Q}_c = \mathbf{Q}$ gives for the first part of the cycle

$$\mathbf{V}'_a = \frac{\mathbf{Q}}{2\pi\epsilon_o}\left[\log_e \frac{2h_1}{r} + \log_e \frac{d_{12'}}{d_{12}} + \log_e \frac{d_{13'}}{d_{13}}\right] + \frac{\mathbf{Q}_e}{2\pi\epsilon_o}\log_e \frac{d_{1e'}}{d_{1e}} \qquad [3.23]$$

$$\mathbf{V}'_e = 0 = \frac{\mathbf{Q}}{2\pi\epsilon_o}\left[\log_e \frac{d_{1e'}}{d_{1e}} + \log_e \frac{d_{2e'}}{d_{2e}} + \log_e \frac{d_{3e'}}{d_{3e}}\right] + \frac{\mathbf{Q}_e}{2\pi\epsilon_o}\log_e \frac{2h_e}{r_e} \qquad [3.24]$$

From [*3.24*]

$$\mathbf{Q}_e = \frac{-\mathbf{Q} \cdot \log_e (d_{1e'}d_{2e'}d_{3e'}/d_{1e}d_{2e}d_{3e})}{\log_e (2h_e/r_e)} \qquad [3.25]$$

Equation [*3.25*] confirms the fact, which should be self-evident, that the same relationship between $\mathbf{Q}_e$ and $\mathbf{Q}$ must exist in all three positions in the transposition cycle.

Substituting from [3.25] for $\mathbf{Q}_e$ in [3.23] gives

$$\mathbf{V}'_a = \frac{\mathbf{Q}}{2\pi\epsilon_o}\left[\log_e \frac{2h_1 d_{12'} d_{13'}}{r d_{12} d_{13}} - \frac{\log_e \dfrac{d_{1e'}}{d_{1e}} \cdot \log_e \dfrac{d_{1e'} d_{2e'} d_{3e'}}{d_{1e} d_{2e} d_{3e}}}{\log_e \dfrac{2h_e}{r_e}}\right]$$

and similarly

$$\mathbf{V}''_a = \frac{\mathbf{Q}}{2\pi\epsilon_o}\left[\log_e \frac{2h_2 d_{23'} d_{12'}}{r d_{23} d_{12}} - \frac{\log_e \dfrac{d_{2e'}}{d_{2e}} \cdot \log_e \dfrac{d_{1e'} d_{2e'} d_{3e'}}{d_{1e} d_{2e} d_{3e}}}{\log_e \dfrac{2h_e}{r_e}}\right]$$

and

$$\mathbf{V}'''_a = \frac{\mathbf{Q}}{2\pi\epsilon_o}\left[\log_e \frac{2h_3 d_{13'} d_{23'}}{r d_{13} d_{23}} - \frac{\log_e \dfrac{d_{3e'}}{d_{3e}} \cdot \log_e \dfrac{d_{1e'} d_{2e'} d_{3e'}}{d_{1e} d_{2e} d_{3e}}}{\log_e \dfrac{2h_e}{r_e}}\right]$$

Making the same approximation as before for the first terms in $\mathbf{V}'_a$, $\mathbf{V}''_a$ and $\mathbf{V}'''_a$, viz. that

$$\sqrt[3]{(8h_1h_2h_3(d_{12'}d_{13'}d_{23'})^2)} \simeq 8h_1h_2h_3$$

$$\mathbf{V}_a = \tfrac{1}{3}(\mathbf{V}'_a + \mathbf{V}''_a + \mathbf{V}'''_a) \simeq \frac{\mathbf{Q}}{2\pi\epsilon_o}\left[\log_e \frac{8h_1h_2h_3}{rD_m^2} - \frac{1}{3\log_e(2h_e/r_e)}\left(\log_e \frac{d_{1e'} d_{2e'} d_{3e'}}{d_{1e} d_{2e} d_{3e}}\right)^2\right]$$

so that

$$C_o \simeq 2\pi\epsilon_o \frac{1}{\log_e \dfrac{8h_1h_2h_3}{rD_m^2} - \dfrac{1}{3\log_e \dfrac{2h_e}{r_e}}\left(\log_e \dfrac{d_{1e'} d_{2e'} d_{3e'}}{d_{1e} d_{2e} d_{3e}}\right)^2} \text{ F/m.} \qquad [3.26]$$

3.8 Corona

If the alternating voltage on a line conductor system is gradually raised, it will reach a value known as the visual critical voltage V_v, at which a faint violet glow known as corona appears, together with a slight hissing noise and smell of ozone. At a voltage V_o somewhat below this voltage V_v, an additional power loss begins to occur in the system. This loss P is given approximately by

$$P = K(V - V_o)^2 \qquad [3.27]$$

where K and V_o the critical disruptive voltage, are constant for a given conductor arrangement, frequency and supply waveform. This loss can be very considerably increased by fog and rain. V, the operating voltage, V_o and V_v are all r.m.s. voltages and for a 3-phase line they are line-to-neutral, i.e. phase voltages. It is not always sufficient to limit the electric field by the criterion of restricting the power loss to a certain proportion, since radio interference can become intolerable while the corona power loss is relatively slight.

Corona is a partial discharge, i.e. an incomplete failure of the air. If the voltage is increased beyond V_v, then spark breakdown of the whole air space would occur. Corona occurs when the local electric field near the conductor surface is sufficiently high to give enough energy to free electrons, arising for example from cosmic rays, for them to ionize gas molecules on collision with them. Thus more electrons and positive ions appear which in turn are accelerated by the electric field and cause further ionizing collisions.

If the electric field at the conductor surface E_r is used instead of the potential difference V, then the values of E_r which give the disruptive (E_o) and visual (E_v) critical points, are independent of spacing. For very large diameter smooth conductors in air at normal temperature and pressure, corona power loss begins at the critical disruptive stress when

$$E_r = E_o \simeq 30 \text{ kV (peak)/cm.}$$

For 50-Hz applied voltage, and for a conductor of radius r cm which is in the normal range of values, corona becomes visible when the peak gradient reaches

$$E_v = E_o(1 + 0{\cdot}3r^{-\frac{1}{2}}). \qquad [3.28]$$

This visual critical voltage can therefore be found by substituting the particular value E_v given by [3.28] in place of the general value E_r.

The stranding of the line conductors causes corona to appear at a lower electric field and voltage than that given by [3.28]: a typical reduction in voltage is 25 to 30%. This may be allowed for by a roughness factor M as shown below (equation [3.32]).

E_o rises in proportion to the air density δ. If δ is taken as 1·0 at 20°C and 76 cm Hg then for a barometric pressure b cm and θ°C,

$$\delta = \frac{3{\cdot}86b}{(273+\theta)}. \qquad [3.29]$$

The student should note that [*3.29*] follows from the characteristic equation for perfect gases. This equation is frequently stated in different symbols from those in [*3.29*], viz. $pV = mRT$ where p is gas pressure, V volume, m is the mass of gas, T its absolute temperature, and R is a universal gas constant. Putting $\delta = m/V$, changing the symbols and eliminating R by considering the new conditions in terms of the standard conditions gives

$$\frac{b}{\delta(273+\theta)} = \frac{76}{1\times 293}.$$

The constant given as 0·3 in [*3.28*] is in fact variable if the density changes, and is proportional to $\delta^{-\frac{1}{2}}$, and may be written as $0{\cdot}30\delta^{-\frac{1}{2}}$, again for a conductor radius r measured in centimetres.

From [*3.1*] the stress at the surface of one of two conductors of a single-phase line, each of radius r is

$$E_r = \frac{Q}{2\pi\epsilon_o r}$$

if the inter-conductor spacing d is large enough for the charge on the other conductor to have negligible effect, and this is true for normal values of d/r. From [*3.7*] if V is the p.d. between the two conductors, then

$$V = \frac{Q}{\pi\epsilon_o}\log_e\frac{d}{r}$$

so that

$$V = 2rE_r \log_e (d/r). \qquad [3.30]$$

Taking into account the effect of pressure, temperature and conductor stranding, and using the nominal value of E_o of 30 kV (peak)/cm at N.T.P., [*3.28*] becomes

$$E_v = \frac{30}{\sqrt{2}}\delta M\left(1+\frac{0{\cdot}30}{\sqrt{(\delta r)}}\right) \text{kV(r.m.s.)/cm.} \qquad [3.31]$$

When in [*3.30*] the stress at the conductor reaches the critical stress E_v for visual corona, the p.d. between the two conductors reaches the critical value V_v.

$$V_v = \frac{60}{\sqrt{2}} r\delta M\left(1+\frac{0{\cdot}30}{\sqrt{(\delta r)}}\right)\log_e \frac{d}{r} \quad \text{kV (r.m.s.).} \qquad [3.32]$$

In the case of a 3-phase line with an r.m.s. phase (line-to-neutral) voltage V, then from [*3.20*]

$$V = \frac{Q}{2\pi\epsilon_o}\log_e \frac{D_m}{r}$$

$$= rE_r \log_e \frac{D_m}{r}$$

so that the critical value V_v of the phase voltage at which corona arises is given by

$$V_v = \frac{30}{\sqrt{2}} r\delta M\left(1+\frac{0{\cdot}30}{\sqrt{(\delta r)}}\right)\log_e \frac{D_m}{r} \quad \text{kV (r.m.s.).} \qquad [3.33]$$

It can be seen from the following example that for normal values of D_m/r encountered in 3-phase lines, the maximum stress is given very closely by

$$E_r = \frac{Q}{2\pi\epsilon_o r}$$

as also applies in the single-phase line given above.

Example 3.1

If the onset of corona in air occurs when the electric stress at any point reaches 30 kV/cm, calculate the maximum voltage at which a 3-phase overhead line can operate without corona if it has conductors of 1·5 cm diameter, with equal spacing between phases of 2·5 m.

Solution

From [*3.1*] we can see that maximum stress will occur at the surface of a conductor at the instant when it has maximum charge on it, i.e. when it is at peak phase voltage V_m with respect to earth, where $V_m = \sqrt{2}\, V$ and V is the r.m.s. phase voltage. $V = (1/\sqrt{3})V_L$ where V_L is the r.m.s. line voltage.

If the charge on conductor a is $\hat{Q}$ C/m at the instant when the p.d. between a and neutral is V_m, then at that instant since the voltages of conductors b and c to neutral are both $-V_m/2$, then their charges/unit length are each $-\hat{Q}/2$ (see Fig. 3.8).

The charge $\hat{Q}$ on conductor a gives an electric stress at p, the point on its surface on a line joining its centre to the centre of the equilateral triangle z, of

$$\frac{\hat{Q}}{2\pi\epsilon_o r}\ \text{V/m} \quad \text{from [3.1]}$$

and this stress is along the line zp.

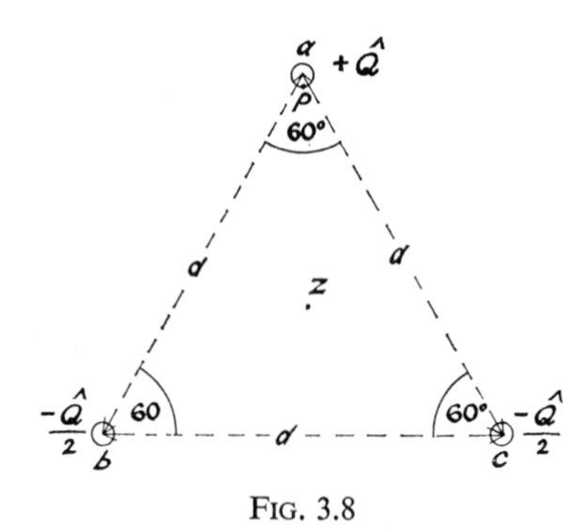

FIG. 3.8

The electric stresses due to charges $-\hat{Q}/2$ on b and c, cancel in the direction normal to zp, but along zp together they give

$$\frac{\hat{Q}}{2\pi\epsilon_o d}\sin 60^\circ = \frac{\sqrt{3}}{2}\hat{Q}\frac{1}{2\pi\epsilon_o d}$$

The total stress at p directed along zp, which is the maximum value, is therefore

$$\frac{\hat{Q}}{2\pi\epsilon_o}\left(\frac{1}{r}+\frac{\sqrt{3}}{2d}\right). \qquad [3.34]$$

V_m and $\hat{Q}$ are related by the capacitance to neutral

$$C = \frac{2\pi\epsilon_o}{\log_e(d/r)} \quad \text{(equation [3.20])}$$

since $D_m = d$ here. Thus

$$\hat{Q} = \frac{2\pi\epsilon_o V_m}{\log_e(d/r)}. \qquad [3.35]$$

Substituting the value of $\hat{Q}$ from [*3.35*] into [*3.34*] gives

$$\text{the maximum stress} = \frac{V_m}{\log_e (d/r)}\left(\frac{1}{r}+\frac{\sqrt{3}}{2d}\right)$$

$$3\times 10^6 = \frac{V_m}{\log_e (250/0{\cdot}75)}\left(\frac{100}{0{\cdot}75}+\frac{\sqrt{3}}{5}\right)$$

$$V_m \simeq 3\times 10^4 \times 0{\cdot}75 \log_e \frac{250}{0{\cdot}75} = 131\ 000 \text{ volts.}$$

The greatest (r.m.s.) line voltage which should be used is therefore $131\sqrt{3}/\sqrt{2} = 160{\cdot}5$ kV.

3.9 Bundle or Multiple Conductors

This section refers to the use of more than one line conductor/phase and single-circuit 3-phase lines are considered.

If a solid conductor were replaced by a multiple conductor set, consisting of n conductors each of radius r arranged on the circumference of a circle, with the aggregate of the cross-sections of these conductors equal to that of the single conductor (i.e. the latter has a radius of $r\sqrt{n}$), then the visual critical voltage V_v would be increased for two reasons:

(*a*) For the same line voltage, the maximum surface electric field is less for the multiple conductors than for the single conductor. The multiple conductors could tend towards a thin cylinder of much larger diameter than that of the single conductor as n rises, and [*3.1*] shows that this reduces the surface field.

(*b*) As [*3.28*] shows, the surface field has to reach a greater value for corona to be visible for a small conductor than for a large one, so that the system voltage can be raised when bundle conductors are employed.

The order of magnitude of these factors (*a*) and (*b*) will now be considered by approximate calculations, with (*a*) taken first.

Fig. 3.9 shows a 3-phase line in air with two conductors/phase each of radius r metres, all in one plane, and the conductors of each phase S metres apart, with a distance d metres between the corresponding conductors of two adjacent phases.

It will be assumed that the conductors of one phase all have the same total charge on them at any instant, and that these charges are distributed uniformly over the surfaces. These assumptions do not invalidate the general conclusions of the following simplified analysis (Parton and Wright, 1965). The analysis applies to positive- or negative-sequence system voltages, so that the charges sum to zero at every instant with a distribution which depends upon the capacitances to earth, to earth wire(s) and between conductors. It will be

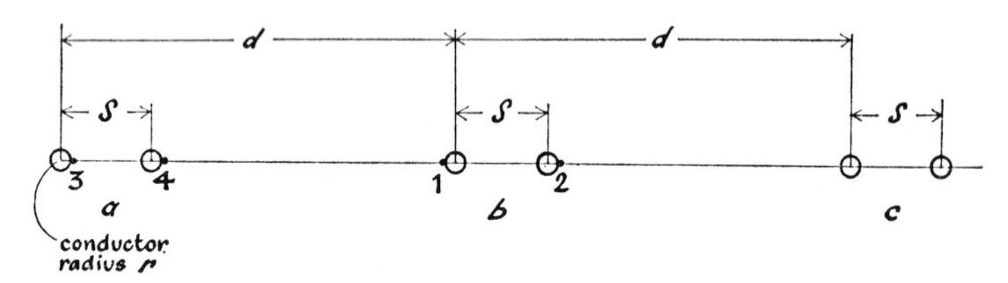

FIG. 3.9. *Single-circuit line with two conductors/phase in one horizontal plane.*

assumed here that the charge distribution depends upon balanced capacitances to earth, so that the charges are of equal amplitude and in phase with the respective voltage to earth. This approximation, whilst it conflicts with the neglect of earth and earth wire(s) in the following analysis, does not significantly affect the comparison between two conductors per phase and a single conductor per phase, and it simplifies the analysis sufficiently for it to be included here within a reasonable space.

Maximum stress will then occur on the surface of a conductor on the line joining the conductor centres, at the instant when the voltage of that phase is a maximum with respect to the neutral. If the expressions for the maximum surface fields at points 1, 2, 3 and 4 are compared, it is seen by inspection that $E_4 > E_1 (= E_2) > E_3$.

When the potential of the two conductors of phase a has its maximum value $\hat{V}_{an}$ with respect to neutral, the potential difference v_{ab} between conductors a and b is 0·866 times its maximum value, i.e. $v_{ab} = 0{\cdot}866\ \hat{V}_{ab}$. If each of the conductors of phase a has a charge $\hat{Q}$ at this instant, then the conductors of phases b and c each have a charge $-\hat{Q}/2$.

$$E_4 = \frac{\hat{Q}}{2\pi\epsilon_o}\left[\frac{1}{r}+\frac{1}{(r+S)}+\frac{1}{2(d-S-r)}+\frac{1}{2(d-r)}+\frac{1}{2(2d-S-r)}+\frac{1}{2(2d-r)}\right]. \quad [3.36]$$

The p.d. between conductors of phases a and b at this instant is

$$v_{ab} = 0{\cdot}866\, \hat{V}_{ab} = \int_{r}^{(d-S-r)} E_x dx$$

where E_x may be obtained from [3.36] by writing x for r.

$$v_{ab} = \frac{\hat{Q}}{4\pi\epsilon_o}\left[3 \log_e\left(\frac{d-S-r}{r}\right)+3 \log_e\left(\frac{d-r}{r+S}\right)- \log_e\left(\frac{d+r}{2d-S-r}\right)-\log_e\left(\frac{d+S+r}{2d-r}\right)\right]. \quad [3.37]$$

Substituting the value of $\hat{Q}$ given by [3.37] into [3.36] gives

$$E_4 = \frac{\sqrt{3}\,\hat{V}_{ab}}{3\log_e\left(\frac{d-S-r}{r}\right)+3\log_e\left(\frac{d-r}{r+S}\right)-\log_e\left(\frac{d+r}{2d-S-r}\right)-\log_e\left(\frac{d+S+r}{2d-r}\right)} \times\left[\frac{1}{r}+\frac{1}{(r+S)}+\frac{1}{2(d-S-r)}+\frac{1}{2(d-r)}+\frac{1}{2(2d-S-r)}+\frac{1}{2(2d-r)}\right] \quad [3.38]$$

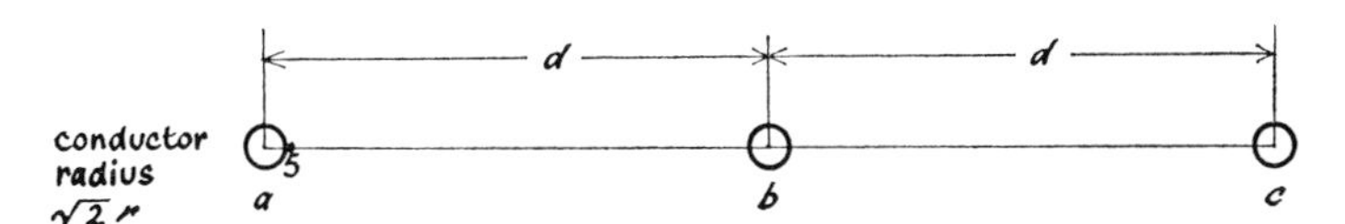

FIG. 3.10 *Single-circuit line with one conductor/phase in one horizontal plane.*

If the alternative single-conductor/phase system of Fig. 3.10 is considered with conductor radius $\sqrt{2}r$, then, making similar assumptions, the maximum stress occurs at point 5 at the instant when $v_{an} = \hat{V}_{an}$ and $v_{ab} = 0{\cdot}866\, \hat{V}_{ab}$.

If the charge on conductor a at this instant is $\hat{Q}_1$ and it is $-\hat{Q}_1/2$ on conductors b and c, then

$$E_5 = \frac{\hat{Q}_1}{2\pi\epsilon_o}\left[\frac{1}{\sqrt{2}\,r}+\frac{1}{2(d-\sqrt{2}\,r)}+\frac{1}{2(2d-\sqrt{2}\,r)}\right] \quad [3.39]$$

and

$$0{\cdot}866\, \hat{V}_{ab} = \int_{\sqrt{2}\,r}^{(d-\sqrt{2}\,r)} E_x dx$$

so that

$$0{\cdot}866\,\hat{V}_{ab} = \frac{\hat{Q}_1}{4\pi\epsilon_o}\left[3\log_e\left(\frac{d-\sqrt{2}\,r}{\sqrt{2}\,r}\right)+\log_e\left(\frac{2d-\sqrt{2}\,r}{d+\sqrt{2}\,r}\right)\right] \quad [3.40]$$

Substituting for $\hat{Q}_1$ in [*3.39*] from [*3.40*]

$$E_5 = \frac{\sqrt{3}\,\hat{V}_{ab}}{3\log_e\left(\frac{d-\sqrt{2}\,r}{\sqrt{2}\,r}\right)+\log_e\left(\frac{2d-\sqrt{2}\,r}{d+\sqrt{2}\,r}\right)} \times\left[\frac{1}{\sqrt{2}\,r}+\frac{1}{2(d-\sqrt{2}\,r)}+\frac{1}{2(2d-\sqrt{2}\,r)}\right]. \quad [3.41]$$

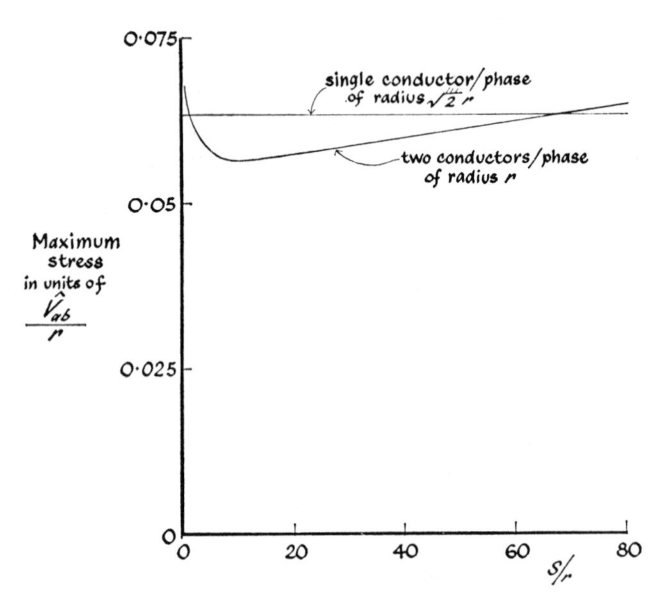

FIG. 3.11. *Variation of maximum electric stress with ratio S/r for $d/r = 700$.*

From [*3.38*] and [*3.41*], it is possible to compare the maximum stress of a 2-conductor/phase 3-phase line with that for a single-conductor/phase line. This is done in Fig. 3.11 for $d/r = 700$, which is of the order of the ratio used for some E.H.V. lines, and it shows that there is a range of values of S/r where a reduction in maximum stress, which in this case is up to about 11%, can be obtained.

If the corona point for each of the smaller conductors in a multiple set is similar to that for a solitary conductor, then the effect of the second factor (*b*) can be seen very approximately from [*3.28*] as raising the stress for visual corona E_v by about

$$\frac{1+0{\cdot}3r^{-\frac{1}{2}}}{1+0{\cdot}3(\sqrt{2}\,r)^{-\frac{1}{2}}},$$

when a single-conductor system is replaced by one with two conductors/phase, and

$$\frac{1+0{\cdot}3r^{-\frac{1}{2}}}{1+0{\cdot}3(2r)^{-\frac{1}{2}}}$$

for a four-conductor/phase system. For $r = 1$ cm, the increase in E_v is respectively 4 and 13%.

Measurements on bundle conductor systems indicate voltage gains rather larger than those given by factors (*a*) and (*b*) (Butterworth, 1954), so that if for the 2-conductor system considered, S/r is chosen so that the voltage gain is say 1·09, the actual gain may exceed $1{\cdot}09 \times 1{\cdot}04$.

It is therefore possible, by replacing a single conductor by two or more conductors with the same total cross-section, to obtain an increase in the corona voltage, i.e. to use a higher working voltage, or for a given system voltage to reduce the radio interference and corona losses. This becomes important when transmission voltages above 132 kV are used; e.g. two conductors spaced 30 cm apart are used on the 275-kV supergrid system, and the 400-kV system frequently has four conductors/phase spaced 30 cm apart.

There is an additional advantage that since the inductance is decreased by using bundle conductors as shown in the previous chapter, and the capacitance is increased, the 'natural load' of the line is raised (see Chapter 1).

It the internal magnetic flux linkages within the conductors are neglected, the inductance/conductor L and capacitance to neutral C of a transmission line in air, are related by $LC = \mu_0\epsilon_0$ (see equations [*3.20*] and [*3.23*] with r_m replaced by r, and see also Volume 2 where surges on transmission lines are considered).

The capacitance must therefore be increased in approximately the same proportions as the inductance is decreased, when two or more conductors/phase are used instead of one; i.e. about 20% for a 275-kV line (see Chapter 2.) The positive-sequence capacitance to neutral can be calculated using [3.20], with the conductor outside radius r replaced by an equivalent radius r'. The ratio r'/r has been calculated and tabulated for various values of the ratio S/r where S is the spacing between the adjacent conductors constituting one phase (Butterworth, 1954). If however an effective radius r_e, which is somewhat similar to the self-geometric mean distance or G.M.R. is used for calculating inductance (see chapter 2), in place of r', then the error is not great. It is found that even for S/r as low as 10, r_e is only about 2% below r' for two, three or four conductors/phase, and the error is less for larger values of S/r. For two conductors/phase $r_e = \sqrt{(rS)}$, for three conductors/phase spaced S apart $r_e = \sqrt[3]{(rS^2)}$, and for four conductors/phase at the corners of a square $r_e = \sqrt[4]{(\sqrt{2}\, rS^3)}$.

REFERENCES

BUTTERWORTH, S. 1954. *Electrical characteristics of overhead lines.* E.R.A., Report O/T4.

CARTER, G. W. 1967. *The electromagnetic field in its engineering aspects.* Longmans, London.

MORTLOCK, J. R. & HUMPHREY DAVIES, M. W. 1952. *Power system analysis.* Chapman and Hall, London.

PARTON, J. E. & WRIGHT, A. 1965. 'Electric stresses associated with bundle conductors.' *International Journal of Electrical Engineering Education*, **3**, pp. 357–367.

STEVENS, R. A. & GERMAN, D. M. 1964. 'The capacitance and inductance of overhead transmission lines.' *ibid.*, **2**, pp. 71–81.

STEVENSON, W. D. 1962. *Elements of power system analysis.* McGraw-Hill (International Students Edition), 2nd edition.

Examples

1. Deduce from first principles, making suitable approximations, the maximum electric stress for a 132-kV, 3-phase overhead transmission line which has one conductor/phase, with the conductors all at the same height above earth and symmetrically positioned about the centre phase. The conductor radii are 1·41 cm, and the distance between adjacent phases is 3 metres. Indicate briefly the factors involved in corona and comment upon whether or not corona is likely to occur on this line under normal operating conditions.

Determine the positive-sequence inductance and capacitance per unit length for this transmission line.

(University of Leeds)(13·67kV/cm; $1{\cdot}17 \times 10^{-6}$H/m, $9{\cdot}89 \times 10^{-12}$F/m)

2. Derive an expression for the capacitance/unit length of a two-wire transmission line h metres above perfectly conducting ground. The conductors are of radius r metres, d metres apart and equidistant from the ground. Assume that r is very much less than d and h.

$$\left(C = 2\pi\varepsilon_0 \Big/ \log_e \frac{d^2}{r^2\left(1+\dfrac{d^2}{4h^2}\right)} \text{ F/m} \right)$$

3. Determine for the following transmission line (a) the ratio of zero-sequence capacitance to positive-sequence capacitance to earth of each conductor if there is no earth wire and (b) the order of magnitude percentage change in zero-sequence capacitance caused by adding the earth wire.

The transmission line conductors each have a radius of 1·3 cm and are at heights above earth of 12, 13·5 and 15 m. The distances between phase conductors are 5, 6 and 7 m. When the earth wire is added it has a radius of 0·9 cm and it is 17 m above earth.

(University of Leeds) (0·575, +6%)

4. A long, straight cylindrical wire of radius r, in a medium of permittivity ϵ is parallel to a horizontal plane conducting sheet. The axis of the wire is at a distance h above the sheet. Derive an expression for the capacitance/unit length between the wire and the plane, stating any assumptions made.

The p.d. between the wire and the sheet is 5 kV, with $r = 0{\cdot}3$ cm and $h = 12$ cm.

Calculate

(1) the capacitance/unit length;

(2) the electric stress in the medium at the upper surface of the sheet at a point 20 cm from the axis of the wire.

(L.C.T.) $\left(C = \dfrac{2\pi\epsilon}{\log_e \dfrac{(2h-r)}{r}}\right.$ F/m, 0·0127 μF/km

6·85 kV/m acting vertically downwards)

5. Calculate the disruptive critical voltage between conductors for a single-phase line with the following details: conductor diameter 1·35 cm; spacing between conductors 4·25 m; coefficient of roughness M = 0·83; temperature 20°C; barometer reading 760 mm Hg (torr); disruptive critical voltage gradient of air = 21·1 kV (r.m.s.)/cm at 25°C and 760 mm Hg.

(H.N.C.) (154·6 kV)

6. If for a single-phase line the critical voltage between conductors is 120 kV for conductors of 1·27 cm diameter spaced 1·22 m apart, find the critical voltage with 1·09 cm diameter conductors spaced 2·44 m apart if all other conditions remain unchanged.

(H.N.C.) (120 kV)

Chapter 4

OVERHEAD LINES

4.1 Insulators for Overhead Lines

Overhead line conductors are not themselves insulated so insulators mounted on a suitable cross-arm are required to give the necessary clearances between conductors, and between conductors and earth against the highest voltage and worst atmospheric conditions to which the line is likely to be subjected. The insulator must also provide the necessary mechanical support for the conductor against the worst likely mechanical loading conditions.

Only two materials are in general use for insulating bare overhead line conductors—porcelain and toughened glass. Neither material is perfect, e.g. both are easily damaged by careless handling and by stone throwing. Porcelain is made from a fine, homogeneous mix of wet plastic clay which is shaped, covered with glaze and then fired in a kiln. This produces porcelain with a very hard, very smooth, brown glazed skin. Glass is described as toughened when its surface is in compression and its interior in tension. This state is achieved by rapidly cooling the glass insulator after shaping and allowing the interior to cool slowly. Toughened glass can withstand greater tension than can annealed glass.

Insulators may be classified by the rated line voltage of the system in which they are to be used but the working voltage is that line voltage divided by $\sqrt{3}$. In service, insulators are also subjected to overvoltage surges due to lightning and to switching (see Volume 2). Impulse voltages used to simulate lightning surges are obtained from impulse generators which produce a unidirectional voltage rising almost uniformly to a peak voltage in (typically) 1 μs and falling exponentially to half peak voltage in 50 μs. Impulse voltages are specified by their peak voltage, which can be positive or negative, and their wave shape, which would be specified as a 1/50 μs wave for the example quoted.

Neglecting the small shackle type insulators used to support 415/240-V lines between houses, insulators can be classified as follows: pin type, post type, string unit type (also called disc, or suspension or cap and pin type).

A pin type insulator is small, simple in construction, cheap, used up to and including 33 kV and is usually of porcelain since toughened glass is more expensive. The conductor is bound into a groove on the top of the insulator, which is cemented on to a galvanised steel pin attached to the crossarm of the pole.

The discussion to follow is applicable to all types of insulators, although the pin type is given as a specific example. If an insulator is clean and dry and in a dry atmosphere it can fail for several reasons.

(*a*) It can puncture. The puncture voltage of an insulator is the voltage which will break through the porcelain between the conductor and the support pin, causing the insulator to be destroyed. The puncture voltage is measured in an overvoltage test by immersing the insulator in insulating oil and applying a 50-Hz voltage across the insulator. B.S. 137 specifies that an 11-kV pin insulator must withstand 95 kV r.m.s. (about 15 times working voltage) without puncture—hence the expression 'the withstand voltage'. Insulators should be so designed that they never puncture.

(*b*) It can flashover. The flashover voltage is the voltage which will cause an arc through the air surrounding the insulator. The voltage used can be either power frequency or impulse, either positive or negative: values are usually given for all three. The dry flashover distance is the minimum distance, measured by a taut tape, from the pin to the conductor. For an 11-kV pin insulator, the withstand (peak) impulse voltage is about 10·6 times the peak working voltage, and the withstand power frequency one minute dry voltage (r.m.s.) is about 7·9 times the working voltage. The reference to one minute means that the test voltage should be raised as quickly as possible to the appropriate value and held for one minute. The actual flashover voltages should be greater than the withstand values. In service, if an insulator flashes over due to a surge voltage, the surrounding air is so ionised that there is usually a follow-through power arc, due to the working voltage, after the surge transient has passed. For this reason insulators are often fitted with arcing horns to keep this power arc away from the insulator.

In service the upper surfaces of an insulator can become damp and polluted by industrial dirt, salt-laden coastal spray, frost and fog to such an extent that these upper surfaces offer greatly reduced resistance to surface leakage current due to the working voltage. The effective surface resistance is due to the protected creepage distance which is the sum of all the distances measured along the curved under surfaces of the insulator which would remain dry to a horizontal spray directed at the insulator with its pin vertical. B.S. 137 gives minimum creepage distances for insulators to be used in moderately-, and heavily-polluted atmospheres: and in the latter case it gives both total and protected creepage distances. Large creepage distances are obtained by designing an insulator with as many sheds or skirts as possible.

The wet flashover distance of an insulator is the sum of all the distances, measured from the outer edge of a shed, perpendicularly to the surface of the shed below (or to the pin or conductor whichever is the lesser). Insulators are usually subjected to the artificial rain test specified in B.S. 137. For an 11-kV pin insulator the power frequency, one minute, wet withstand voltage is about 4·7 times its working voltage: its wet flashover voltage would also be measured. These rain tests are very difficult to perform with any reasonable accuracy: the tendency is to rely on creepage and wet flashover distances based on experience of insulators in service, rather than on rain tests. Clearly the surface resistance depends not only on the creepage distance but also on the surface width, thickness and composition of the conducting film. The minimisation of surface leakage current is a major problem in the design and maintenance of insulators. A well-designed insulator with a hard, smooth glaze free from cracks or scratches, used in an area with little pollution, will require little maintenance, especially if the rainfall is sufficient to keep the insulator clean.

The second type of insulator mentioned was the post type. This is similar to a pin-type insulator but has a metal base, and frequently also has a metal cap so that more than one unit can be mounted in series. The creepage distance is made large by using many identical sheds, especially for an outdoor insulator. A post-type insulator has a low capacitance and a low voltage gradient at the live conductor so that its corona voltage is high (see section 3.8). Post insulators of small diameter (about 15 cm or less) are solid, but larger diameters are often hollow and filled with an insulating medium, usually moisture-free oil. Post insulators supporting an overhead line conductor are called line post insulators. Post insulators are sometimes fitted with arcing horns, and can be used up to the highest voltages.

The third type of insulator is the string insulator unit which consists of a single shed of porcelain or toughened glass with several skirts underneath, the number and creepage distance of the skirts increasing for units to be used in heavily-polluted, fog-prone areas. Each insulator is cemented into a metal cap above and a metal pin (or ball) below, such that several units can be joined together in ball and socket joints to form an insulator string. This string is flexible, and hangs almost vertical, and the number of units in the string is chosen to suit the system voltage (usually 66 kV and higher), and the pollution likely to be experienced. Since the string is flexible, allowances must be made to ensure adequate clearances between live conductors and between live conductors and earthed metal when the string swings, in the wind, transversely to the line. String units are also used for the lowest system voltages and upwards to attach the two ends of the line to the terminal towers, and also to attach the two ends of the line to an angle tower where the line makes an appreciable change in direction. Such insulators are called tension strings, and their axes are almost horizontal.

A toughened glass insulator unit, when damaged, shatters into small pieces, and is therefore easily detected: but it is essential that the cement bond between the cap and pin should hold or the line will drop and may give a clearance height to earth less than the statutory minimum.

4.1.1 VOLTAGE DISTRIBUTION OVER AN INSULATOR STRING

A major problem with string insulators is the voltage distribution over the string. A string of identical, clean, dry units with no metal nearby (other than thin conductors to apply a voltage across the string) would share the applied voltage equally amongst the units, since they would act as a capacitor voltage divider. In use, however, each metal cap-and-pin joint has a capacitance to the earthed tower, see Fig. 4.1. It is usual to assume that these capacitances are all equal and given by kC where C is the capacitance of each string unit and k usually lies between 0·1 and 0·2. The numerical relation between the voltage across, and the current in, a capacitor is given by

$$I = 2\pi fCV \qquad [4.1a]$$

and hence

$$I \propto C \text{ for fixed } V \qquad [4.1b]$$

$$V \propto I \text{ for fixed } C. \qquad [4.1c]$$

The voltage distribution can be calculated by assuming that the tower-end unit takes one arbitary unit of current (denoted by 1 *a*) and

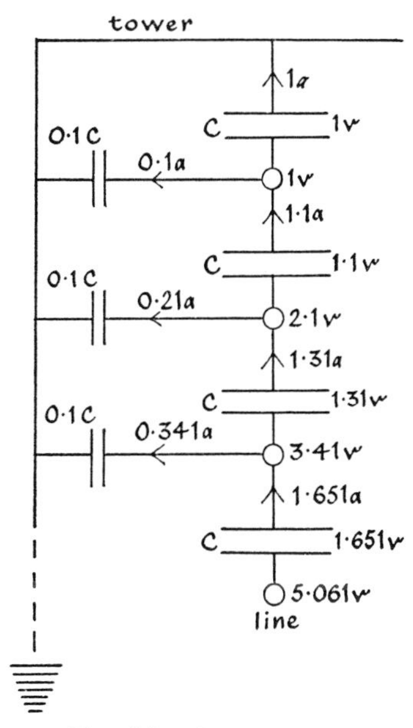

FIG. 4.1. *String efficiency.*

one arbitrary unit of voltage (1 *v*). If, for example $k = 0{\cdot}1$, then by [*4.1b*], the current in the uppermost shunting capacitor is 0·1 *a*. Thus the current in the second unit is 1·1 *a*, and the voltage across it, by [*4.1c*], is 1·1 *v*. The relative voltage across each unit and across the whole string can thus be found. Clearly the line-end unit takes the largest share of the total voltage. Assuming that this unit is about to flashover, then the whole unit is about to flashover.

The string efficiency is defined (in per unit) as the ratio

$$\frac{\text{flashover voltage of the string}}{n \times \text{flashover voltage of one unit}} \qquad [4.2]$$

where n is the number of units in the string. In the above example string efficiency is

$$\frac{5{\cdot}061 \times 100}{4 \times 1{\cdot}651} = 76{\cdot}8\%.$$

The string efficiency decreases as the number of units increases.

The string efficiency could be improved by reducing the shunt capacitance relative to the capacitance of each unit. Another method would be to increase the capacitance of each unit from the tower end towards the line end, but this is seldom done as it would involve carrying spares of different types of units, which is contrary to the tendency to standardise on as few types as possible. An increase in the string efficiency can be achieved by changing the line-end arcing horn to a grading ring. This has the effect of increasing the effective capacitance of the line-end units, but its effectiveness diminishes as the number of units in the string increases. Calculations based on this method usually involve changing a delta-connected set of capacitors to a star-connected set, using the formula given in Appendix 2.

The voltage distribution over a very long insulator string could be analysed using the theory of the uniform long line (see section 1.3.2) by replacing per km by per insulator unit, but the student should consider whether, in practice, the shunting capacitances between the cap-and-pin metal and the tower are uniform, especially near the tower end of the string. Since these shunting capacitances are the cause of the non-uniform voltage distribution over a clean, dry insulator string, the removal of the metal between the insulator units would appear to be a possible solution to the problem. Such single piece, or long-rod, insulators have not been an unqualified success. They are, in effect, a post type insulator suspended from the top cap.

The above method of calculation assumes that the voltage distribution is entirely dependent on capacitance effects, i.e. that the insulators are clean and dry. If however the units are badly polluted, the surface leakage (resistance) currents could be greater than the capacitance currents, and the extent of the pollution could vary from unit to unit, giving an unpredictable voltage distribution. It is thus important, that all types of insulators should be cleaned regularly.

Another method of improving the voltage distribution is to use a semi-conducting (or stabilising) high resistance glaze on the units, in an attempt to give a resistor voltage-divider effect. (The description semi-conducting does not have the meaning usually associated with electronic devices.) In an attempt to overcome the problem of surface pollution, some insulators have been covered with a thin film of silicone grease which absorbs the dirt and makes the surface water form into droplets rather than a thin film.

4.2 Conductor Materials for Overhead Lines

The following conductor materials are in general use for overhead lines: hard-drawn (HD), high-conductivity (HC) copper, hard-drawn cadmium-copper, hard-drawn aluminium and aluminium alloy, steel-cored aluminium (SCA). Standard data relating to these materials and to standard conductors made from these materials are given in the appropriate British Standard Specifications.

Possibly the most important property of an overhead line conductor material is high tensile strength (high breaking load) so that the spans between towers (or poles) can be as long as possible and the sag as small as possible (see section 4.3), thus reducing the number and height of these towers and the number of insulators required. The conductor should have a low resistivity to reduce the I^2R losses and the voltage drop. Its capital, installation and maintenance cost should be low: an important point here is the expected life of the conductor. The final choice of material is often a compromise.

Overhead line conductors are invariably stranded to make them more flexible during erection and while in service. Most conductors have a single central wire, the first layer has 6 wires, the next layer 12, the next 18, each layer increasing by 6 wires (this assumes the usual case that all the wires are of equal diameter). Each layer is twisted in the sense opposite to the layer below (the outside layer being always right-handed) in order to increase the total mechanical strength of the conductor.

Conductors can corrode due to chemical interaction with the polluted atmosphere, or due to oxidation or due to electro-chemical action between two different materials in an electrolyte due to salt spray or sulphuric acid due to sulphur dioxide in the atmosphere.

When conductors have been in service their surface tends to become rough due to corrosion and pollution and, in a turbulent wind, they tend to vibrate or flutter at relatively high frequency (10-Hz) with low amplitude. Occasionally conductors dance or gallop at low frequency (1-Hz) but high amplitude, due possibly to a coating of ice. A conductor, on suddenly shedding a covering of ice, jumps upwards because of its elasticity. Factors such as these must be taken into account in determining the spacing between conductors. The conductors must not come so close together that flashover takes place between them. Conductor spacing and overall diameter affect the inductance, capacitance and corona inception voltage, and their final choice is subject to many considerations, both technical and economic. Spacing is kept to a minimum, consistent with the above considerations, especially for the lower voltage lines: for the higher voltage lines, a large overall diameter (e.g. bundle conductors, see section 3.9) is advantageous.

Most overhead line materials are hard-drawn. The cyclic current loading of a conductor tends to anneal the wire and this, together with the corresponding effect on conductor joints, affects the current rating of a line. The ability of the conductor to dissipate heat depends on its circumference, and on the ambient temperature, wind and solar radiation. Lines have a cold, normal and hot weather current rating, a 10 and 20 minute overload rating, and a short-time fault current rating.

In the data below, the standard temperature is taken as 20°C.

High-conductivity annealed copper

Although this electrolytically refined material is not used for overhead lines, its electrical properties are often taken as a standard of reference.

Resistivity $= 1/58 = 0{\cdot}01724\Omega\text{-mm}^2/\text{m}$ or $\mu\Omega$-m
$= 1{\cdot}724\ \mu\Omega$-cm

Such a material is said to have a conductivity of 100%.

Temperature coefficient of resistance at 20°C (for constant mass) $= 1/254{\cdot}5 = 0{\cdot}00393$ per °C.

Density at 20°C $= 8{\cdot}89\ \text{g/cm}^3 = 8890\ \text{kg/m}^3$

Coefficient of linear expansion $\alpha = 17 \times 10^{-6}$ per °C.

(Note: Throughout this chapter α is the symbol for the coefficient of linear expansion and is not the temperature coefficient of resistance. BS. 1991 gives α as the symbol for both.)

Hard-drawn high-conductivity copper

Hard-drawn wires have a higher mechanical strength but also a higher resistivity than an annealed wire.

$$\text{Density} = 8{\cdot}89 \text{ g/cm}^3$$
$$\text{Coefficient of linear expansion } \alpha = 17\times10^{-6} \text{ per } ^\circ\text{C}$$
$$\text{Young's modulus of elasticity } E = 1{\cdot}24\times10^{11} \text{ N/m}^2$$

Relative to alternative materials, copper has a high conductivity and a small overall diameter for a given current rating. Thus the transverse wind and ice load on the conductor is small but the corona inception voltage is also low (see Section 3.8) so copper is not suitable for very high voltage lines. Also it has a relatively low tensile strength so that it could only be used for short spans. Copper has a very long life since it is not subject to any form of corrosion.

Hard-drawn cadmium copper

The addition of up to 1% cadmium to copper raises the tensile strength but at the expense of conductivity.

$$\text{Resistivity (maximum)} = 2{\cdot}177\ \mu\Omega\text{-cm}$$
$$\text{Conductivity (minimum)} = (1{\cdot}724/2{\cdot}177)100 = 79{\cdot}2\%$$
$$\text{Density} = 8{\cdot}945 \text{ g/cm}^3$$
$$\text{Coefficient of linear expansion } \alpha = 17\times10^{-6} \text{ per } ^\circ\text{C}$$
$$\text{Young's modulus of elasticity } E = 1{\cdot}24\times10^{11} \text{ N/m}^2$$

Cadmium copper is used for medium spans but, like copper, it is not suitable for the higher voltages, and is more expensive than copper. In order to give a quick comparison between different types of conductors, the equivalent copper area (E.C.A.) is defined as that area of solid hard-drawn copper having the same resistance per unit length as the conductor. Thus a cadmium copper conductor described as 1·13 cm^2 (E.C.A.) 19/0·3 cm, has an actual section of 1·34 cm^2. E.C.A. should not be confused with calculated area (given in the B.S.S. tables) which is the area of solid cadmium copper, of the same conductivity, having the same resistance per unit length as the stranded conductor.

Hard-drawn aluminium

Resistivity (maximum) = 2·826 $\mu\Omega$-cm
Conductivity (minimum) = 61·1%
Density = 2·703 g/cm^3
Coefficient of linear expansion α = 23×10^{-6} per °C
Young's modulus of elasticity $E = 6{\cdot}6 \times 10^{10}$ N/m^2.

The data for aluminium varies depending on its purity and previous physical history. Aluminium tends to be used (more especially for underground cables) when the price of copper is high. The comparison of aluminium with copper depends on the basis of the comparison, e.g. equal weight, equal resistance or equal current-carrying capacity from the point of view of heat dissipation. Aluminium has a low tensile strength so it is only used for short spans. Since it generally has a larger section, it is subject to higher wind and ice loading. Aluminium is subject to atmospheric corrosion.

Hard-drawn aluminium alloy

The addition of about $\frac{1}{2}$% each of silicon and magnesium increases the resistance to corrosion and also the tensile strength at the expense of conductivity. Aluminium alloy can be used for slightly longer spans than could hard-drawn copper.

Resistivity (maximum) = 3·25 $\mu\Omega$-cm
Conductivity (minimum) = 53%
Density = 2·70 g/cm^3
Coefficient of linear expansion α = 23×10^{-6} per °C
Young's modulus of elasticity E = $6{\cdot}8 \times 10^{10}$ N/m^2

Steel-cored aluminium (S.C.A.)

This tends to be the standard conductor in use at 132 kV and above. For example, a 2·6 cm^2 (E.C.A.), 7/54/0·32 cm S.C.A. conductor has a central steel strand surrounded by a layer of 6 steel strands, then three layers of aluminium of 12, 18 and 24 strands respectively, all of 0·32 cm diameter. The steel provides the necessary mechanical strength. The aluminium provides the electrical conductivity while that of the steel is neglected.

The data for galvanised steel is as follows

Density = 7·8 g/cm^3
Coefficient of linear expansion α = $11{\cdot}5 \times 10^{-6}$ per °C
Young's modulus of elasticity E = $1{\cdot}93 \times 10^{11}$ N/m^2

Both steel and aluminium oxidise due to atmospheric oxygen and electrolytic action (corrosion) can take place between the zinc and aluminium especially in areas of industrial pollution and salt spray. For this reason a layer of grease is put between the steel and the aluminium during the manufacture of the conductor.

Because of its strength, S.C.A. is used for all long spans, in order to limit the sag and decrease the number of towers needed. Its large diameter raises its wind and ice loading but also raises its corona inception voltage.

The current-carrying layers of aluminium are effectively long-drawn-out solenoids so there is a longitudinal field inside the conductor which magnetises the steel core. Since alternate layers of aluminium are twisted in opposite senses, the magnetic field of each layer tends to oppose (but not to cancel) those of adjacent layers. The internal flux-linkages and inductance of a solid conductor were discussed in section 2.2.1 on the basis of a uniform distribution of current over the conductor section. Clearly this simple theory is not valid for an S.C.A. conductor.

4.2.1 VIRTUAL MODULUS OF ELASTICITY OF S.C.A.

Young's modulus of elasticity E is defined as E = stress/strain, and applies up to the elastic limit of the material, i.e. while strain is proportional to stress. Thus E also equals change in stress/corresponding change in strain. Values of E for steel and aluminium have already been given: the problem now is to find a virtual value E_v for a composite conductor such as S.C.A.

Assume that at each end of a length of conductor the steel and aluminium are firmly clamped together so that no slip can take place between the two materials. Then any strain must be the same for the steel, the aluminium and the composite conductor. If the conductor, initially free of any tension, has a force F applied to it then

$$F = f_s A_s + f_a A_a$$

where f = stress and A = cross sectional area. The average stress is

$$(f_s A_s + f_a A_a)/(A_s + A_a)$$

The strain $= f_s/E_s = f_a/E_a$ and the virtual modulus is defined as

$$E_v = \frac{\text{average stress}}{\text{strain}}$$

$$= \frac{(f_s A_s + f_a A_a)E_s}{(A_s + A_a)f_s}$$

$$= \frac{A_s E_s + A_a E_a}{A_s + A_a}. \qquad [4.3]$$

4.2.2 VIRTUAL COEFFICIENT OF LINEAR EXPANSION OF S.C.A.

The coefficient of linear expansion α of a material is the increase in length per unit length per unit rise in temperature. If a length l increases by δl when the temperature rises $\delta\theta$ then

$$\alpha = \delta l/(l\,.\,\delta\theta).$$

For simplicity assume $l = 1$ and $\delta\theta = 1$, then $\alpha = \delta l$. Consider unit length of S.C.A. conductor, lying on a horizontal frictionless surface, the conductor being fixed at one end and subjected to a tensile force F at the other end: as before the ends are clamped so that at each end the steel and aluminium cannot slip relative to each other. Assume the temperature rises one degree. Then the steel should expand by an amount α_s and the aluminium by α_a. But any change in length will result in a change in stress. Let the tensile stress in the steel and aluminium increase by δf_s and δf_a respectively, while the applied force is unchanged so that $\delta f_s A_s + \delta f_a A_a = 0$. Due to the increase in tensile stress the corresponding increases in length are $\delta f_s/E_s$ and $\delta f_a/E_a$. The actual changes in length of the steel, aluminium and S.C.A. conductor must all be equal so

$$\alpha_s + \delta f_s/E_s = \alpha_a + \delta f_a/E_a = \alpha_v$$

where α_v is the change in length of the S.C.A. per unit length per unit rise in temperature, i.e. its virtual coefficient of linear expansion. By eliminating δf_s and δf_a the student should show that

$$\alpha_v = \frac{\alpha_s A_s E_s + \alpha_a A_a E_a}{A_s E_s + A_a E_a}. \qquad [4.4]$$

4.3 Sag and Stress Calculations

4.3.1 PARABOLA

Fig. 4.2(a) shows an overhead line strung between two fixed points A and B which are on the same level. The line is assumed to be

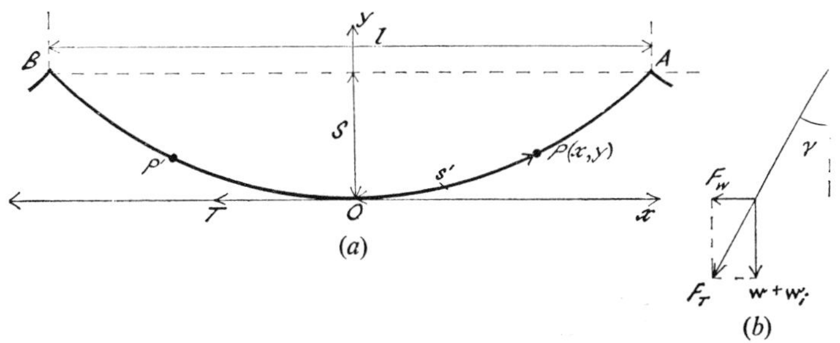

FIG. 4.2. *(a) Parabolic form (P′OP arc length = s = 2s′); (b) Effect of wind force* F_W.

perfectly flexible and to sag below the level AB due to its own uniform weight. The shape of the line will therefore be that of a catenary (see mathematical text books and section 4.3.4). Except possibly for supergrid lines of long span and large sag, it is usually sufficiently accurate to assume that the line is so taut that its shape is that of a parabola $y = ax^2$ where a is a constant for a given line, and O is the origin.

Let l = span = horizontal distance between supports, metres

S = sag at mid-span, metres

Then $y = S$ when $x = l/2$ and

$$a = 4S/l^2 \qquad [4.5a]$$

and

$$y = 4S(x/l)^2. \qquad [4.5b]$$

Consider now the equilibrium of the half line OA.

Let T = tension (N) at O (assumed constant over the whole span)

w = conductor weight, N/m

Assuming that the conductor is almost horizontal, taking moments about A gives $TS = (wl/2)(l/4)$ and hence

$$S = wl^2/8T \qquad [4.6]$$

Eliminating S between [*4.5b*] and [*4.6*] gives

$$y = wx^2/2T \qquad [4.7a]$$

and

$$a = w/2T. \qquad [4.7b]$$

The Overhead Line Regulations lay down a minimum clearance height for the line above ground, and if to this is added the sag, the height of the insulator support points A and B can be found. The Regulations state that the sag must be calculated under the specified worst conditions, namely: that the line is at a temperature of 22°F, covered with ice to a radial thickness of 0·96 cm with ice weighing 8900 N/m³ and simultaneously subjected to a 80 km/hour wind, at right angles to the line, such that the pressure on the line is 378 N/m² of total projected area, and that the maximum tension in the line under these conditions is not greater than half the breaking tension on the line. The variation in tension along a line will be discussed in section 4.3.4 but for the present it will be assumed that the tension is constant at all points along a taut line. The same Regulation applies to medium/low voltage lines except that the ice thickness is reduced to 0·48 cm.

The practical complication is that the lines will be erected under warmer and nearly still-air conditions and yet must comply with the worst conditions stated above. Consider 1 metre length of conductor.

Let d = diameter of conductor, metres.

t = radial thickness of ice, metres.

D = overall diameter of ice-covered conductor = $d+2t$ metres.

The wind force on the conductor is (see Fig. 4.2.(b))

$$F_W = 378\,D \text{ N/m} \qquad [4.8]$$

The weight of ice on the conductor is

$$\begin{aligned} w_i &= (D^2-d^2)\ \ 2225\pi \\ &= 2{\cdot}8\times10^4 t\,(d+t) \text{ N/m} \end{aligned} \qquad [4.9]$$

The conductor is deflected from a vertical plane by an angle γ.

Let F_T = total force acting on the conductor per metre run

$$= \{F_W^2+(w+w_i)^2\}^{\frac{1}{2}} \text{ N/m} \qquad [4.10]$$

F_T must lie in the new plane of the conductor, and

$$\tan\gamma = F_W/(w+w_i). \qquad [4.11]$$

Equations [*4.5a*] to [*4.7b*] still hold providing that all the measurements are made in the new plane of the conductor and w is replaced by F_T. If F_T is the total force per metre on the conductor under the worst conditions, and T is the limiting tension defined by the Regulations, then the sag in the new plane is, from [*4.6*].

$$S = F_T l^2/8T \quad [4.12]$$

and the sag measured vertically is $S\cos\gamma$.

It is now necessary to obtain an approximate expression for the actual (curved) length of the half line (s'), assuming a parabolic shape. Consider a small right-angled triangle on the parabola $y = ax^2$. Then

$$(\mathrm{d}s')^2 = (\mathrm{d}x)^2+(\mathrm{d}y)^2$$

$$(\mathrm{d}s'/\mathrm{d}x)^2 = 1+(\mathrm{d}y/\mathrm{d}x)^2 = 1+4a^2x^2$$

$$\mathrm{d}s' = (1+4a^2x^2)^{\frac{1}{2}}\mathrm{d}x \simeq (1+2a^2x^2)\mathrm{d}x.$$

since $a = 4S/l^2$ is small. On integrating

$$s'_x = x+(2a^2x^3/3)+c$$

where $c = 0$ since $s' = 0$ when $x = 0$. Thus

$$s'_x = x+2a^2x^3/3. \quad [4.13]$$

Note that s'_x is the curved distance from O to P(x,y). If s is the double curved distance, $2s'_x$, then the total curved length of the parabola is, from [*4.5a*],

$$s = l+8S^2/3l. \quad [4.14]$$

Substituting from [*4.12*], the total curved length is

$$s = l(1+F_T^2 l^2/24\,T^2). \quad [4.15]$$

If a line is subjected to one set of physical conditions, denoted by subscript 1, and then to a changed set of physical conditions, denoted by subscript 2, a relationship between them can be derived as follows. If T_1 increases to T_2 then the conductor elongation will be $s_1(T_2-T_1)/AE$ where A in^2 is the cross-sectional area of the conductor. If the temperature increases from θ_1 to θ_2 the conductor expansion will be $\alpha s_1(\theta_2-\theta_1)$. Thus

$$s_1+s_1(T_2-T_1)/AE+\alpha s_1(\theta_2-\theta_1) = s_2 \quad [4.16a]$$

Substituting for s_1 and s_2 from [4.15], using the binomial expansion and neglecting second-order terms, [4.16a] reduces to

$$(l^2/24)(F_{T2}^2/T_2^2-F_{T1}^2/T_1^2) = (T_2-T_1)/AE+\alpha(\theta_2-\theta_1) \quad [4.16b]$$

A practical problem is to solve [4.16b] for T_2), by re-arranging it as

$$T_2^2[T_2-\{T_1-\alpha AE(\theta_2-\theta_1)-AEF_{T1}^2 l^2/(24T_1^2)\}] = AEF_{T2}^2 l^2/24 \quad [4.17]$$

Equation [*4.17*] gives the erection tension T_2 such that should the line be subjected to the worst conditions (as specified in the Regulations), the tension on the line will not exceed T_1 which is half the breaking load of the line. The sag and height of the line above ground (clearance height) at the time of erection must be such that the clearance height will not fall below the statutory minimum at any future time.

Equation [*4.17*] can be written in the form

$$T_2^2(T_2-a) = b. \quad [4.18]$$

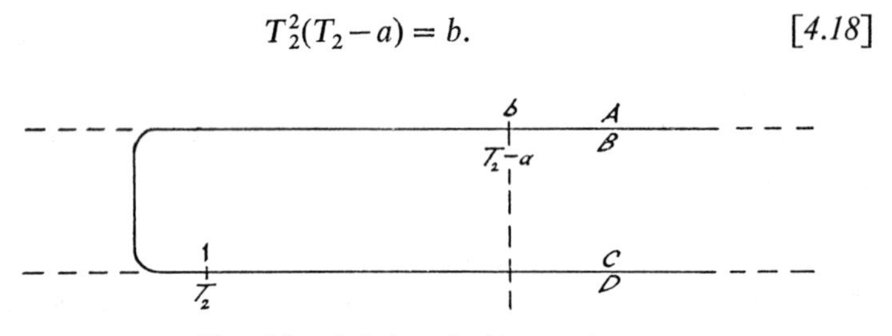

FIG. 4.3. *Solution of cubic equation.*

This cubic equation can be solved using a slide rule. Calling the scales A, B, C and D (see Fig. 4.3), set the cursor to b on the A scale and adjust the slide by trial and error until the number on the D scale opposite 1 or 10 on the C scale (T_2) is such that the number under the cursor on the B scale is T_2-a. If, for example, $b = 75$, it will be found that this method gives the same significant figures for T_2 whether b is taken as 75 or 7·5.

4.3.2 SPANS OF UNEQUAL LENGTH: EQUIVALENT SPAN

When a line consisting of several sections of unequal length is strung on its supporting insulators, it is very desirable that these insulators should hang vertically. Thus at each support point, the horizontal components of the tensions, at the support points, of the adjacent sections should be equal. If each section is strung according to the formulae developed in the previous section, the resulting tensions would not necessarily be equal.

The tensions are designed to be equal by defining an equivalent span l_e and basing all the calculations on this equivalent span. The two conditions used are that the sum total of the horizontal spans should be equal and that the sum total of the curved lengths should be equal for the actual line and the equivalent line.

If there are n sections then

$$nl_e = \Sigma l \qquad [4.19]$$

and, by [4.15]

$$n(l_e + w^2 l_e^3/24T^2) = (\Sigma l) + w^2(\Sigma l^3)/24T^2. \qquad [4.20]$$

Substituting [4.19] in [4.20] gives

$$l_e^2 = (\Sigma l^3)/nl_e = \Sigma l^3/\Sigma l$$

$$l_e = \{(l_a^3 + l_b^3 + \ldots)/(l_a + l_b + \ldots)\}^{\frac{1}{2}} \qquad [4.21]$$

where the sections are $a, b, c, \ldots$

Knowing l_e the tension T, to which all the sections will be erected, can be found.

4.3.3 SUPPORTS AT DIFFERENT LEVELS

Fig. 4.4 shows a line strung between two support points B and C which are at different levels. Let BOCA be the complete parabola

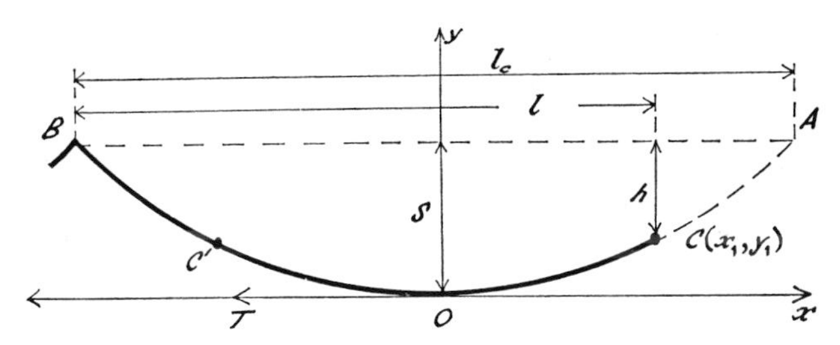

FIG. 4.4. *Supports at different levels.*

with A and B at the same level, of which the actual line is assumed to be a part.

Let l_c = span of the complete parabola (AB)
l = span of the actual line

then

$$x_1 = l - l_c/2 \qquad [4.22]$$

The equation of the parabola, $y = ax^2$, holds for BOC and BOA, so [4.7a] still holds.

Thus

$$S = wl_c^2/8T. \qquad [4.23]$$

and from [*4.5b*]

$$x_1^2/(S-h) = l_c^2/4S. \qquad [4.24]$$

Eliminating S gives

$$l_c = l+2Th/wl. \qquad [4.25]$$

The above theory is still valid even if the curved line length BC is less than BO, i.e. the ground level is falling so fast that the lowest point of the parabola is outside the actual section of line (e.g. BC′). Knowing l_c, section 4.3.1 can now be applied.

4.3.4 CATENARY

Formulae based on the catenary should be used for very high voltage lines whenever sag>span/10 approximately. When these formulae

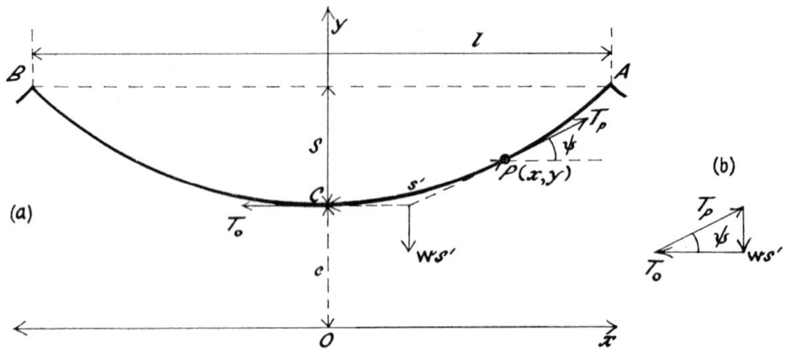

FIG. 4.5. *Catenary.*

are expanded as infinite series, an estimate can then be made of the errors involved in using the approximate formulae based on the parabola.

Referring to Fig. 4.5(a), the part CP is in equilibrium under three forces, T_o the horizontal tension at the lowest point C, T_p the tension at P and ws' the weight of the part acting vertically downwards. These three forces must meet at a point which determines the line of action of ws', and the relationships between these forces can be determined from Fig. 4.5(b). The length of a part of the catenary is s' while the corresponding length of the part of the catenary suspended from two points on the same level is $s = 2s'$.

Considering vertical and horizontal components,

$$T_p \sin \psi = ws' = ws/2 \quad [4.26a]$$

$$T_p \cos \psi = T_o. \quad [4.26b]$$

It is convenient to write

$$T_o = wc \quad [4.26c]$$

so that [*4.26b*] can be rewritten as

$$T_p \cos \psi = wc \quad [4.26d]$$

where c is called the parameter of the catenary. From [*4.26a*] and [*4.26d*]

$$s' = c \tan \psi = s/2. \quad [4.27]$$

The curved length of the catenary can be found by the method used for the parabola (section 4.3.1) and is given by

$$s' = c \sinh (x/c) = s/2 \quad [4.28a]$$

which can be rewritten, using [*4.26c*] as

$$s' = (T_o/w) \sinh (wx/T_o) = s/2. \quad [4.28b]$$

Substituting dy/dx for $\tan \psi$ in [*4.27*] and integrating gives

$$y = c \cosh (x/c) \quad [4.29a]$$

$$y = (T_o/w) \cosh (wx/T_o) \quad [4.29b]$$

$$y - c = (T_o/w)\{\cosh (wx/T_o) - 1\}$$

$$= \frac{wx^2}{2T_o}\left\{1 + \left(\frac{x^2}{12}\right)\left(\frac{w}{T_o}\right)^3 + \ldots\right\}. \quad [4.29c]$$

Equation [*4.29c*] gives a comparison between the equations of the catenary and the equivalent parabola for which $c = 0$. During worst loading conditions $T_o = T_1$ and $w = F_T$: during erection conditions $T_o = T_2$.

From [*4.28*] and [*4.29*] we get

$$y^2 = c^2 + (s')^2 = c^2 + (s/2)^2. \quad [4.30]$$

From [*4.26a*], [*4.26d*] and [*4.30*],

$$T_p = wy. \quad [4.31]$$

From [*4.26c*], [*4.29*] and [*4.31*]

$$T_p = T_o \cosh (x/c) \quad [4.32a]$$

$$= T_o \cosh (wx/T_o). \quad [4.32b]$$

As x increases, T_p increases for a given T_o : T_o is the minimum tension in the line.

Considering now the complete catenary of span l, putting $x = l/2$ and expanding as infinite series, the following formulae are obtained. Conductor length s is given by

$$s = (2T_o/w)\sinh(wl/2T_o) \qquad [4.33a]$$

$$= l\left\{1+\left(\frac{l^2}{24}\right)\left(\frac{w}{T_o}\right)^2+\left(\frac{l^4}{1920}\right)\left(\frac{w}{T_o}\right)^4+\ldots\right\}. \qquad [4.33b]$$

The sag S is given by

$$S = (T_o/w)\{\cosh(wl/2T_o)-1\} \qquad [4.34a]$$

$$= \frac{wl^2}{8T_o}\left\{1+\left(\frac{l^2}{48}\right)\left(\frac{w}{T_o}\right)^2+\ldots\right\}. \qquad [4.34b]$$

The tension at the support point A is

$$T_A = T_o\cosh(wl/2T_o) \qquad [4.35a]$$

$$= T_o+wS. \qquad [4.36b]$$

If the line were to be drawn so tight as to be a horizontal straight line, then $s' = x$ and [4.28a] becomes

$$(s'/c) = \sinh(x/c).$$

Since this equation has to be true for all values of x, as well as for $s' = x = 0$, then $c \to \infty$, and $w/T_o = 1/c \to 0$.

The student should check that if the sag is $0{\cdot}1 \times$ span, then $l/c \simeq 0{\cdot}8$: and that if $l = c$ then the sag $\simeq 0{\cdot}1276 \times$ span.

REFERENCES

BRADBURY, J. et al., 'Long-term creep assessment for overhead-line conductors', 1975, *Proc. I.E.E.*, **122**, pp 1147–1151.
ELECTRICITY COMMISSION, 1937, *Electricity supply regulations.*
GRACEY, G. C. 1963. *Overhead electric power lines.* Benn, London.
GRIDLY, J. H. (Ed.) 1961. *High voltage distribution practice.* Benn, London.
MCCOMBE, J. & HAIGH, F. R. 1966. *Overhead line practice.* Macdonald, London.
WADDICOR, H. 1964. *The principles of electric power transmission.* Chapman and Hall, London.
WOOD, A. B., TAYLOR, J. V. and LIPTROT, F. J., 'Design, testing and construction of the 500-kV overhead lines for the El Chocon to Buenos Aires transmission system', 1976, *Proc. I.E.E.*, **123**, pp 51–59.

BRITISH STANDARD SPECIFICATIONS:

125. Hard-drawn Copper Conductors.
215. Part 1. Aluminium Conductors.
Part 2. Steel-cored Aluminium Conductors.
672. Hard-drawn Cadmium Copper Conductors.
1989. Memorandum on Values for the Properties of High-conductivity Copper.
3242. Aluminium Alloy Conductors.
137. Porcelain and Toughened Glass Insulators for Overhead Power Lines.
923. Impulse Voltage Testing.

Examples

1. A suspension insulator string comprises four similar units, the self-capacitance of each of which is four times that between each link-pin and earth. If the conductor is energised at 76 kV, measured with respect to earth, find the voltage across each unit and the string efficiency.

(I.E.E.) (11, 13·8, 20, 21·2 kV, 60·8 %)

2. Determine the voltage distribution and the string efficiency of a three-unit suspension insulator if the capacitance of the link-pins to earth and to the line is respectively 15 % and 5 % of the self-capacitance of each unit.

(L.C.T.) (30·2, 31·1, 38·7, 86·3 %)

3. In a string of three disc units, each unit has a self-capacitance C and from each cap-and-pin joint to the tower the capacitance is $0{\cdot}2\,C$. The capacitance from the grading ring to the lowest joint is $0{\cdot}4\,C$ and to the second lowest joint $0{\cdot}1\,C$. Calculate the string efficiency.

(L.C.T.) (100 %)

4. In a 3-unit suspension insulator string, the capacitance between each link-pin and earth is 0·2 of the capacitance of the unit. Determine by how much the capacitance of the lowest unit should be increased to achieve a string efficiency of 90 %, the other two units being left unchanged.

(I.E.E.) (26·5 %)

5. An insulator string containing three identical units is to have equal voltage across each unit by a suitable choice of the capacitance between each cap-and-pin and the grading ring. If the tower-end

unit has unit capacitance and the cap-and-pin to tower capacitance is 0·2 unit, calculate the relative capacitance of the upper and the lower cap-and-pin to the grading ring.

(L.C.T.) (0·1, 0·4)

6. A steel-cored aluminium line has an equivalent copper area of 1·15 cm^2, and has a resistance of 0·158 ohm/km at 20°C. Hence deduce the percentage conductivity of hard-drawn copper, referred to standard annealed copper as 100%, given that the latter has a resistivity at 20°C of 1/58 ohm-mm^2/m.

(L.C.T.) (96·4%)

7. Calculate the virtual modulus of elasticity and the virtual coefficient of linear expansion of a 1·15 cm^2, 7/30/0·28 cm diameter steel-cored aluminium conductor assuming the data given in the text.

(L.C.T.) ($9{\cdot}18 \times 10^{10}$ N/m^2, $18{\cdot}42 \times 10^{-6}$ per °C)

8. An overhead line conductor is supported at a water crossing from two towers, the heights of the supports being 29 m and 33·6 m respectively above water level, with a horizontal span of 336 metres. If the conductor weighs 8·33 N/m and its tension is not to exceed $3{\cdot}34 \times 10^4$N, calculate (i) the clearance between the lowest point of the conductor and the water and (ii) the horizontal distance of this point from the lower support.

(L.C.T.) (27·4, 113 metres)

9. A transmission line has a span of 183 metres between level supports and the vertical sag is 6·1 metres. The conductor has a cross-sectional area of 1·29 cm^2 and weighs 11·4 N/m. Allowing a breaking stress of 41 400 N/cm^2, find the working factor of safety when the wind pressure at right angles to the span is 1195 N/m^2 of projected area.

(H.N.C.) (5·33)

10. Determine the sag for a 183-metre span of galvanised steel conductor having 37 strands, each 0·259 cm diameter. The weight of conductor is 7·15 N/m and the breaking strength is 67 700 N. The factor of safety to be used is 2·5. The wind loading is 382 N/m^2 of

projected area (coated with ice). The weight of ice is 8920 N/m^3, and a radial thickness of ice of 0·96 cm may be assumed.

(University of London) (3·2 metres)

11. The conductors of an overhead line are 1·15 cm diameter and span 122 metres. Calculate the sag at mid-span when the conductors have an ice loading of 0·96 cm radial thickness, and are subjected to a horizontal wind pressure of 382 N/m^2. The permissible conductor tension is $3·56 \times 10^4$ N. If the suspension insulator strings are 1·22 m long and the minimum clearance from the supporting tower is 0·458 m, calculate the least distance from the tower at which the insulator string may be attached to its cross-arm. The weight of conductor is 5·83 N/m and the weight of ice 8920 N/m^3.

(University of London) (0·865, 1·32 metres)

12. Estimate the sag of a copper conductor in still air at 72°F, such that when the temperature falls to 22°F and the wind loading is 11·2 N/m and ice loading 5·27 N/m, the stress in the conductor does not exceed 17 200 N/cm^2. The area of cross-section of the conductor is 0·628 cm^2 and the weight of conductor/m = 5·6 N. Young's modulus (E) = $12·4 \times 10^6$ N/cm^2. The coefficient of linear expansion/degree F = $9·22 \times 10^{-6}$.

(H.N.C.) (2·28 m)

13.. An overhead line is erected on a gradient of 1 in 45. The height of the supports is 7·62 m and the span length measured horizontally is 137 m. If the conductor equivalent cross-sectional area is 0·893 cm^2 and the maximum allowable stress is 6900 N/cm^2, calculate the maximum sag relative to the lower support. Assume that the conductor with ice loading weighs 5·1 N/m and wind pressure is 382 N/m^2.

(H.N.C.) (1·2 m)

14. The steel-cored aluminium conductors of an overhead line are 1·95 cm overall diameter, and are carried by suspension insulator strings 1·43 metres long. Calculate the height of the lowest cross-arm above ground level, if the minimum clearance between conductor and ground is 7·62 m when there is an ice load 0·96 cm thick and a horizontal wind pressure of 382 N/m^2 on span lengths of 244 m. The allowable tension is $3·56 \times 10^4$ N, the conductor weighs 8·31 N/m and ice weighs 8920 N/m^3.

(H.N.C.) (12·5 m)

Chapter 5

UNDERGROUND CABLES

5.1 Introduction

Paper-insulated, metal-sheathed cables which are used for most underground power transmission and distribution, may be either

(*a*) solid-type, in which the pressure within the oil-impregnated paper is not boosted above atmospheric, and may even fall below it locally, e.g. in voids. This can lead to breakdown at high electric stresses, and this type is not generally used above 33 kV (i.e. $33/\sqrt{3}$ kV to earth);

(*b*) pressurised cables, in which pressure is maintained above atmospheric, either by gas in gas-pressure cables, or by oil in oil-filled cables. Voids are then respectively either filled with gas at relatively high pressure, or are prevented from forming. In compression cables which are available up to 275 kV, voids are prevented by high-pressure gas holding a thin diaphragm sheath against the core insulation. The oil-filled designs are currently in service in Great Britain up to 275 kV with ratings of over 700 MVA, and 400 kV, 3200 A, i.e. 2200-MVA cables are available with forced cooling. Oil-filled cables have been made for use at 550 kV, and this type has now emerged as the dominant type, supplanting those systems relying upon gas pressurisation, such as sulphur hexafluoride or nitrogen.

Either of these types (*a*) or (*b*) when used in a 3-phase system, may consist of three single-core cables or of one 3-core cable. The latter may, if used below 22 kV, be of the belted type (illustrated in Fig. 5.2), or for higher voltages it will be screened, i.e. each core then has a metallised screen around it to enable higher voltages to be used, as discussed later. Also each core may have a separate sheath of lead (type S.L. or H.S.L.) or aluminium (type S.A.).

The lapped paper insulation has a 50-Hz breakdown strength for single unimpregnated sheets in a test lasting a few minutes of about

60 kV/cm, but impregnation with a suitable oil mixture raises this to about 500 kV/cm at 50 Hz, though the breakdown strength of the actual multi-layer insulation of an oil-filled cable is of the order of 300 kV/cm. The actual values of working stress which can be permitted, when long life is required under working conditions, which include repeated cycles of expansion and contraction from particular working temperatures, vary from about 40 kV/cm for a screened solid-type cable, to between 100 and 150 kV/cm for most present-day pressurised cables. All the above values are r.m.s. kV/cm. Attempts are being made to use synthetic papers and films in place of the paper insulation even at the highest voltages. Superconducting cables seem as though they would be cheaper than oil-filled cables above about 5 GVA but the need for such ratings is some way off, at least in the UK.

5.2 Capacitance of Single-core Cable

Since a single-core cable has an earthed metallic sheath, there is only an electric field between the conductor and sheath to consider.

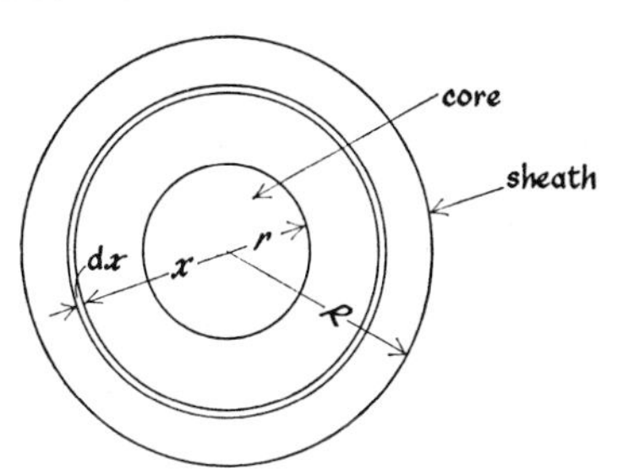

FIG. 5.1. *Cross-section of a single-core cable.*

If conductor and sheath radii are r and R metres respectively, as shown in Fig. 5.1 and the charge on the conductor surface is Q coulomb/metre of length, then Coulomb's law gives the electric flux density at a radius of x metres ($r<x<R$) as

$$D_x = \frac{Q}{2\pi x} \text{ coulomb/m}^2$$

The electric field or potential gradient at that radius is

$$E_x = \frac{Q}{2\pi\epsilon_o\epsilon_r x} \text{ V/m} \qquad [5.1]$$

where ϵ_r is the relative permittivity of the cable insulation and $\epsilon_0 = 8{\cdot}85 \times 10^{-12}$ F/m. Q, D, E and V are all the r.m.s. values of quantities varying sinuisoidally in time. V and Q are here italicised because only their magnitude and not their phase is of concern at this point.

Similarly D and E are shown italicised here although they vary not only in time but also in space.

The p.d. V between core and sheath is given by

$$V = \int_r^R E_x \, \mathrm{d}x$$

$$V = \frac{Q}{2\pi\epsilon_0\epsilon_r} \log_e \frac{R}{r} \text{ volts} \qquad [5.2]$$

and the capacitance between core and sheath is

$$C = \frac{2\pi\epsilon_0\epsilon_r}{\log_e (R/r)} \text{ F/m.} \qquad [5.3]$$

Equation [*5.3*] applies also to the majority of 3-core cables used at voltages above 11 kV, since they have earthed metal screens around each core and/or have separate sheaths for each core. Where the conductor cross-section is not circular, e.g. if it is oval, the radius r used in [*5.3*] may be taken as the actual conductor periphery $\div 2\pi$.

Substituting for Q in [*5.1*] from [*5.2*] give the potential gradient at a radius x as

$$E_x = \frac{V}{x \log_e (R/r)} \text{ V/m.} \qquad [5.4]$$

The maximum electric stress in the cable dielectric which is shown by [*5.4*] to occur at the surface of the conductor is given by

$$E_r = \frac{V}{r \log_e (R/r)}$$

and it is increased by about 15 to 25% above this value by the stranding of the conductor (unless the conductor surface has a conducting smooth screen applied to it).

Equating $\mathrm{d}E_r/\mathrm{d}r$ to zero, gives the minimum value of the stress at the conductor surface as V/r when $\log_e R/r = 1$, i.e. when $R = 2{\cdot}718\, r$. The application of this result is limited, but it does indicate that for given V and R there can be a limit to how small the conductor should be made, particularly at the higher voltages.

Since the insulation can only be stressed to its limiting operating voltage gradient at the conductor surface and it is under-stressed further from the conductor, it is attractive to attempt to reduce the amount of insulation by redistribution of stress, so as to increase it in outer layers without increase at the conductor. Two methods which would do this are:

(*a*) capacitance grading, where two or more insulating materials are used with those having the larger permittivities nearer to the conductor (see [*5.1*]);

(*b*) intersheath grading, where a single insulating material is separated into two or more layers by thin metallic intersheaths maintained at the appropriate potentials by being connected to tappings on the winding of the transformer supplying the cable.

Since neither of these methods is employed in practice, they are not discussed further, but some examples on them have been included at the end of the chapter, as exercises in the principles involved. Thinner papers are however applied near the conductor to increase the electric strength in this region since it reduces the 'butt-gaps'.

Example 5.1

A 3-core cable 6·43 km long has each core separately sheathed. The diameter of each conductor is 2·5 cm and the radial thickness of its insulation is 0·5 cm. The dielectric has a relative permittivity of 3·2. Calculate the maximum stress in the dielectric, and the charging kVA when the cable is connected to a 3-phase 50-Hz 33-kV supply.

SOLUTION

The maximum stress occurs at the surface of each conductor and is given by $E_r = V/(r \log_e R/r)$ where V is the voltage between each core and its earthed screen, i.e. V is the phase voltage of the system $= 33\,000/\sqrt{3}$ volts. $r = 1{\cdot}25$ cm and $R = 1{\cdot}25 + 0{\cdot}5$ cm

$$E_r = \frac{33\,000}{\sqrt{3} \times 1{\cdot}25 \log_e (1{\cdot}75/1{\cdot}25)} = 45{\cdot}4 \text{ kV/cm.}$$

The capacitance/metre for each core

$$= 2\pi\epsilon_0\epsilon_r/(\log_e R/r) \text{ F}$$

$$= \frac{2\pi \times 8{\cdot}85 \times 10^{-12} \times 3{\cdot}2}{\log_e (1{\cdot}75/1{\cdot}25)} \text{ F}$$

$$= 5{\cdot}31 \times 10^{-4}\ \mu\text{F.}$$

The total capacitance for each core

$$= 5{\cdot}31 \times 10^{-4} \times 6{\cdot}43 \times 10^{3}\ \mu\text{F}$$
$$= 3{\cdot}41\ \mu\text{F}.$$

The charging current flowing in each core is therefore

$$I = 2\pi \times 50 \times 3{\cdot}41 \times 10^{-6} \times 33\,000/\sqrt{3} = 20{\cdot}4\ \text{A}.$$

The total charging kVA for all three cores

$$= 3VI \times 10^{-3} = \sqrt{(3)} \times 33\,000 \times 20{\cdot}4 \times 10^{-3}$$
$$= 1168\ kVA.$$

5.3 Capacitance of 3-core Belted Cable

As discussed later, this variety of solid-type cable is only used up to about 11 kV. As shown in Fig. 5.2, each conductor is lapped with

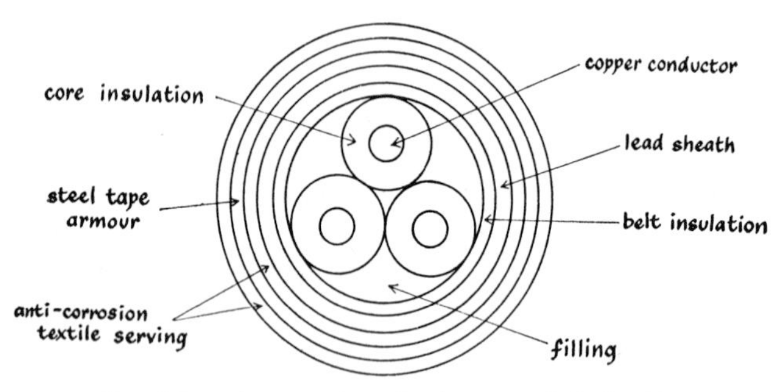

FIG. 5.2. *Cross-section of a three-core belted cable.*

paper, the spaces are then filled to form an overall circular section, and a further belt of paper insulation surrounds the three cores. This belt insulation is needed because with line voltage V_L between cores, the core insulation is only sufficient for $V_L/2$, but the voltage between each core and earth is $V_L/\sqrt{3}$.

Since the conductor section is frequently not circular, and the conductors are not surrounded by isotropic homogeneous insulation of one known permittivity, the capacitances C_c between cores, and C_s between a core and sheath, are not easily calculated and are generally obtained by measurement. The effective capacitors representing these capacitances/m are shown in Fig. 5.3. The mesh-connected capacitors C_c may be replaced by three capacitors each $3C_c$ connected in star (see Appendix 2). The effective capacitance of each core to the earthed neutral $C = (C_s+3C_c)$.

In determining the capacitances of this type of cable, two common measurements are:

(*a*) the capacitance C_a between conductor 1 and the other two conductors both joined to the sheath, where $C_a = C_s + 2C_c$;

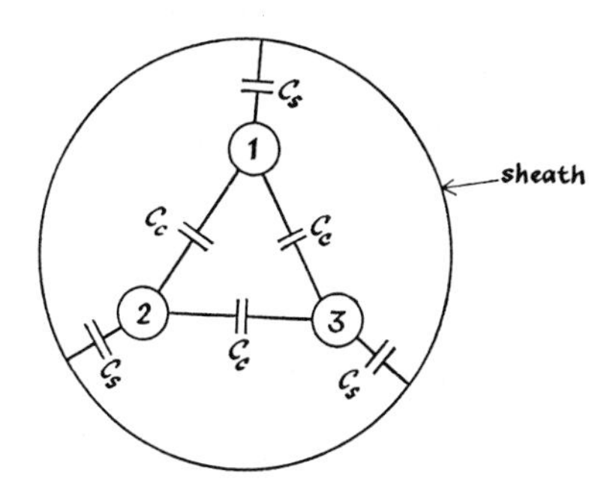

FIG. 5.3. *Capacitances between cores and to sheath of a three-core belted cable.*

(*b*) the capacitance C_b between the sheath and all three conductors joined together, where $C_b = 3C_s$.

From these values, the effective capacitance to neutral is

$$C = \frac{9C_a - C_b}{6} \qquad [5.5]$$

5.4 3-core Cables

5.4.1 SOLID-TYPE CABLES

These cables are mainly mass-impregnated, i.e. the insulation is dried and impregnated with oil after the paper has been lapped on to the conductors. In a 3-core belted cable, the electric field varies from instant to instant and is in fact rotating at constant speed. The stresses are particularly high at the surface of the conductor which is at that moment at the highest potential, at points nearest to the other conductors, to the sheath and to the centre of the cable. These stresses are mainly radial, and can be kept to values which the paper insulation can withstand normal to its surface. When these cables were used at 33 kV however, failures occurred, and these were found to be mainly in or near the filling material in the centre of the cable, and particularly where the conductor insulation met the filler. In these regions the electric field has a considerable component tangential to the paper surface, and the electric strength of paper is less along the fibre direction than normal to it. Furthermore voids are more likely to form in or at the boundary of the filler material, and ionisation of the gas in these voids can lead to breakdown, as discussed later.

The problem of extending cables to voltages above 33 kV was solved partly by the application, following work by Höchstädter, of an earthed metallic screen around the insulation of each conductor. This screen can consist of perforated foil, or tape with a space between turns, so that the impregnating oil or gas can pass freely. In this case or in the separately-sheathed cable which also has individual screening, there is no belt insulation, and each core has its own insulation to earth and the electric field in it is virtually entirely radial. Using screens it was possible to make a few single-core solid-type cables for 66 kV, but pressurised cables of various types were developed to meet the need for cables above 33 kV, and above this voltage solid-type cables are not economic because of the large insulation thickness needed.

A 33-kV solid-type cable has a radial thickness of insulation of about 0·75 cm (this becomes slightly less the larger the conductor), and this gives a stress at the conductor surface of about 40 kV/cm. The impulse strength of this insulation is about 1000 kV/cm, so that the cable impulse breakdown voltage greatly exceeds the impulse withstand voltage test for 33 kV cable, which is 194 kV, and the thickness of insulation is governed by the power frequency performance.

5.4.2 OIL-FILLED CABLES

In these cables, a low viscosity oil is maintained at sufficient pressure to prevent void formation, by connecting ducts within the cable to oil reservoirs. In single-core cables, the duct is formed within the conductor by winding the conductor strands on a tube of steel strip spiral, or by self-supporting segmental construction. In 3-core cables, three ducts are formed, one in each filler space, by steel or aluminium spiral. Alternatively, where aluminium sheaths are used, fillers may be omitted from the inter-core spaces and these are then fully available to oil flow without the need for separate spiral ducts. The system must prevent the pressure in the cable falling below atmospheric, and a lower limit of about 21·7 kN/m^2 gauge is set. It must be designed to prevent transient pressures, developed when load is suddenly increased, exceeding a limit which is about $8{\cdot}63 \times 10^5$ N/m^2 gauge. Oil expansion can take place by compression of flexible compartments containing gas within the reservoirs.

The discharge inception stress is 300 to 400 kV/cm which is much greater than that for a solid-type cable, so that the normal operating stresses are raised very considerably. Thus a 132-kV cable with normal working stress at the conductor of 110 kV/cm needs only about 1 cm radial thickness of insulation, but since the pressurising and elimination of voids give little if any increase in impulse strength, this thickness (as with gas-pressure cables) is largely governed by impulse breakdown requirements. Since void formation is largely prevented the normal maximum conductor temperature can exceed that for a solid-type cable, and is 85°C in Britain. These cables are now increasingly being used at 33 kV as well as at the higher voltages.

5.4.3. GAS-PRESSURE CABLES

The three main types are gas-filled, impregnated-pressure (now only rarely used) and compression cables.

In gas-filled cables, the paper dielectric is impregnated with petroleum jelly before lapping on the conductors so that it has no free compound. The interstices between the layers of paper are then filled with gas, usually dry nitrogen, at a pressure up to $1{\cdot}38 \times 10^6$ N/m^2 forming a composite dielectric. The gas flows throughout the cable via the filler spaces, butt-gaps in the dielectric and conductor strands, or by leaving a small clearance to sheath in the case of a single-core cable.

Impregnated-pressure cables which are not used now unless specifically requested for modification of existing systems, have a lead pipe in the filler space for easy gas flow to the mass-impregnated dielectric. The impregnating compound used is suitable for higher electric stresses, and has a low dielectric loss with a relatively small change in power factor with temperature.

Both oil and gas cables have variants of these types, including those with oil and gas respectively at a pressure of about $1{\cdot}38$–$1{\cdot}72 \times 10^6$ N/m^2 within a steel pipe. In the former, the oil-pressure pipe-type cable, the oil permeates the insulation of the three screened cores, whereas in the pipe-type gas compression cable used up to 275 kV, the pressure acts on the thin diaphragm lead sheaths of the three solid-type cores in order to prevent void formation. The compression cable, though in service at 275 kV is restricted in its rating to about 400 MVA, due to mutual heating of the three cores in one pipe. If instead, a self-contained cable is used in which each core has its own reinforced outer sheath to contain the gas pressure, there is a limitation in size due to the cost.

5.5 Conductor Inductive Reactance

For single- and 3-core cables, [*2.9*] and [*2.23*] respectively of Chapter 2 may basically be used to calculate the inductance which applies effectively to each conductor. These can be modified if necessary for such effects as:

(*a*) reduction due to mutual coupling with the sheath. This is generally small except in certain cases such as large single-core cables with aluminium sheath;

(*b*) increase due to mutual coupling with the armour of 3-core cables. This can be up to 10 or 20%;

(*c*) reduction due to the lower effective spacing between cables with conductors of shaped cross-section as compared with corresponding round conductors.

If [*2.23*] of Chapter 2 is applied to a 3-core belted cable with 37/0·238 cm conductors with conductor insulation 0·444 cm thick, then D_m and r_m may be found as follows. A 37-strand conductor has a central strand surrounded by three layers containing respectively 6, 12 and 18 strands so that the overall conductor radius = 3·5× 0·238 cm = 0·834 cm; r_m is therefore 0·778×0·834 = 0·643 cm. The distance between conductor centres is uniform and D_m = 2(0·444+0·834) = 2·54 cm

The inductance of each conductor is therefore

$$\frac{\mu_o}{2\pi}\log_e\frac{D_m}{r_m} = 2\times10^{-7}\log_e\frac{2\cdot54}{0\cdot643}\text{ H/m}$$

before making allowance for factors such as (*a*), (*b*) and (*c*) mentioned above. Where a 3-phase circuit consists of three single-core cables [*2.23*] may again be used, and for two such cables in a single-phase circuit [*2.29*] applies.

5.6 Effective Conductor Resistance

The effective resistance R_{eff} of each conductor of a cable must take into account the following five factors, the d.c. resistance at the operating temperature, skin effect, proximity effect, sheath losses and armour losses, and these are now considered in turn.

D.C. resistance. The maximum conductor temperature has to be fixed after taking account of such factors as: loss of electrical and mechanical strength of the paper and of the impregnating fluid; the expansion of conductor and sheath; the possibility of void formation due to differential expansions; and the method of laying the cable. For pressurised cables the conductor temperature must not exceed 85°C, and for other types of cables, temperatures between 50°C and 80°C are specified. For a copper conductor, the d.c. resistance at maximum temperature may therefore exceed the value at 20°C by between 12 and 26%.

Skin effect. The increase in resistance due to skin effect at a given frequency depends on the cross-sectional area, and to a lesser extent on the temperature. For 3 cm^2, 6 cm^2 and 12 cm^2 conductors, the increase at 50 Hz would be about 2%, 7% and 24% respectively, so it is significant above about 6 cm^2. Above about 9 cm^2 a special construction may be used to reduce this effect, and the increase in resistance for 12 cm^2 would then be about 6%.

Proximity effect. This increase in resistance being due to non-uniformity in current density over the conductor section caused by the magnetic field of current in the other phase conductors, is greater for a given size in single-core cables, than in 3-core belted ones. For the latter it may in general be of about the same magnitude as that due to skin effect.

When the d.c. resistance at the operating temperature is increased by factors (which may be obtained from tables) for skin and proximity effects a conductor a.c. resistance R_{ac} results.

Since appreciable power may be lost in the cable sheath and armour due to magnetically-induced currents in them, this a.c. resistance R_{ac} must be further increased to give the effective value R_{eff}. It is usual to express the sheath and armour losses as a function of conductor loss since they arise from the magnetic field of the conductor current.

If sheath loss $= \lambda_1 \times$ conductor loss and armour loss $= \lambda_2 \times$ conductor loss then $R_{eff} = R_{ac}(1+\lambda_1+\lambda_2)$.

Sheath losses. The sheath losses are due either to circulating currents or to eddy currents, but the latter are generally small except in very large cables. The losses due to currents circulating between sheaths are greater for larger conductor sizes, and since aluminium has a lower resistivity than lead, they are greater in aluminium sheathed cables. Current will be induced due to the resultant magnetic field of current in the other two phases, in the sheath of one single-core cable, or in the sheath of one core of a 3-core cable, which has each core separately sheathed with lead or aluminium (S.L. or S.A. type). For 3-core belted or 3-core screened cables, with only a single sheath enclosing all three phases, the sheath losses are less for otherwise comparable conditions. The closer the three phase conductors are together the smaller the loss, since neglecting screens or S.L. sheaths, the eddy losses in the outer sheath arise from the rotating magnetic field which would be zero if the three cores were in exactly the same position in space.

If the sheaths of three single-core cables are not bonded electrically together, induction between conductors and sheaths produces voltages between one sheath and another which depend upon the spacing but can sometimes be of the order of 100 volts/km for each 1000 amperes flowing in the cores. In order to avoid these voltages, sheaths are generally bonded. If this is done at both ends then currents circulate between sheaths, and although this introduces loss, it tends to cancel the loss due to the currents flowing within a single sheath outlined above. The voltage, and thus the loss, is greater the further the cores are laid apart. Methods of cross-bonding at joints have, however, been developed, by which circulating current sheath losses can be virtually eliminated and inter-sheath voltages kept small, and this may be necessary where spacings between cores are large. For 275 kV, 6 to 20 cm^2 cables at 18 cm centres, the sheath circulating loss would be equal to the conductor loss and it is therefore essential to have cross-bonding or some other alternative. If the transfer of power from one phase to another, and the resultant unbalanced voltage drop due to three single-core cables being laid in the same plane, is too large, they may be transposed.

Armour losses. Eddy current and hysteresis losses in the steel armour may exceed the losses in the lead sheath, and may reach as much as 20% for some 3-core cables. Steel armouring is generally unacceptable in single-core cables because of the large losses which would occur.

5.7 Breakdown

Although impregnation of the individual sheets of the paper insulation of a cable, raises the nominal 50-Hz breakdown stress at the conductor surface to about 500 kV/cm if the voltage is applied for a few minutes, (and 2000 kV/cm for a 1/50 microsecond impulse test), breakdown will occur at less than 200 kV/cm if the 50-Hz voltage is maintained for a few days. The breakdown voltage during a short-time test is not a reliable guide as to the breakdown voltage that will occur after a long time, e.g. a cable which has a higher initial value may have a lower long-time value. Breakdown is particularly affected by conditions in the butt-gaps between paper tapes where stress concentrations occur.

Breakdown in cable insulation may occur due to:

(*a*) puncture, i.e. a clean hole through the insulation. This will commonly occur in a laboratory test of a specimen of insulation when the voltage is raised quickly, but is very rare in service;

(*b*) thermal instability;

(*c*) tracking.

These latter two processes have caused some failures in service in the past, but should not occur if certain temperature limits are not exceeded, and failures in modern pressurised cables due to these electrical mechanisms, which will now be discussed briefly, are exceedingly rare. The location of faults in cables and some of their causes are discussed by Gooding (1966).

5.7.1 THERMAL INSTABILITY

The charging current of a cable, as shown in Fig. 5.4, will lead the applied voltage by an angle ϕ_d which is very slightly less than 90°, due to a very small power loss arising partly from a conduction current, since the bulk and surface leakage resistances are not infinite. The bulk insulation resistance may be seen by analogy with [*5.3*] to be given by

$$\frac{\rho \log_e (R/r)}{2\pi l} \text{ ohms} \qquad [5.6]$$

for a length of l metres where ρ is the dielectric resistivity. When an r.m.s. alternating phase-to-neutral voltage V is applied across the effective capacitance C (given by [5.3]), the power loss $\simeq \omega CV^2 \cos\phi_d$ and this exceeds that for an applied direct voltage. The increased power loss occurs in the bulk of the insulation due to dielectric hysteresis, and is generally much greater than the leakage loss.

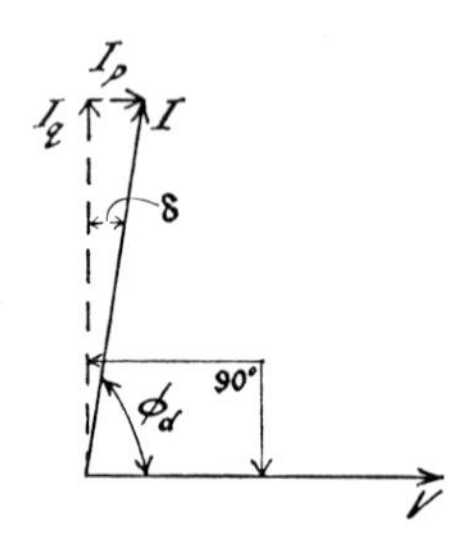

FIG. 5.4. *Phasor diagram for cable dielectric.*

$\cos\phi_d = \sin\delta \simeq \tan\delta \simeq \delta$ (radians), since $\delta = (90-\phi_d)$ and $\delta < 0{\cdot}5°$ for most cables. Cos ϕ_d should be kept very small under all operating conditions, since, if it is large, the power loss is large, and

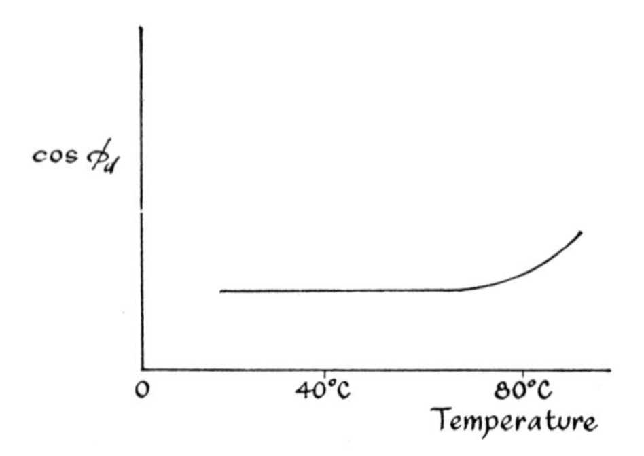

FIG. 5.5. *Variation of cos ϕ_d with temperature.*

the insulation temperature T rises appreciably. If the cable were to operate under conditions where $\partial(\cos\phi_d)/\partial T$ was sufficiently large (see Fig. 5.5), then the temperature could continue to increase until the insulation was damaged. This is because the increasing temperature causes increased power dissipation in the dielectric, which again gives further temperature rise, and the effect is cumulative unless the rate of heat dissipation equals or exceeds the rate of generation.

Modern cable dielectrics have such a low power factor, that thermal instability will not occur under normal operating conditions, except perhaps after the start of the tracking process. Whereas with thermal instability no permanent damage occurs if the temperature rise is halted before a certain maximum is reached, once ionisation has led to the formation of a carbonised core, the following mechanism is likely to continue until breakdown occurs, perhaps after a very long time.

5.7.2 TRACKING FOLLOWING VOID IONISATION

When insulation failures have occurred in service, e.g. in the early solid-type cables used at 33 kV, it was frequently due to a carbonised track, starting from the conductor surface, where the stress was highest, and gradually extending outwards, with 'tree-burning' patterns on the surface of the paper tapes, until breakdown occurred.

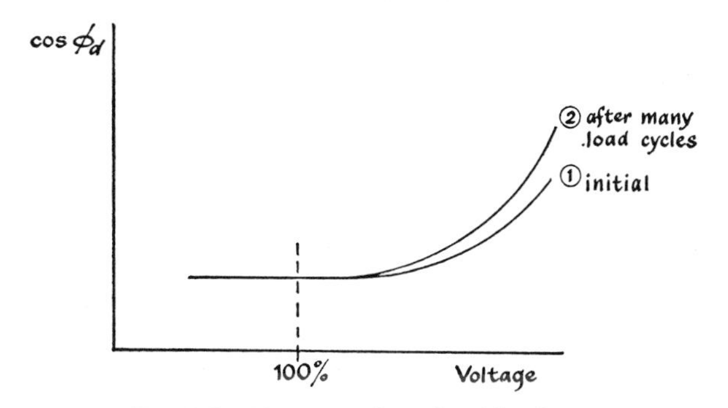

FIG. 5.6. *Variation of cos ϕ_d with voltage.*

This is initiated by ionisation occurring in air or other gas, possibly below atmospheric pressure, in voids which can occur between layers of paper perhaps during load cycles or due to compound drainage. ϕ_d does not vary appreciably with increasing voltage until the stresses are sufficient to cause ionisation to begin, after which cos ϕ_d would rise markedly as shown in curve (1) of Fig. 5.6, and in Fig. 5.5, if more than 100% voltage were applied. In many cables there would be little rise in ϕ_d up to 200% voltage. When the load on the cable is varied, insulating compound expansion, while the temperature is rising, could cause the sheath to become slightly distended, and since this distortion would remain after the compound contracts, voids might form in the insulation. After repeated load cycles, the power factor could therefore increase (see curve (2) in Fig. 5.6), if the temperature were allowed to exceed a certain limit. Under normal operating conditions cos ϕ_d will not vary much either with temperature or voltage, and is likely to be less than 0·005.

The problems involved in determining the maximum permissible temperatures which can be permitted, for continuous normal operation at full rated current, for specified amounts, frequency and duration of emergency loading and during the interval whilst faults are cleared, are of course very complex, and involve far more factors then the simple consideration of thermal instability and tracking above might suggest. Under short-circuit conditions, for example, the permissible temperature rise is limited by mechanical considerations, such as buckling of joints due to conductor expansion, weakening of joint ferrules due to overheating solder, cracking of lead sheaths, and bursting of belt insulation by electromagnetic forces. The continuous loading and the consequent temperature conditions, must be set so that the ageing of the paper does not lead to a failure of its mechanical properties. A reduction of 10°C in ageing temperature increases the time taken to reach the same deterioration by a factor of 2 to 2·5. The maximum permissible temperature of the conductor depends upon the way in which the cable is laid, not only because the dissipation of heat can vary, but also because if a cable is free to contract and expand in a duct, there is a greater possibility of mechanical deterioration of the paper, and more sheath movement than if it is buried directly in the ground. For example, the maximum conductor temperature of a 33-kV solid-type cable in Britain, which is limited rather by avoidance of void formation, than by deterioration of the dielectric materials, is reduced from 65°C for cables buried directly in the ground, to 50°C for plain lead-covered cables installed in ducts, and this is because of lead sheath movement. In the case of an armoured cable, the conductor temperature in ducts is maintained at 65°C because the armour protects the sheath from abrasion. A pressurised cable, where void formation is prevented, is operated at 85°C conductor temperature in ducts, because it has a tough outer sheath and in the event of sheath failure a fall in pressure would indicate damage which can be repaired at convenience, without an electrical failure having occurred due to ingress of moisture.

5.8 Thermal Considerations and Current Rating

Cables may either be directly laid in the ground or may be drawn into ducts or pipes. Wherever possible, the first system is used because it is simpler and cheaper, and the heat dissipation is better. If there is more than one feeder in the same trench the cables of the two circuits are laid a foot or more apart so as to limit mutual heating and avoid a fault on one cable circuit affecting another. They are laid at the bottom of an open trench and are covered with soil or with a special sand. Where three single-core solidly-bonded cables form a 3-phase circuit, they are laid as close together as possible to avoid excessive sheath losses and the inductance is reduced. If cross-bonding is used then the current rating can be substantially increased because of the elimination of sheath circulating currents, and because the spacing between centres may be increased to 25 cm or more.

For a given maximum permissible conductor temperature, the maximum continuous current-carrying capacity, i.e. the rated current, may be determined from the heat dissipation. The heat generated in the conductors in a length of 1 metre of a 3-core cable, when each core carries a current I, is $P = 3\,I^2R_{ac}$ watts. When steady state has been reached this is equal to the heat dissipated through the cable insulation, and away through the soil in the case of a directly-laid cable.

If the steady-state temperature rise of the conductor above ambient temperature is T_c °C and G_c and G_g thermal ohms/metre length of cable respectively represent the thermal resistance between conductor and cable outer surface, and between the latter and the ground surface, then

$$3I^2R_{ac} = \frac{T_c}{G_c+G_g}. \qquad [5.7]$$

A thermal ohm is that thermal resistance which needs a temperature difference of 1°C across it to cause a heat flow of 1 watt. Since the flow of heat and flow of current are analogous, then [5.7] corresponds to Ohm's law. The thermal ohm has a unit of °C/watt. In [5.7], the power and the thermal resistances are those of 1 metre length of cable. The thermal resistivity g is analogous to electrical resistivity, and therefore has the unit of °C m/watt or more commonly is given in °C cm/watt, and this is generally used instead of thermal ohm cm.

Equation [5.7], modified if necessary to include other heat sources such as sheath, armour and dielectric, and the corresponding parts of the thermal resistances through which this heat flows, can be used to calculate the current rating if the thermal resistances are estimated.

For a single-core cable, [5.6] gives by analogy,

$$G_c = \frac{g_1}{2\pi l} \log_e \frac{R}{r} \text{ thermal ohms}/l \text{ metres} \qquad [5.8]$$

where g_1 is the thermal resistivity of the cable dielectric and this may be taken as about 5·5 thermal ohm-m (i.e. 550°C cm/watt) for most high-voltage cables. In the case of a 3-core cable $G_c = 1/3 \times (g_1/2\pi)$ multiplied by an empirical geometrical factor.

In order to calculate G_g, the ground is sometimes assumed to have a constant thermal resistivity g_2 of about 120°C cm/watt, but g_2 varies greatly from one soil to another, e.g. from 60° to 300°C cm/watt from wet clay to dry sand, and this change halves the current rating of a cable for an 85°C maximum temperature. It also varies with time, since it depends markedly on moisture, and the heat of the cable tending to dry out the soil near it, raises g_2 locally.

As with many calculations outlined in a short summary of this kind, the actual physical processes cannot closely be represented by the simple idealised models given, and it is a matter of experience which of the more detailed calculations existing in specialist books or papers must be used. The equations given are therefore intended only to illustrate principles.

If the ideal case of a single cable with external radius R_e buried d metres below the ground surface in a soil of constant thermal resistivity g_2 is considered, then the thermal field is exactly analogous to the electric field between a single overhead line conductor and the earth plane. Comparison with [3.13] of Chapter 3 therefore shows that for a length of l metres

$$G_g = \frac{g_2}{2\pi l} \log_e \frac{2d}{R_e} \text{ thermal ohms.} \qquad [5.9]$$

A change in g_2 of 50% may change the current rating by 10 to 30%, whereas increasing d from 1 m to 2 m may only reduce the rating by about 5%.

If more than one cable is laid in a trench the effects of mutual heating must be taken into account. This may be done in an ideal case by superposition, and the use of image heat sinks, as shown in

Fig. 5.7, where P_A denotes that cable A is a source of P_A watts/metre, and A' is its image sink of $-P_A$ watts/metre.

If heat is considered to flow from a metre length of A alone in an infinite medium of constant resistivity g_2 (which is considered to extend on either side of the ground surface) then the temperature

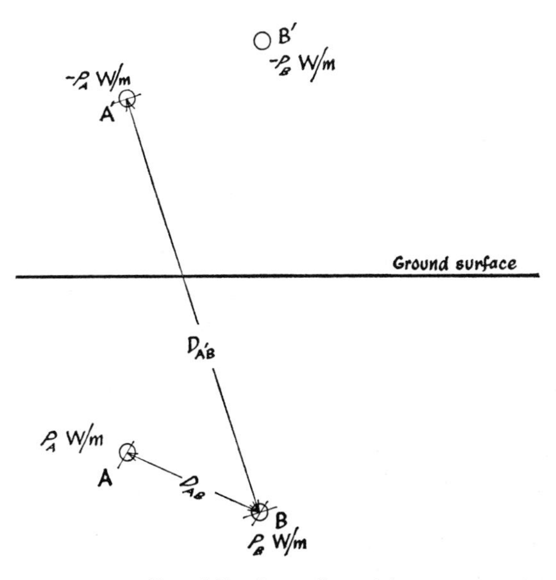

FIG. 5.7. *Image heat sinks.*

difference between B and a point X distant x_1 from A and x_2 from A' is

$$T_1 = P_A\frac{g_2}{2\pi}\log_e\frac{x_1}{D_{AB}}$$

and similarly considering only heat to flow into A', the corresponding temperature difference is

$$T_2 = \frac{-P_A g_2}{2\pi}\log_e\frac{x_2}{D_{A'B}}.$$

The difference in temperature between B and X due to both A and A' is

$$T_1+T_2 = \frac{P_A g_2}{2\pi}\log_e\left(\frac{x_1}{x_2}\cdot\frac{D_{A'B}}{D_{AB}}\right).$$

As x_1 and $x_2 \rightarrow \infty$, $x_1/x_2 \rightarrow 1$ and the temperature at X approaches the ambient temperature. Thus the temperature rise of B above ambient due to cable A is

$$\frac{P_A g_2}{2\pi} \log_e \frac{D_{A'B}}{D_{AB}}. \qquad [5.10]$$

Similarly the temperature rise of cable 1 due to other cables in a total group of n cables may be added, giving the result

$$\frac{g_2}{2\pi} \sum_{A=2}^{A=n} P_A \log_e \frac{D_{A'1}}{D_{A1}}. \qquad [5.11]$$

The total temperature rise of the outer surface of cable 1 buried at depth d with a power generated of P_1 watts/m is given from [5.9] and [5.11] as

$$T_1 = \frac{g_2}{2\pi}\left[P_1 \log_e \frac{2d}{R_e} + \sum_{A=2}^{A=n} P_A \log_e \frac{D_{A'1}}{D_{A1}}\right] \qquad [5.12]$$

The current rating of cables laid in ducts is more complex, because many other factors, including convection losses, must be taken into account, but results are available in tables, charts and factors. Similarly, data are available to determine the currents that can be carried other than continuously, e.g. for intermittent or emergency loading, or short-circuit ratings, when higher than normal operating temperatures may be permitted for very limited periods.

A number of methods of improving the cooling of underground cables in order to increase their current ratings, are under investigation. On a number of 275-kV cable circuits, separate pipes carrying water have been buried near the cables. It has the further advantages of allowing cables to be laid closer together and for two circuits to use one trench, so that the cost of trenches is reduced, and the selection of the route may be easier. Other possible schemes include using pipes in which the cable is immersed in water, and superconducting cables with virtually no resistance.

5.9 D.C. Cables

The losses in a d.c. cable are less than those with a.c., because with a d.c. cable:

(*a*) there is no skin effect so that special construction is avoided;

(*b*) power loss in the dielectric is small since it is only due to leakage current;

(*c*) sheath losses are small as they are only due to leakage and ripple currents, so that cross-bonding equipment is not needed and avoidance of sheath voltages would reduce jointing as longer lengths could be used;

(*d*) there is no continuous charging current so that shunt reactors at the supply end of the cable are no longer needed to take compensating lagging current. Also the difficulty does not arise of the load current of an a.c. cable being restricted by its charging current, which can approach the load current in a relatively few miles of E.H.V. cable, thus necessitating for some lengths further reactive compensation at the load end (see section 1.5). (For 132, 275 and 400-kV cables, the load and charging kVA become about equal in 70, 50 and 40 km respectively.)

Due to these factors, the current-carrying capacity is increased for d.c., by amounts varying from about 10% at 6 cm^2 cross-section to about 18% at 20 cm^2 cross-section.

The great advantage of d.c. working is, however, that the d.c. breakdown stress of oil-filled cable dielectric of about 1000 kV/cm is some three times that for 50 Hz breakdown, so that if the working stress can also be raised by a factor of 3 to about 400 kV/cm, then two d.c. cables can transmit rather more than 2·5 times as much power as three a.c. cables which would have the same capital cost. This increase in working stress involves control of voltage transients so that they do not reach 2·5 times rated voltage, since the impulse strength of oil-filled cable dielectric is about 1000 kV/cm. D.C. cables have now been fully developed up to ±600 kV.

In a d.c. cable the voltage distribution depends upon the insulation resistance instead of upon the capacitance as it does for a.c. operation. From [*5.6*], the leakage current flowing/metre length for a direct voltage V between conductor and sheath is

$$\frac{2\pi V}{\rho \log_e (R/r)} \text{ A/metre.}$$

The voltage across a radial thickness dx at radius $x(r<x<R)$ is

$$\frac{2\pi V}{\rho \log_e (R/r)} \cdot \frac{\rho \, dx}{2\pi x} = \frac{V \, dx}{x \log_e (R/r)}$$

and the stress at this radius is given by

$$E_x = \frac{V}{x \log_e (R/r)} \text{ V/m}$$

which is the same value as for a.c. operation (equation [*5.4*]), and the maximum stress again occurs at the conductor surface, and is of value

$$\frac{V}{r \log_e (R/r)} \text{ V/m.}$$

This, however, has assumed that the dielectric electrical resistivity is everywhere constant, which cannot be true since it is a function of temperature. At a temperature of $T°$

$$\rho_T = \rho_o \, e^{-\alpha T}$$

where ρ_o is the resistivity at a given standard temperature, and α is a constant for a given dielectric. Since after some time on a given load, the resistivity will vary from a minimum at the conductor surface where the temperature is greatest, to a maximum at the sheath or screen, it is possible for the stress at the screen to exceed that at the conductor. Care must be taken to ensure that it does not exceed the designed working stress. The d.c. cable should therefore only have a maximum temperature of 85°C like the a.c. cable, if the temperature gradient has not then already been reached at which this stress inversion occurs. This latter limit is reached at lower conductor temperatures, the thicker the insulation, i.e. the higher the voltage. There is however some compensating effect of electric stress on resistivity, since its value for a stress E, ρ_E is believed to be given by

$$\rho_E = \rho_o \, e^{-\beta E}.$$

In addition to these factors, further difficulties arise due to space charges which do not respond at once to transient temperature changes, and to migration of moisture and contamination in the electric field.

REFERENCES

BARNES, C. C. 1966. *Power cables: their design and installation.* Chapman and Hall, London, 2nd edition.

CHERRY, D. M. 'Containing the cost of undergrounding', 1975, *Proc. I.E.E.*, **122,** pp 293–300.

ENDACOTT, J. D. and GOSLING, C. H. 1966. 'D.C. transmission: overhead lines and cables', *Electrical Review*, **178,** No. 8, pp. 301–304.

GOLDENBERG, H. 1957. 'The calculation of cyclic rating factors for cable laid direct in the ground or in ducts', *E.R.A. Report F/T186*; 1958. 'The calculation of continuous current ratings and rating factors for transmission and distribution cables', *E.R.A. Report F/T187*; and 1958. 'The calculation of cyclic rating factors and emergency loading for one or more cables laid direct in the ground or in ducts', *Proc. I.E.E.*, **105,** C, pp. 46–56.

GOODING, H. T. 1966. 'Cable-fault location on power systems'. *Proc. I.E.E.*, **113,** No. 1, pp. 111–119.

GOSLAND, L. & PARR, R. G. 1960. 'A basis for short-circuit ratings for paper-insulated lead-sheathed cables up to 11 kV (copper conductors, unscreened)', *E.R.A. Report F/T195.*

MIRANDA, F. J. & GAZZANA PRIAROGGIA P., 1976. 'Self-contained oil-filled cables. A review of progress'. *Proc. I.E.E.*, **123,** pp 229–238, and 882–886.

ROSS, A., 'Cable practice in electricity-board distribution networks', 1974, *Proc. I.E.E.*, **121** (IIR), pp 1307–1344.

Examples

1. A 3-phase 3-core belted cable has a capacitance between the three conductors connected together and sheath of 0·6 μF. The capacitance between one conductor and the other two connected to the sheath is 0·7 μF. Calculate the charging current supplied to each conductor when connected to a 6·6 kV, 3-phase, 50 Hz supply.

(H.N.D.) (1·14 A)

2. A cable composed of a conductor of 2 mm diameter covered with 1 mm of insulating material is found to have an insulation resistance of 480 megohm/km. What thickness of a similar material would be required for a 3 mm diameter conductor in order to have an insulation resistance of 960 megohm/km?

(H.N.C.) (4·5 mm)

3. A central conductor radius r is surrounded by a concentric earthed metallic tube of inner radius R. The space between them is filled by insulating materials of differing relative permittivities. From radius r to r_1 it is ϵ_{r_1}, from r_1 to r_2 it is ϵ_{r_2} etc. If a p.d. V is applied between conductor and tube, obtain an expression for the voltage gradient at any radius x from the tube centre.

If two materials fill the space, the inner having a relative permittivity of 5 and the outer 2·5, calculate at what value of r_1 the two materials will withstand equal voltages if $R = 5$ cm and $r = 1$ cm.

(University of London)

$$\left(E_x = \frac{V}{\epsilon_{r\,x} \cdot x\left[\frac{1}{\epsilon_{r_1}}\log_e\frac{r_1}{r} + \frac{1}{\epsilon_{r_2}}\log_e\frac{r_2}{r_1} + \ldots\right]};\ 2{\cdot}9\text{ cm}\right)$$

4. A single core cable with homogeneous dielectric has a conductor radius r cm and sheath radius R cm. A single intersheath radius x cm is to be employed. The voltage gradient at the conductor surface is E_1 and that at the intersheath surface nearest the external sheath is E_2. Obtain an expression from which can be determined the value of x which will enable the cable to withstand the highest voltage between core and sheath when $E_1 > E_2$.

(University of London)

$$\left(\frac{E_1 r}{x} + E_2 \log_e \frac{R}{x} = E_2\right)$$

5. A 3-core belted cable is connected to an 11-kV 50-Hz 3-phase supply. Each core takes a charging current of 2 A. C_s is 0·4 μF. Calculate the angle of phase displacement between the charging current entering core 1 and the voltage between cores 2 and 3. Calculate also the magnitudes and phase relationships of the component currents which make up the total charging current of core 1.

(University of London)

(180°, $0{\cdot}8\underline{/0°}$ A, $0{\cdot}69\underline{/30°}$ A and $0{\cdot}69\underline{/-30°}$A)

6. Two concentric cylinders of radii r and R metres, are separated by a homogeneous dielectric of maximum working stress E_m V/m. A third concentric cylinder of radius x and negligible thickness is inserted between the two.

Find expressions (*a*) from which x can be determined for given values of r and R, and (*b*) from which the p.d. between the third cylinder and the outer one can be found; in order that the highest possible voltage may be maintained between inner and outer cylinders.

(University of London)

$(R/x = e^{(1-r/x)};\ V_{x\,R} = E_m(x-r))$

7. A single-core cable laid in the ground has a conductor of diameter 2·5 cm, a dielectric thickness of 4 cm, a thermal resistivity of the dielectric of 5 thermal ohm metre, an electrical core resistance of 100 microhm metre and a current rating of 500 A. Determine the thermal resistance between the sheath and the ground surface, if

the temperature difference between the conductor and the ground surface is 50°C at rated current.

(L.C.T.) (0·855 thermal ohm)

8. A single-phase concentric cable 2·4 km long is connected to 50-Hz, 6·6-kV busbars. The inner conductor has a diameter of 1·25 cm and the radial thickness of insulation is 0·8 cm. The relative permittivity of the dielectric is 3·5. Determine the charging kVA.

(H.N.C.) (7·75 kVA)

9. A single-phase voltage of 10 kV at a frequency of 50 Hz is applied between two of the cores of a 3-phase belted cable. The core-to-core capacitance is 0·15 μF and the core-to-sheath capacitance is 0·1 μF.

Calculate (*a*) the potential difference between the third core and the sheath and (*b*) the total charging current taken by the cable.

(University of London) (0, 0·86 A)

10. A 3-phase 11-kV 50-Hz belted cable has a capacitance between any core and sheath of 0·5 μF and a core-to-core capacitance of 0·3 μF. Calculate the instantaneous value of the current between the sheath and any selected core at the instant when the charging current carried by that core has its maximum value.

(University of London) (1·41 A)

11. Calculate the charging current taken by a belted-type 3-core cable 32 km long when connected to a 3-phase 50-Hz supply with 20 kV between lines.

The capacitance between two cores with the third core and sheath insulated is 0·0625 μF/km.

If a supply is given at the far end of the cable to a factory which takes 857 kVA at 0·7 power factor lagging, calculate the power factor and current at the sending end of the cable. Neglect cable losses and inductance.

(H.N.C.) (14·5 A, 0·98, 17·8 A)

12. A coaxial cable has a cylindrical inner conductor of radius *a* and an outer conductor of inner radius *b*. The insulant has a resistivity ρ and a relative permittivity ϵ_r. Find an expression for the capacitance/metre length of the cable, and from it obtain an expression for the insulation resistance/metre length, assuming the resistivity to be constant and uniform.

Due to the temperature-gradient in service, the resistivity of the insulating material of a high voltage d.c. coaxial cable working at a voltage V between inner and outer conductors is related to the radius x by the expression $\rho_x = kx^2$. Find the distribution of voltage gradient through the insulant. State the values of the gradient at the inner and outer radii if $V = 300$ kV, $a = 2{\cdot}5$ cm and $b = 7{\cdot}0$ cm. Compare these values with what they would be had the resistivity been uniform throughout, and comment on the implications.

(University of London) (35 kV/cm, 98 kV/cm, 117 kV/cm, 42 kV/cm)

13. Three $16{\cdot}1$ cm^2 single-core lead-sheathed oil-filled paper-insulated copper cables are laid together in a duct, through which circulating water flows. They carry a 3-phase 400-kV, 50-Hz supply of 1100 MVA over a distance of 3·2 km. The outer and inner diameters of the insulation are 10 cm and 5·45 cm respectively. The three lead sheaths are cross-bonded at intervals.

If the mean copper temperature is 80°C and the insulation has a relative permittivity of 2·25 and a power factor of 0·012, determine (*i*) the total copper loss, (*ii*) the total dielectric loss, (*iii*) the peak electric stress in the insulation and (*iv*) the water flow in litres per minute. The sheath losses are equal to 75% of the copper loss value and the temperature of the cooling water rises by 6°C during its passage through the duct.

[For copper at 0°C, resistivity $\rho = 0{\cdot}66\ \mu\Omega$-in, temperature coefficient $\alpha = 0{\cdot}00427$ per °C.]

(University of London) (340 kW, 398 kW, 197 kV/cm, 2370)

Chapter 6

TRANSFORMERS

6.1 Equivalent Circuit of a Two-winding Transformer

If a 3-phase transformer is supplied with balanced 3-phase voltages and the load on it is balanced, then conditions in each phase are similar except for the phase displacement. A single-phase equivalent

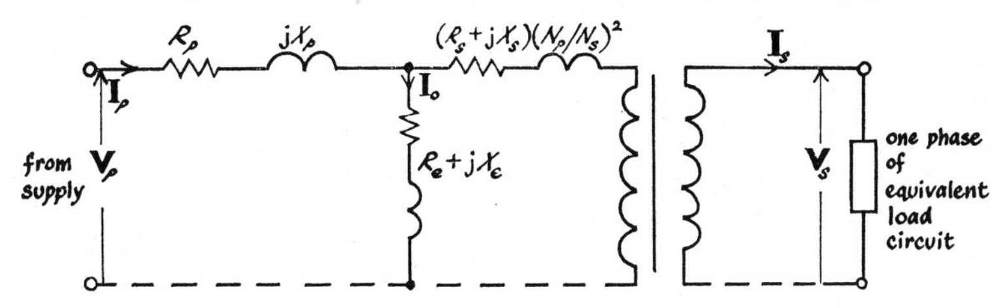

FIG. 6.1.

$\mathbf{V}_p$ = *line to neutral, i.e., phase, voltage applied to primary winding.* $\mathbf{V}_s$ = *line to neutral, i.e., phase, voltage, at the secondary terminals.* $\mathbf{I}_p$ = *current flowing from the supply into each phase winding of the primary.* $\mathbf{I}_s$ = *current in each phase winding of the secondary flowing to the load circuit.* $\mathbf{I}_o$ = *no-load current flowing in each phase of primary.* R_p = *primary resistance/phase, ohms.* X_p = *primary leakage reactance/phase, ohms.* R_s = *secondary resistance/phase, ohms.* X_s = *secondary leakage reactance/phase, ohms.* R_e+jX_e = *effective exciting impedance in which the no-load current* $\mathbf{I}_o$ *flows.* $\mathbf{I}_o$ *has components which (a) magnetise the core and (b) supply the power lost in the iron due to the alternating flux.* N_p = *number of primary turns/phase.* N_s = *number of secondary turns/phase.*

circuit such as that shown in Fig. 6.1, where the secondary impedance has been referred into the primary, may be used for steady-state operation of a star/star connected transformer.

It is less justifiable to consider the exciting impedance to be a linear element, i.e. constant, than it is in the case of the series impedances, since core loss and magnetising components of current vary non-linearly with core flux density. However, the power transformer normally operates at almost constant voltage and frequency, so that the exciting impedance is approximately constant over the normal working range.

The no-load current $\mathbf{I}_o$ is, however, very small compared with the rated current. The latter is frequently referred to as 1·0 per unit current, and $\mathbf{I}_o$, though varying with the type and size of transformer, is generally between 0·02 and 0·04 per unit (see Appendix 1). Since also, most power system calculations concern a network of which each transformer is only one piece of equipment among many, it is generally acceptable to neglect $\mathbf{I}_o$, so that the exciting impedance $R_e + \mathrm{j}\, X_e$ can be omitted from the equivalent circuit, in which case the short-line formulae of section 1.2 apply. There are of course, cases where this cannot be done, e.g. in considering the harmonics (section 6.5), magnetising-inrush current (section 6.6), or the zero-sequence network required in unbalanced 3-phase calculations (section 6.10), or in switching over-voltages which may arise in interrupting the no-load current of a transformer (see Volume 2).

If the exciting impedance is neglected, the transformer may then be represented for steady-state conditions, by a series resistance and leakage reactance together with an ideal transformer.

These ohmic values can be referred to the primary side as in Fig. 6.1, or to the secondary side. In the latter case an impedance of

$$\left[R_s + R_p\left(\frac{N_s}{N_p}\right)^2\right] + \mathrm{j}\left[X_s + X_p\left(\frac{N_s}{N_p}\right)^2\right] \quad \text{ohms}$$

may be placed in the secondary circuit, with no impedance on the primary side. It may be noted that for efficient design conditions R_s is nearly equal to $R_p(N_s/N_p)^2$ and X_s is nearly equal to $X_p(N_s/N_p)^2$, and similar conditions apply when impedances are referred to the primary winding.

It is more usual and convenient to use per unit values of impedance, where the per unit resistance at rated MVA

$$= \left[R_p + R_s\left(\frac{N_p}{N_s}\right)^2\right] \times \frac{\text{rated primary phase current}}{\text{rated primary phase voltage}}$$

and the per unit reactance is similar. It is shown in Appendix 1 that for a given MVA base and for operation at rated voltages on both sides of the transformer, there can only be one value of per unit impedance. Exactly the same value is given if secondary quantities are used in the above expression for per unit resistance or reactance. The series impedance of a power transformer is generally between about 0·05 and 0·15 per unit, and tends to the higher values at the higher voltages and ratings. The exciting impedance to positive-sequence current, on the other hand is of the order of 25 per unit.

In the case of a tap-changing transformer, where the turns ratio can vary from the nominal value which is the ratio between the nominal rated voltages of the two systems to which it is connected, two points may arise:

(*a*) these two systems may be connected by two or more transformers of differing ratios; this can be dealt with by referring all ohmic impedances to a single voltage level and putting ideal transformers in series with those with off-nominal taps to allow for the difference in voltage; and

(*b*) the impedance may vary with tap position, although one of the aims in the design of tap-changers is to minimise this variation.

If one of the transformer windings is delta-connected, then ohmic impedances are still referred into the other winding by multiplying by the square of the ratio of the turns/phase. If the impedances are referred in this way from the star-connected winding into the delta-connected winding, then these impedances and those of the latter winding are connected in delta. These three equal delta-connected impedances **Z** can then be replaced by an equivalent star-connected set of impedances **Z**/3 by using the delta-star transformation (see Appendix 2).

It is then possible to consider a star-connected winding with the new impedances giving the same line voltage and line current. If impedances are stated in per unit then they are independent of whether the windings are connected in delta or in star. The method of connection of the transformer windings is immaterial in determining system currents, until the actual currents within the transformer phases need to be calculated.

6.2 3-phase Two-winding Transformer Types

3-phase transformers are generally either of the 3-limb core-type, or consist of a bank of three separate single-phase units. The latter arrangement reduces the weight and size of individual units, which eases transport problems, but is less efficient and more expensive except in so far as the provision of a spare unit costs less if it is single-phase.

Fig. 6.2 indicates the arrangement of a 3-limb core-type transformer with star/delta connected windings. This figure is merely diagrammatic, e.g. it shows primary and secondary windings side by

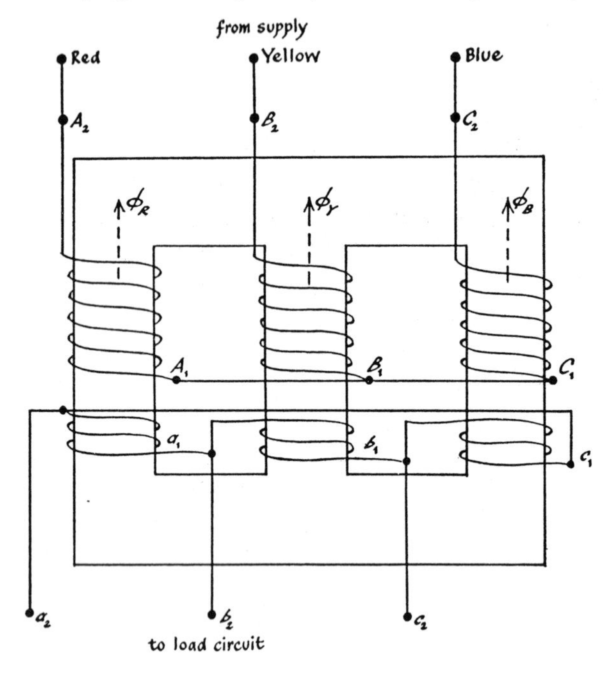

FIG. 6.2. *3-limb core-type transformer with star/delta connected windings, shown diagrammatically.*

side on a limb which would give a high leakage reactance. Some 5-limb transformers are made when it is vital to obtain the lowest height, e.g. for certain very large transformers such as 600-MVA, 400-kV units.

6.3 Phase Shift

If both primary and secondary windings are connected in the same way either in star or in delta, with corresponding terminals joined, then there is no phase shift between the corresponding voltage of any phase to neutral on either side (although a phase shift of 180° could occur if for example A_1, B_1 and C_1 terminals were joined to form a star point on the primary side, and a_2, b_2 and c_2 were joined on the secondary side).

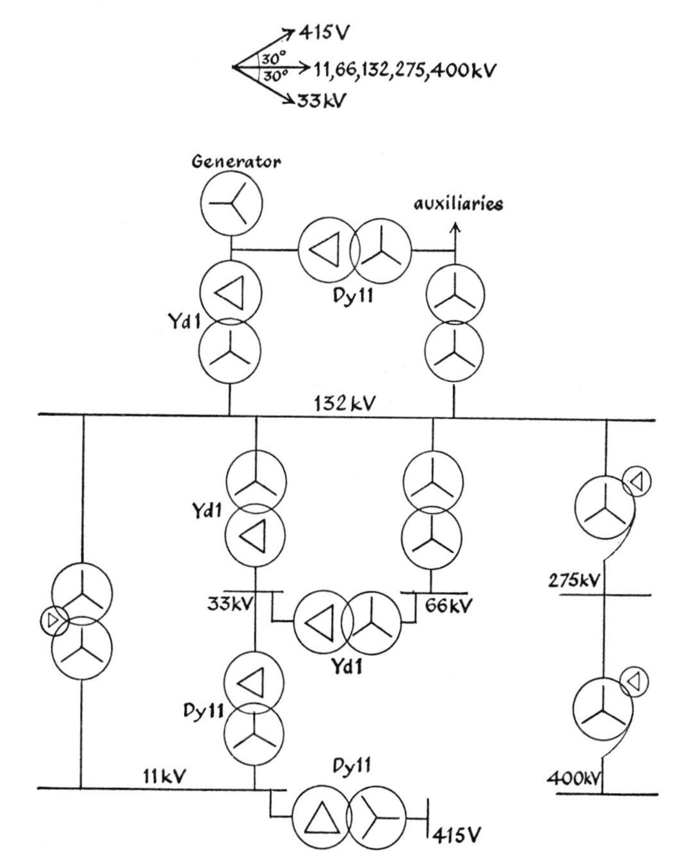

FIG. 6.3. *Typical power transformer connections.*

The following conventions are adopted in drawing connection and phasor diagrams for transformers (B.S. 171). The H.V. (high-voltage) winding is given capital letters and the L.V. winding small letters. Subscript 1 is the neutral end (if any); subscript 2 (or higher if there are tappings) is the line end. The polarity markings are such that e.m.f.s are instantaneously in time phase between terminals with corresponding markings, i.e. at the instant when the p.d. of

A_1 with respect to A_2 is at a positive maximum value, that of a_1 with respect to a_2 is also at positive maximum value.

For reasons of harmonic reduction which are discussed in section 6.5, a delta-connected winding is desirable in many power transformers, and it is only for some larger sizes that it is economical to make a third winding (tertiary) which can be delta-connected.

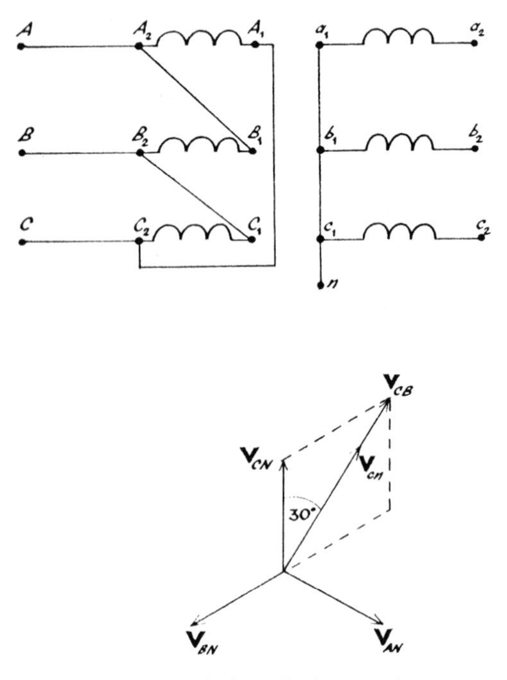

FIG. 6.4. *Delta/star Dy 1 connection.*

In the C.E.G.B. and Area Board Systems, there are therefore different transformer connections as illustrated in Fig. 6.3. It can be seen that there is no phase shift between the 400-kV and 11-kV systems, but the 415-volt system has phase voltages which lead those of the 11-kV system by 30°. Since a 30° phase shift is introduced between the 132-kV and 33-kV bars, the 33/11-kV transformers must give the opposite 30° shift. The reason for the forward or backward 30° shift introduced by a delta/star or star/delta transformer can be seen from Figs. 6.4 and 6.5.

In the case of the Dy 1 connection shown in Fig. 6.4, the voltage $\mathbf{V}_{CB} = \mathbf{V}_{CN} + \mathbf{V}_{NB} = \mathbf{V}_{CN} - \mathbf{V}_{BN}$ is across the C phase of the primary, and this voltage will be in phase with the voltage $\mathbf{V}_{cn}$ across the c phase of the secondary since these windings are wound on the same limb and are linked by the same flux. Hence the latter voltage $\mathbf{V}_{cn}$ lags by 30° behind the corresponding phase voltage $\mathbf{V}_{CN}$ on the primary side. (The neutral point N on the primary side may or may not be physically available. If it is available it may be at the star

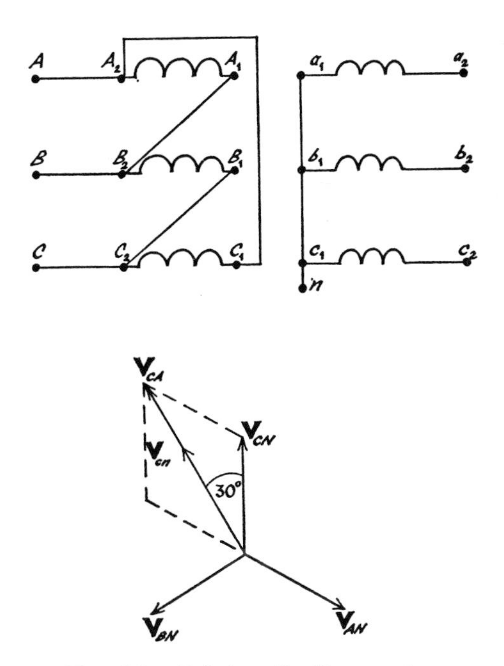

FIG. 6.5. *Delta/star Dy 11 connection.*

point of a star-connected alternator stator or transformer secondary supplying the delta/star transformer, or it may be obtained by means of a zig-zag earthing transformer.)

In the case of the Dy 11 connection shown in Fig. 6.5 the voltage across the C phase winding of the primary is now $\mathbf{V}_{CA} = \mathbf{V}_{CN} - \mathbf{V}_{AN}$. There is a 30° forward phase shift between the phase voltages on the secondary side, and the corresponding phase voltages on the primary side.

6.4 Parallel Operation of 3-phase Transformers

Satisfactory operation of 3-phase transformers in parallel requires fulfilment of the following conditions:

(*a*) All transformers must belong to the same group as regards phase shift: i.e. all star/star with corresponding phase ends joined to form neutral points, or all Dy 11, etc. It follows that all terminals must be correctly marked and the appropriate ones connected. This condition is absolutely essential.

(*b*) All transformers should have the same voltage ratios. If the secondary e.m.f.s differ, a circulating current will flow between the transformers, and since the windings have a low impedance, only a very small difference in ratio can be permitted without undue circulating current.

(*c*) The per unit impedances of all transformers expressed at their full-load MVA rating should be the same. (An alternative way of expressing this condition is that their ohmic impedances should be inversely proportional to their full-load MVA ratings.)

It will be seen below that this condition should be met if the transformers are to share the load in proportion to their ratings, and are thus able to give a total load equal to the sum of their ratings.

(*d*) The X/R ratios (leakage reactance/resistance for both windings together) should be equal for all transformers. This is shown below to result in all transformers having the same power factor, which gives the most economical condition. This is because it minimises the currents in the transformers for a given total load, and the transformers can then supply a greater load without exceeding their individual MVA ratings.

6.4.1 EQUAL RATIOS

Two transformers A and B with equal turns ratios, are in parallel supplying a balanced 3-phase load. The impedance of the primary when referred to the secondary and added to the impedance of the secondary, gives a total impedance of $\mathbf{Z}_A = R_A + \text{j}\, X_A$ ohms and $\mathbf{Z}_B = R_B + \text{j}\, X_B$ ohms respectively as shown in Fig. 6.6.

Impedances are shown bold-face, viz. $\mathbf{Z}$ because although they do not vary sinusoidally in time as phasors do, they are complex operators, i.e. $\mathbf{Z} = R + \text{j}\, X = Z/\theta$ where $Z = (R^2 + X^2)^{\frac{1}{2}}$ and $\theta =$

$\tan^{-1} X/R$. It is important that the complex form of $\mathbf{Z}$ is used and not merely its modulus $|\mathbf{Z}|$ or Z, and this is true in virtually all power system calculations.

Fig. 6.6 is drawn for a star/star transformer for simplicity. A delta-connected winding can be replaced by an equivalent star-connected winding. The current in one phase of the load is given by

$$\mathbf{I} = \mathbf{I}_A + \mathbf{I}_B \qquad [6.1]$$

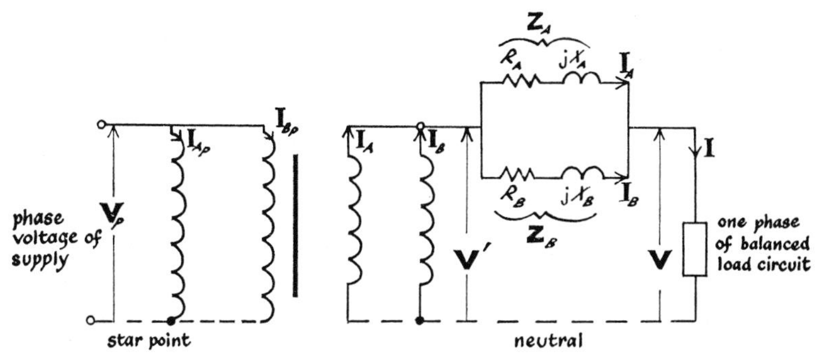

FIG. 6.6. *Transformers in parallel.*

and the two impedance voltages are equal so that

$$\mathbf{I}_A \mathbf{Z}_A = \mathbf{I}_B \mathbf{Z}_B. \qquad [6.2]$$

Equations [*6.1*] and [*6.2*] give

$$\mathbf{I}_A = \mathbf{I} \, . \, \frac{\mathbf{Z}_B}{\mathbf{Z}_A + \mathbf{Z}_B} \qquad [6.3]$$

$$\mathbf{I}_B = \mathbf{I} \, . \, \frac{\mathbf{Z}_A}{\mathbf{Z}_A + \mathbf{Z}_B}. \qquad [6.4]$$

If transformers A and B supply power of P_A and P_B watts respectively and lagging reactive power Q_A and Q_B then (see Appendix 4)

$$P_A + \mathrm{j}\, Q_A = 3\mathbf{V}\mathbf{I}_A^*$$

$$P_B + \mathrm{j}\, Q_B = 3\mathbf{V}\mathbf{I}_B^*$$

where $\mathbf{V}$ is the phase terminal voltage across the load and $\mathbf{I}_A^*$ and $\mathbf{I}_B^*$ are the conjugates of $\mathbf{I}_A$ and $\mathbf{I}_B$. From [*6.3*] and [*6.4*]

$$P_A + \mathrm{j}\, Q_A = (P + \mathrm{j}\, Q) \, . \left(\frac{\mathbf{Z}_B}{\mathbf{Z}_A + \mathbf{Z}_B} \right)^* \qquad [6.5]$$

and

$$P_B + \mathrm{j}\,Q_B = (P + \mathrm{j}\,Q)\,.\left(\frac{\mathbf{Z}_A}{\mathbf{Z}_A + \mathbf{Z}_B}\right)^* \qquad [6.6]$$

where $(P+\mathrm{j}\ Q) = 3\mathbf{VI}^*$ is the total load supplied by the two transformers. Dividing [6.5] by [6.6] gives

$$\frac{P_A + \mathrm{j}\,Q_A}{P_B + \mathrm{j}\,Q_B} = \left(\frac{\mathbf{Z}_B}{\mathbf{Z}_A}\right)^*. \qquad [6.7]$$

In order to be able to supply the maximum possible load volt-amperes the transformers must share the total load in proportion to their ratings; e.g. suppose that a factory load is rising and an existing 1000-kVA transformer A is to be reinforced by connecting a new 200-kVA transformer B in parallel with it. If $Z_B/Z_A = 4$ instead of the correct ratio of 5, then [6.7] shows that the new transformer will take 1/5 of the total load, so that when the latter is 1000 kVA, the new transformer is fully loaded. There would therefore be no increase at all in the load that could be supplied after installing the new transformer. This indicates that what might appear to be a relatively small deviation from the correct impedance ratio can be serious. It is therefore necessary that the ratio of the loads which must also be the ratio of the ratings, gives from [6.7]

$$\frac{|P_A + \mathrm{j}\,Q_A|}{|P_B + \mathrm{j}\,Q_B|} = \frac{Z_B}{Z_A} = \frac{\text{full-load rating of } A}{\text{full-load rating of } B} = \frac{3VI'_A}{3VI'_B} \qquad [6.8]$$

where I'_A and I'_B are the rated or full-load currents on the secondary sides. Equation [6.8] shows that the ohmic impedances must be inversely proportional to the ratings. It is more convenient, however, in most cases to use per unit impedances. The per unit impedance of transformer A at its own volt-ampere rating is $I'_A\,\mathbf{Z}_A/V$ and that of B at its own rating is $I'_B\mathbf{Z}_B/V$. It must be noted of course that these per unit impedances are complex quantities in exactly the same way as ohmic impedances.

Equation [6.8] shows from these definitions of per unit impedance, that the transformers should have equal magnitudes of per unit impedances (each being based on its own rating).

Fig. 6.7 shows that if the transformers are operated with each of their secondary power factors equal to that of the load, viz. $\cos\phi$, then for a given total load, smaller currents flow through them than if they were to operate on different power factors $\cos\phi_A$ and $\cos\phi_B$.

Thus if the power factors are equal the transformers can supply a greater total load without exceeding their ratings (see section 1.2.5). From Fig. 6.7(b) the power delivered to the load is given by

$$P_A = 3VI_p$$
$$Q_A = 3VI_q$$

and

$$\tan \phi_A = \frac{I_q}{I_p} = \frac{Q_A}{P_A} \qquad [6.9]$$

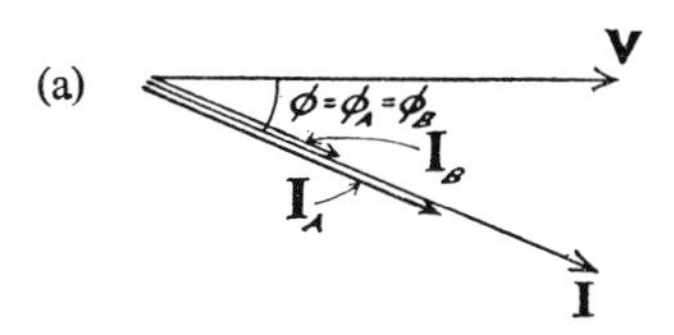

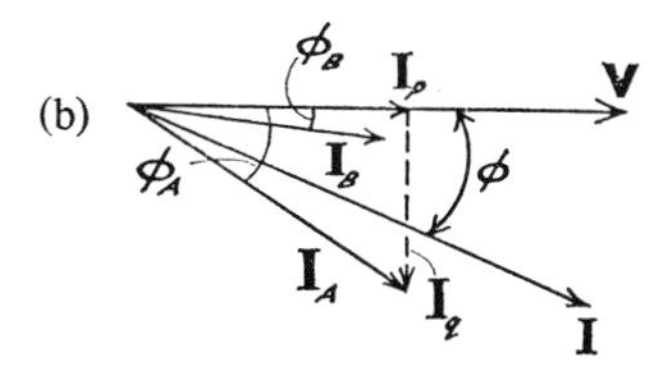

FIG. 6.7. *Secondary phasor diagram (a) for equal power factors, (b) for unequal power factors.*

and a similar expression can be written for $\tan \phi_B$. Hence for equal power factors

$$\frac{Q_A}{P_A} = \frac{Q_B}{P_B}$$

and the ratios given by [6.7]

$$\frac{P_A + jQ_A}{P_B + jQ_B} = \left(\frac{\mathbf{Z}_B}{\mathbf{Z}_A}\right)^*$$

must be 'real', and this is satisfied if

$$\frac{X_A}{R_A} = \frac{X_B}{R_B}. \qquad [6.10]$$

It should be noted that the power factor of the load, cos ϕ, which is the cosine of the angle between secondary phase current and secondary phase terminal voltage, is not the same as the cosine of the angle between primary phase current and primary phase terminal voltage, because of the impedance voltage of the transformer. This can be seen from the phasor diagram shown in Fig. 6.8, which

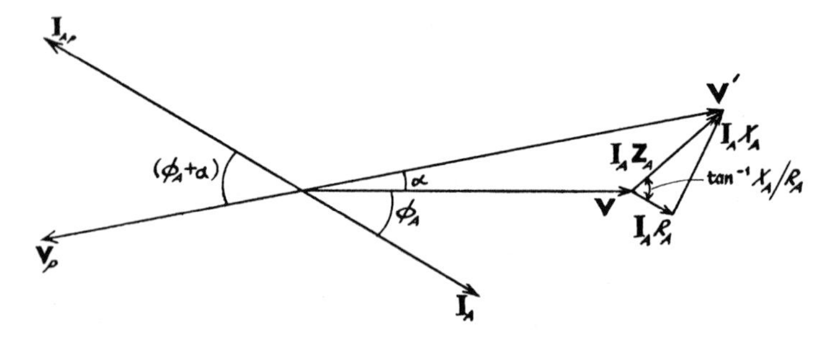

FIG. 6.8. *Phasor diagram for one transformer.*

represents conditions for transformer A in Fig. 6.6. The primary power factor of transformer A is cos $(\phi_A+\alpha)$. If the conditions of [*6.8*] and [*6.10*] are satisfied, $\mathbf{I}_A$ and $\mathbf{I}_B$ are in phase so that both primary power factors are equal and similarly both secondary power factors are equal.

Parallel operation of transformers with equal ratios will now be illustrated with an example. In order to show that it does not matter whether impedances are referred to primary or secondary sides, the problem is solved on the primary side, instead of considering the secondary side as has been done in this section. The student should note that the symbols $\mathbf{I}_A$ and $\mathbf{I}_B$ are used (for convenience in avoiding double suffixes), as the primary currents, whereas they have been used for secondary currents in this section.

Worked Example 6.1

Two 3-phase transformers A and B with equal ratios of transformation, and each star-connected on both primary and secondary sides, operate in parallel to supply a load. Transformer A has a primary current of 100 A and its power factor on the primary side is 0·7 lagging.

Calculate the power factor and current on the primary side of transformer B. Calculate also the power factor of the load and the total load current. Both transformers have three times as many turns/phase in the secondary winding as in the primary. $R_A = 0{\cdot}5\ \Omega$, $X_A = 1{\cdot}5\ \Omega$, $R_B = 0{\cdot}6\ \Omega$ and $X_B = 1{\cdot}8\ \Omega$, where these equivalent resistances and leakage reactances are all referred to the primary and are those for each phase.

If the primary line voltage is 10 kV, calculate the resistance and reactance per phase of a single star-connected load which would be equivalent to the actual load.

If for transformer A, an inductive reactance of $0{\cdot}5\ \Omega$ with negligible resistance, is connected between each primary terminal and its supply terminal, calculate the new value of primary current and its power factor if the load impedance remains unchanged.

SOLUTION

Fig. 6.9 shows one phase of the transformers and load. If the load is referred to the primary then the equivalent circuit of Fig. 6.10 can

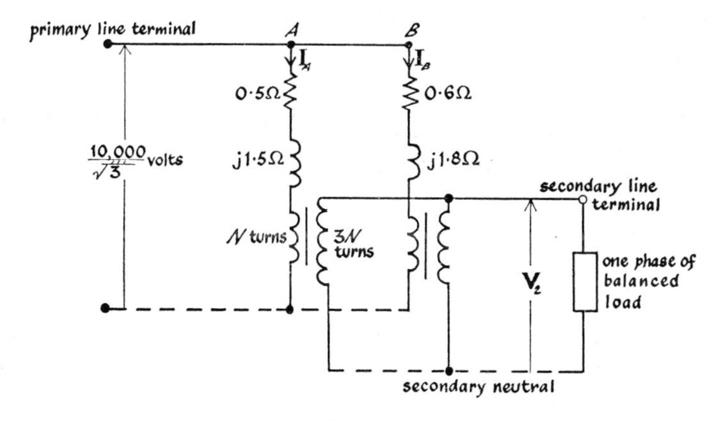

FIG. 6.9.

be drawn, with the impedance-less primary and secondary windings omitted.

The primary current of transformer A is

$$\mathbf{I}_A = 100 \times 0{\cdot}7 - \mathrm{j}\ 100(1-0{\cdot}7^2)^{\frac{1}{2}} = 70 - \mathrm{j}\ 71{\cdot}2\ \mathrm{A}$$

where the reference phasor is the voltage applied to the corresponding primary phase, which is $(10\ 000/\sqrt{3})\underline{/0^\circ}$ volts.

Equating the voltage drops across the impedances of the two transformers,

$$(70 - j\,71{\cdot}2)(0{\cdot}5 + j\,1{\cdot}5) = \mathbf{I}_B(0{\cdot}6 + j\,1{\cdot}8)$$

$$\mathbf{I}_B = \frac{5}{6}(70 - j\,71{\cdot}2) = 58{\cdot}3 - j\,59{\cdot}3\ \text{A}$$

$$I_B = (58{\cdot}3^2 + 59{\cdot}3^2)^{\frac{1}{2}} = 83\ \text{A}$$

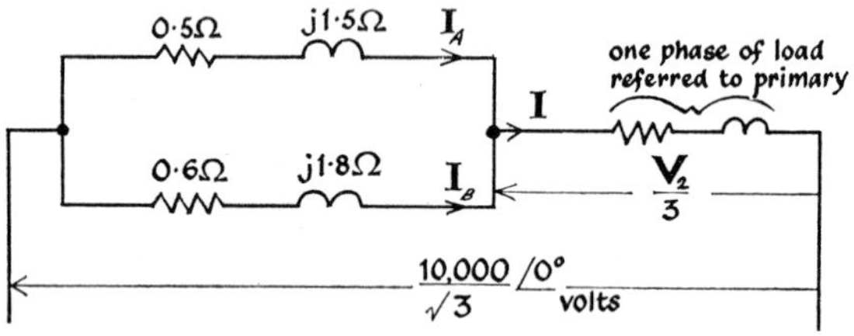

FIG. 6.10.

The primary power factor of B is 0·7 since $X_A/R_A = X_B/R_B$ (see the discussion at the end of section 6.4.1).

Since the power factors are equal, the currents $\mathbf{I}_A$ and $\mathbf{I}_B$ are in phase so that

$$I = I_A + I_B = 183\ \text{A}.$$

This is the load current referred to the primary so that the actual current in the load is 183/3 = 61 A.

The load power factor is the cosine of the angle between 61 A and $\mathbf{V}_2$ or between 183 A and $\mathbf{V}_2/3$. This latter voltage which is across one phase of the load when it is referred to the primary is given by

$$\frac{\mathbf{V}_2}{3} = \frac{10\,000}{\sqrt{3}} - \mathbf{I}_A\mathbf{Z}_A = 5780 - (70 - j\,71{\cdot}2)\,(0{\cdot}5 + j\,1{\cdot}5)$$

$$= 5638 - j\,69\ \text{volts}$$

$$\frac{\mathbf{V}_2}{3}\ \text{lags}\ \frac{10\,000}{\sqrt{3}}\ \text{by}\ \tan^{-1}\frac{69}{5638} = 0^\circ\ 42'$$

$$183\ \text{A lags}\ \frac{10\,000}{\sqrt{3}}\ \text{by}\ \cos^{-1} 0{\cdot}7 = 45^\circ\ 36'$$

$$183\ \text{A lags}\ \frac{\mathbf{V}_2}{3}\ \text{by}\ 44^\circ\ 54'$$

so that the load p.f. = cos 44° 54′ = 0·71 lagging.

The load impedance referred to the primary

$$= \frac{5638-j\,69}{183(0{\cdot}7-j\,0{\cdot}712)}\,\Omega = 20{\cdot}9+j\,21{\cdot}3\,\Omega.$$

Each phase of the equivalent star-connected load at the secondary terminals is therefore

$$9(20{\cdot}9+j\,21{\cdot}3)\,\Omega = 188{\cdot}1+j\,191{\cdot}7\,\Omega.$$

When j 0·5 Ω is connected in series with each primary phase of transformer *A* the equivalent circuit becomes that of Fig. 6.11.

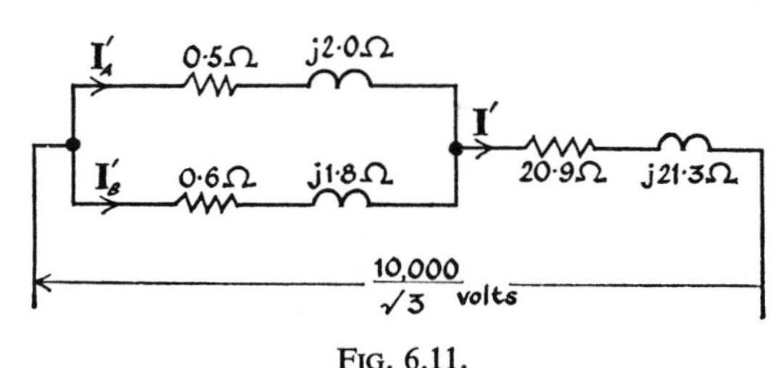

FIG. 6.11.

The total primary input impedance per phase

$$= 20{\cdot}9+j\,21{\cdot}3+\frac{(0{\cdot}5+j\,2)\,(0{\cdot}6+j\,1{\cdot}8)}{1{\cdot}1+j\,3{\cdot}8}\,\Omega$$

$$= 21{\cdot}24+j\,22{\cdot}25\,\Omega$$

$$\mathbf{I}' = \frac{10\,000}{\sqrt{3}\,(21{\cdot}2+j\,22{\cdot}25)} = 187{\cdot}8\underline{/-46°\,20'}\text{ A}$$

$$\mathbf{I}'_A = (129{\cdot}7-j\,135{\cdot}7)\frac{(0{\cdot}6+j\,1{\cdot}8)}{(1{\cdot}1+j\,3{\cdot}8)}$$

$$= 59{\cdot}4-j\,67{\cdot}4\text{ A}$$

$$I'_A = 89{\cdot}7\text{ A}$$

The power factor of transformer *A* is

$$\cos\left(\tan^{-1}\frac{67{\cdot}4}{59{\cdot}4}\right)$$

$$= 0{\cdot}661\text{ lagging.}$$

6.4.2 UNEQUAL RATIOS

If transformers *a* and *b* have slightly different turns ratios, so that in the equivalent circuit with all impedance referred to the secondary side, the e.m.f.s induced in the secondaries are $\mathbf{E}_a$ and $\mathbf{E}_b$ where $E_a > E_b$, then a local circulating current

$$\mathbf{I}_l = \frac{\mathbf{E}_a - \mathbf{E}_b}{\mathbf{Z}_a + \mathbf{Z}_b}$$

will flow in the transformer windings as shown in Fig. 6.12. If the transformers contribute $\mathbf{I}_a'$ and $\mathbf{I}_b'$ towards the total load current $\mathbf{I}$,

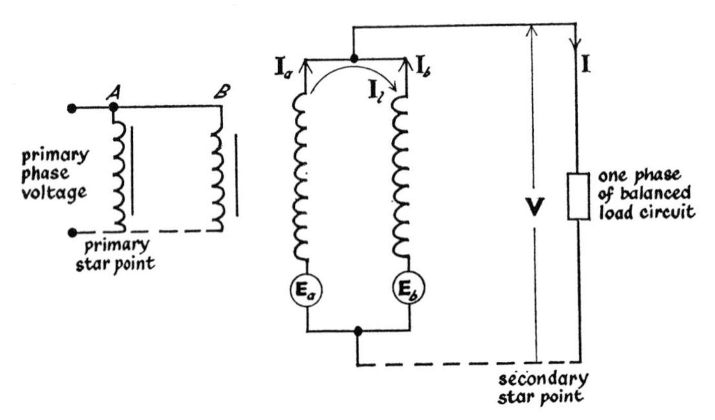

FIG. 6.12. *Transformers with unequal turns ratios in parallel.*

the actual transformer currents are $\mathbf{I}_a = (\mathbf{I}_a' + \mathbf{I}_l)$ and $\mathbf{I}_b = (\mathbf{I}_b' - \mathbf{I}_l)$. Equation [*6.1*] still applies, viz. $\mathbf{I}_a + \mathbf{I}_b = \mathbf{I}$ but [*6.2*] does not, and is replaced by

$$(\mathbf{E}_a - \mathbf{I}_a\mathbf{Z}_a) = (\mathbf{E}_b - \mathbf{I}_b\mathbf{Z}_b) = \mathbf{V}.$$

The general case of a number of transformers in parallel with either equal or unequal ratios, can more conveniently be dealt with by replacing the transformer ohmic impedances by the admittances $\mathbf{Y}_a = 1/\mathbf{Z}_a$, etc. as in Fig. 6.13, and the equivalent load circuit phase impedance $\mathbf{Z}_L = 1/\mathbf{Y}_L$. $\mathbf{E}_a$, $\mathbf{E}_b$, etc. are the e.m.f.s induced in the transformer secondaries if all impedances are referred to that winding.

$$\mathbf{I} = \mathbf{I}_a + \mathbf{I}_b + \mathbf{I}_c + \ldots.$$

$$\mathbf{I}_a = (\mathbf{E}_a - \mathbf{V})\mathbf{Y}_a \text{ etc.} \qquad [6.11]$$

and

$$\mathbf{I} = \mathbf{V}\mathbf{Y}_L$$

so that

$$\mathbf{V}\mathbf{Y}_L = (\mathbf{E}_a - \mathbf{V})\mathbf{Y}_a + (\mathbf{E}_b - \mathbf{V})\mathbf{Y}_b + (\mathbf{E}_c - \mathbf{V})\mathbf{Y}_c + \dots$$

$$\mathbf{V} = \frac{\mathbf{E}_a\mathbf{Y}_a + \mathbf{E}_b\mathbf{Y}_b + \mathbf{E}_c\mathbf{Y}_c + \dots}{\mathbf{Y}_L + \mathbf{Y}_a + \mathbf{Y}_b + \mathbf{Y}_c + \dots} = \frac{\Sigma\mathbf{E}\mathbf{Y}}{\Sigma\mathbf{Y}} \qquad [6.12]$$

Equation [6.12] is known by various names such as the parallel-generator, nodal-voltage or Millman's theorem. It should be noted

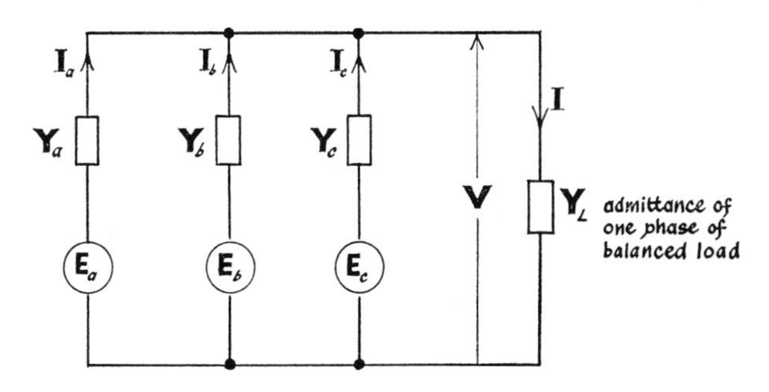

FIG. 6.13. *A number of sources in parallel supplying a common static load.*

that since a static load has been considered here, the term involving $\mathbf{Y}_L$ in the numerator of [*6.12*] has vanished, since it is multiplied by an e.m.f. in the load branch which is zero. When the terminal voltage has been found from [*6.12*], the individual transformer currents can be found from [*6.11*].

If the above analysis is carried out in terms of impedances instead of admittances, it will be seen that, in [*6.12*], the numerator is the total current fed into a short-circuit across the terminals and the denominator is the reciprocal of the impedance looking into the terminals of the passive network. Hence

(voltage between terminals) = (short-circuit current at terminals) $\times$ (impedance looking into the terminals). [*6.12a*]

6.5 Harmonics

Power transformers connected to the 50-Hz system are supplied with a voltage which is virtually free of all harmonics. Odd harmonics in the supply voltage are kept to a very low order, by suitable arrangement of the stator and rotor windings and of the magnetic circuit in the alternators, and even harmonics are not generated (except of very small magnitude under transient conditions).

If the waveform of the 50-Hz supply voltage is assumed to be a pure sine wave, and the transformer per unit impedance is small, then the e.m.f. induced in the windings and the core flux density B must both be nearly sinusoidal. They can only depart from this due to the voltage drop in the transformer impedance caused by harmonics in the current, and these as we now see only arise in the magnetising current which is a very small part of the transformer load current.

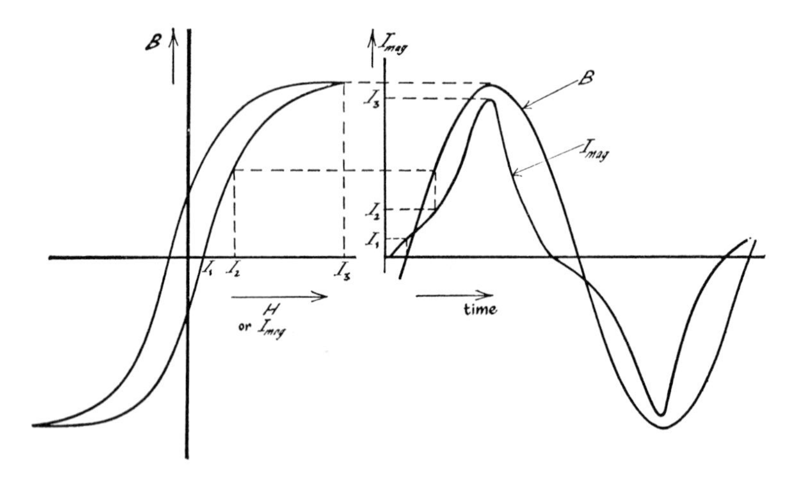

FIG. 6.14. *Distorted magnetising current waveform for a sinusoidal flux density.*

Fig. 6.14 shows that in the iron of the core when it is working in the normal range of the B/H loop, the flux density can only vary sinusoidally in time if the magnetising current contains harmonics. Since the B/H loop is symmetrical, the positive and negative half-cycles of current are symmetrical, indicating that the harmonics, which increase sharply with increasing degree of saturation, are all odd ones, with the 3rd harmonic being the most prominent.

Since the supply voltage contains only exceedingly small harmonics it cannot supply sufficient harmonic current for magnetisation. The harmonic m.m.f. must, however, be provided, and it will be shared between the primary, secondary and tertiary windings. If there is no tertiary winding, the harmonic current will flow mainly in the primary even under load conditions, since the secondary circuit contains a load impedance which is relatively high compared with the low impedance of the supply. The harmonic driving voltages, since they are not present in the supply, are induced in the windings by the flux departing slightly from a sine wave.

If a transformer has a series impedance of 0·1 p.u. at full load; a no-load current which is 4% of rated current; and a 3rd harmonic current which is 25% of the no-load current at rated voltage and frequency, then the 3rd harmonic current is 1% of rated current. The 3rd harmonic voltage drop on the series impedance will therefore be approximately 3% of the full-load impedance drop since this is mainly reactive. It is therefore about 0·3% of rated voltage, so that the 3rd harmonic flux is only about 0·1% of the fundamental flux. This order of magnitude check indicates that the departure of the flux from the sinusoid is likely to be small.

The harmonic fluxes combine with the fundamental flux to give a resultant which is intermediate between a sine wave and that produced by a sinusoidal magnetising current. If the impedance of the supply to the harmonics is much less than that of the transformer, then virtually no harmonic voltage appears across the supply terminals. The smaller the impedance to a given harmonic the smaller the e.m.f. and flux of that frequency. Thus if one of the main windings or a tertiary winding is connected in delta, then since the triplen (3rd, 9th, etc.) harmonics of voltage in each phase winding are in phase with each other, only a small harmonic phase voltage is necessary to drive the current which circulates in the mesh and provides the required triplen m.m.f. Therefore, the core flux, secondary voltage and magnetising current contain much less triplen harmonics than for a transformer with no delta-connected winding. If it is the primary winding which is delta-connected, then no triplen harmonic currents flow from the supply into the transformer, since they circulate around the mesh, and no triplen frequency p.d. appears at the terminals because the triplen harmonic e.m.f. of each phase is absorbed by its own impedance voltage drop. Similarly if the secondary winding is delta-connected, the circulating currents of triple frequency flow within the delta only and not in the load circuit, and the triple frequency distortion in the primary magnetising current, core flux and in secondary voltage is small.

Third and other triple frequency harmonics are less troublesome in 3-limb core-type transformers than in three single-phase units or in a 5-limb transformer. In the former, Fig. 6.2 shows that since the sum of the fundamental fluxes $\mathbf{\Phi}_R+\mathbf{\Phi}_Y+\mathbf{\Phi}_B = 0$, no return flux path is needed, but the triplen harmonic fluxes are in phase in all three limbs, and their return flux path is largely in air and insulation, so that the magnitude of these fluxes is small. This is spatially similar to the conditions for zero-sequence flux discussed in section 6.10, though the latter varies at fundamental frequency (see also Chapter 8).

If the star-point of a star-connected transformer primary winding is connected to the neutral point of the supply, triplen harmonic currents will flow in this neutral connection, and since the supply impedance is small, the core flux need only contain very small triplen harmonic components. If no neutral path exists so that no triplen frequency magnetising current can flow from the supply, and if there is no delta-connected tertiary winding, then the core flux contains a triplen frequency component which can be particularly severe on no load. As an example, the following results on a laboratory demonstration transformer bank of three single-phase units each of under one kVA may be quoted: with the primary star-point not connected to that of the supply and the delta tertiary open, the no-load secondary voltage with rated supply voltage, contained 44% 3rd harmonic. Connecting the primary neutral or closing the tertiary delta reduced the 3rd harmonic to 0·1 and 0·3% respectively. A similar test on a 3-phase core-type transformer of the same size showed a 3rd harmonic secondary voltage component of only 1% with the neutral path open and delta tertiary open. This low 3rd harmonic was due to the high reluctance of the magnetic path to 3rd harmonic flux.

6.6 Magnetising Inrush Current

If a switch is closed connecting an a.c. source to a single-phase circuit consisting of an air-cored coil of inductance L henry and resistance $R\,\Omega$, then the current generally contains a d.c. transient component which dies away with the time constant L/R of the circuit (see section 7.5). The transient has a maximum initial value equal to the peak of the steady-state alternating current, and if the resistance were negligible, this would occur if the switch were closed at voltage zero. In general, the maximum transient occurs if the switch closes at the instant when the voltage is $V_m \cos(\tan^{-1} \omega L/R)$ for a supply voltage $v = V_m \sin \omega t$.

When the coil is iron-cored and the peak flux density is in the saturation region, as in a power transformer operating at normal voltage and frequency, then the transient distortion of magnetising current may be enormously magnified, depending upon the shape of the B/H curve and degree of saturation, the instant of circuit closure and the residual magnetism in the iron core.

If a supply voltage $v = V_m \sin \omega t$ is applied at $t = 0$ to a transformer winding of N turns, which is assumed to have negligible per unit resistance and leakage reactance, and if any other windings are on open circuit, then the instantaneous e.m.f. induced in the winding which is equal and opposite to the applied voltage is given by

$$-N\frac{d\phi}{dt} = -V_m \sin \omega t$$

and the instantaneous flux is

$$\phi = -\frac{V_m}{\omega N} \cos \omega t + A = -\Phi_m \cos \omega t + A.$$

If there were no residual magnetism in the core just prior to the closure of the switch so that $\phi = 0$ at $t = 0$ then $A = \Phi_m$ and

$$\phi = \Phi_m[1 - \cos \omega t]. \qquad [6.13]$$

This suggests that one half-cycle after closing the circuit at voltage zero, the flux would reach twice normal value. This cannot in fact occur since the excursion of flux beyond the normal operating point on the magnetisation curve, causes such a large magnetising current to flow that the impedance voltage drop is no longer negligible. There will in practice be a random point-on-wave circuit-closure, and the worst condition of closure at or very near voltage zero will not always occur. However, there will generally be some residual magnetism in the core and if this flux is in the same direction as that set up by the applied voltage, then the excursion of flux is greater and the magnetising current surge is much increased.

The transient magnetising current decays as shown in Fig. 6.15, but the first peak of current may very considerably exceed the rated current of the transformer. In extreme cases the peak inrush current can reach almost twenty times the normal full-load current, and for transformers larger than 10 MVA, one second or more may elapse before the inrush current falls to 50% of the first peak value (Hudson, 1966). Mal-operation of protective relay circuits on switching a transformer into circuit must consequently be guarded against (see harmonic bias protection of transformers in section 9.5.1). This

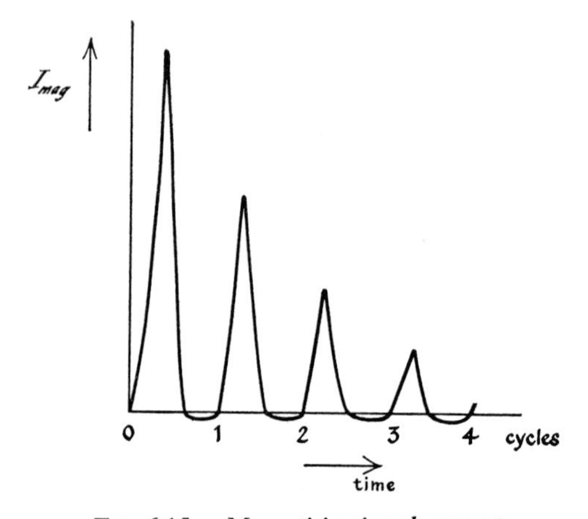

FIG. 6.15. *Magnetising inrush current.*

effect is sometimes observed by students when a fuse blows in the laboratory on switching a Variac, possibly unloaded, into circuit. A 2-kVA Variac, with full load current of 8 A and steady-state magnetising current of 0·4 A can give a current of 60 A in the first cycle even when the residual magnetism is zero.

6.7 Tap Changing

On-load tap changing is required on certain transformers in a power system, in order to regulate the voltage and keep its level at all consumers within limits, despite changes in the amount and distribution of load. Taps will frequently be on the high voltage winding since tappings on very low voltage windings may not give sufficiently close control of voltage. Tap changing, by altering the in-phase component of the system voltage, affects the distribution of vars in a system and may therefore be used to control the flow of reactive power.

The position of the taps depends upon a number of factors, e.g. minimisation of the variation of impedance and mechanical stresses on the winding, leads to taps at the centres of the legs, whereas for a very high voltage transformer, which has graded insulation because

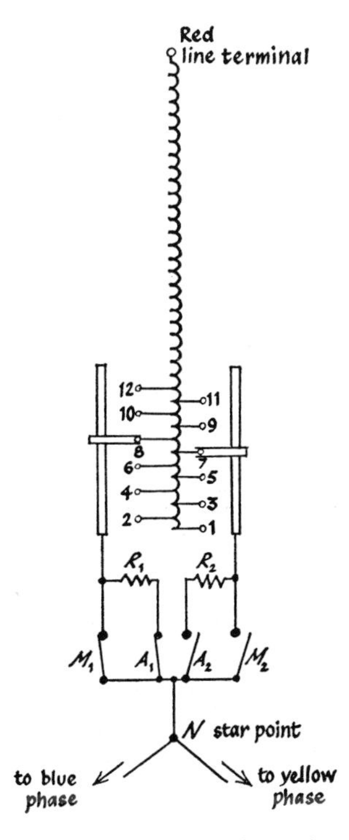

FIG. 6.16. *On-load tap changer, shown diagrammatically.*

it is to be used only on a system with earthed neutral, the taps may be at the neutral end, so that the voltages to earth are low. As a further protection against surge voltages, non-linear resistors may be connected across the taps.

Fig. 6.16 illustrates one phase of an on-load tap changer at the neutral end of a 3-phase high voltage transformer. A selector switch is shown making contact with taps 7 and 8 and the main contactor M_1 connects the neutral to the latter tap. To change to tap 7, M_1 opens and the auxiliary contactor A_2 closes, so that the transition resistors R_1 and R_2 are temporarily in series across the winding between taps 7 and 8 until A_1 opens and M_2 closes. The use of transition resistors in place of centre-tapped bridging reactors reduces arcing at diverter switch contacts.

6.8 Unbalanced Loading

If a 3-phase star/star-connected bank of three single-phase units has its primary star-point connected to that of the supply, then the load on each primary phase is proportional to that on the corresponding secondary phase, and the degree of unbalance is therefore the same on both sides of the transformer. When the star-points are not connected, then the primary phase voltages are not constrained to be balanced, and with a single load across one phase of the secondary, current must flow in the two phases which have no compensating secondary current. These two phases therefore act as chokes of relatively high reactance and their voltages tend towards line voltage, whilst that of the loaded phase is low, so that the secondary phase voltages are also unbalanced. This connection is not used where unbalanced loading is possible, unless there is a delta-connected tertiary which provides an m.m.f. in each transformer, and tends to equalise the voltages. A similar difficulty arises in the case of a 5-limb core-type transformer.

For a 3-limb core-type transformer, the magnetic path on which the unbalanced m.m.f.s on the three limbs act, is of relatively high reluctance, so that the core fluxes and phase voltages are not unbalanced to such an extent. The distribution of current in the primary windings of a 3-limb or 5-limb core-type transformer may be determined as follows. If the primary star point is isolated and a current $\mathbf{I}$ flows in a load connected between the secondary terminal a and the neutral point, the currents $\mathbf{I}_A$, $\mathbf{I}_B$ and $\mathbf{I}_C$ in the primary phases may be found from the equality of primary and secondary ampere-turns, which must exist between each pair of line terminals, since the supply line voltages are balanced. If there are N_p turns in each primary phase winding and N_s in each secondary phase winding, then between terminals A and B, the equality of primary and secondary ampere-turns so that the resultant ampere-turns between line terminals are zero in all cases, gives

$$(\mathbf{I}_A-\mathbf{I}_B)N_p-\mathbf{I}N_s = 0 \qquad [6.14]$$

and between terminals B and C

$$(\mathbf{I}_B-\mathbf{I}_C)N_p = 0$$

so that $\mathbf{I}_B = \mathbf{I}_C$ which is apparent from symmetry, and may be substituted into $\mathbf{I}_A+\mathbf{I}_B+\mathbf{I}_C = 0$ giving

$$\mathbf{I}_A = -2\mathbf{I}_B.$$

$\mathbf{I}_A$ and $\mathbf{I}_B$ are in phase with each other and from [*6.14*]

$$\left.\begin{aligned} \mathbf{I}_A &= \frac{2}{3}\mathbf{I}\frac{N_s}{N_p} \\ \mathbf{I}_B = \mathbf{I}_C &= -\frac{1}{3}\mathbf{I}\frac{N_s}{N_p} \end{aligned}\right\}. \qquad [6.15]$$

Since, as discussed in section 6.10, the high reluctance paths for zero-sequence flux in a 3-phase core-type transformer limit its value, then the unbalance in flux and shift of neutral potential are less for this type of transformer than for three single-phase units.

The presence of a delta-connected tertiary winding (N_t turns/phase) helps to stabilise the potential of the primary neutral, since a (zero-sequence) current of $1/3 \times \mathbf{I}(N_s/N_t)$ circulating within the mesh equalises the m.m.f. on all three limbs. It does not, however, alter [*6.14*] because the zero-sequence m.m.f. on limb *A* exactly cancels that on limb *B*. Similarly the equation following [*6.14*] is unchanged, so that the distribution of primary currents given by [*6.15*] is unaltered by the presence of the delta-connected tertiary winding.

The method of symmetrical components (see Chapter 8) may be used to solve unbalanced loading conditions, as an alternative to using m.m.f. balance as shown here, but these methods do not deal with harmonics which may have appreciable effect.

6.9 Three-winding Transformer Equivalent Circuit

If the no-load current of a 3-winding transformer is neglected, it is possible to draw a simple single-phase equivalent circuit which enables balanced load conditions to be evaluated for all three windings. It is most convenient if the impedances are expressed in per unit and they must then all be given to the same MVA base. This latter consideration is automatically satisfied in 2-winding transformers since both primary and secondary windings have the same MVA rating. In 3-winding transformers, however, although the primary and secondary may or may not have the same MVA rating, that of the tertiary will certainly be less.

If short-circuit tests are carried out on a 3-winding transformer with N_p, N_s and N_t turns/phase on the three windings respectively, the following *ohmic* impedances (composed of resistance and leakage reactance) may be measured:

$\mathbf{z}_{ps}$ = impedance measured in the primary circuit with the secondary short-circuited and the tertiary open.

$\mathbf{z}_{pt}$ = impedance measured in the primary circuit with the tertiary short-circuited and the secondary open.

$\mathbf{z}'_{st}$ = impedance measured in the secondary circuit with the tertiary short-circuited and the primary open.

$\mathbf{z}'_{st}$ may also be referred to the primary winding giving

$$\mathbf{z}_{st} = \left(\frac{N_p}{N_s}\right)^2 \mathbf{z}'_{st}.$$

If $\mathbf{z}_p$, $\mathbf{z}_s$ and $\mathbf{z}_t$ are the ohmic impedances of the three separate windings referred to the primary voltage level, then

$$\left.\begin{aligned} \mathbf{z}_{ps} &= \mathbf{z}_p+\mathbf{z}_s \\ \mathbf{z}_{pt} &= \mathbf{z}_p+\mathbf{z}_t \\ \mathbf{z}_{st} &= \mathbf{z}_s+\mathbf{z}_t \end{aligned}\right\}. \qquad [6.16]$$

Equations [*6.16*] yield

$$\left.\begin{aligned} \mathbf{z}_p &= \tfrac{1}{2}(\mathbf{z}_{ps}+\mathbf{z}_{pt}-\mathbf{z}_{st}) \\ \mathbf{z}_s &= \tfrac{1}{2}(\mathbf{z}_{ps}+\mathbf{z}_{st}-\mathbf{z}_{pt}) \\ \mathbf{z}_t &= \tfrac{1}{2}(\mathbf{z}_{pt}+\mathbf{z}_{st}-\mathbf{z}_{ps}) \end{aligned}\right\}. \qquad [6.17]$$

If $\mathbf{Z}_{ps}$ = per unit impedance corresponding to $\mathbf{z}_{ps}$ based on the MVA and voltage rating of the primary

$\mathbf{Z}_{pt}$ = per unit impedance corresponding to $\mathbf{z}_{pt}$ based on the MVA and voltage rating of the primary

$\mathbf{Z}'_{st}$ = per unit impedance corresponding to $\mathbf{z}'_{st}$ based on the MVA and voltage rating of the secondary, then

$\mathbf{Z}_{st}$, the per unit impedance between the secondary and tertiary windings with the primary open, based upon the MVA and voltage rating of the primary is given by

$$\mathbf{Z}_{st} = \mathbf{Z}'_{st} \cdot \frac{\text{MVA of primary}}{\text{MVA of secondary}}$$

since the ratios of the voltage ratings are the turns ratios, and [*6.17*] become

$$\left.\begin{aligned} \mathbf{Z}_p &= \tfrac{1}{2}(\mathbf{Z}_{ps}+\mathbf{Z}_{pt}-\mathbf{Z}_{st}) \\ \mathbf{Z}_s &= \tfrac{1}{2}(\mathbf{Z}_{ps}+\mathbf{Z}_{st}-\mathbf{Z}_{pt}) \\ \mathbf{Z}_t &= \tfrac{1}{2}(\mathbf{Z}_{pt}+\mathbf{Z}_{st}-\mathbf{Z}_{ps}) \end{aligned}\right\}. \qquad [6.18]$$

$\mathbf{Z}_p$, $\mathbf{Z}_s$ and $\mathbf{Z}_t$ are the per unit impedances of the three windings, each based upon the rated MVA and voltage of the primary winding. These three impedances may now be connected together, in a single-phase equivalent star circuit as shown in Fig. 6.17. The part within

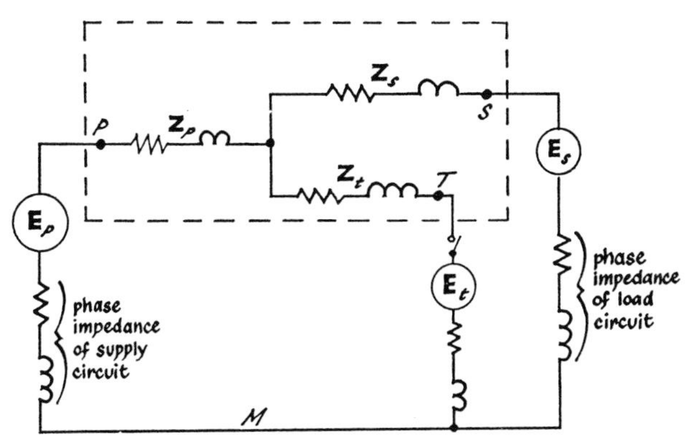

FIG. 6.17. *Equivalent circuit for one phase of three-winding transformer.*

the dotted lines represents the transformer, and that outside shows the external circuit connections to the transformer.

$\mathbf{E}_t$ and the power-frequency current in the tertiary may both be zero if the tertiary consists of a closed delta-connected winding unconnected to any system of supply or load, and in this case terminal *T* is isolated from the system. It is true that triple-frequency current circulates in the tertiary in order to produce the m.m.f. required to maintain the core flux substantially sinusoidal, but Fig. 6.17 relates only to power frequency conditions. If the secondary supplies only static loads, then $\mathbf{E}_s$ is zero, and *S* is connected to *M* via the load circuit impedance. *P* and *M* are connected by the phase supply voltage $\mathbf{E}_p$ and the supply circuit impedance/phase. It must be emphasised that this circuit of Fig. 6.17 applies to balanced conditions, i.e. to positive- or negative-sequence current only, and represents one phase.

Three-winding auto-transformers which are commonly used at the highest transmission voltages can be represented by a similar equivalent circuit, but there are two arrangements possible (Mortlock and Humphrey Davies, 1952).

6.10 Zero-sequence Impedance

The impedances $\mathbf{Z}_1$ and $\mathbf{Z}_2$ presented by the transformer to positive- and negative-sequence currents respectively are equal to the values shown in Fig. 6.1, except for the 30° phase shift for delta/star

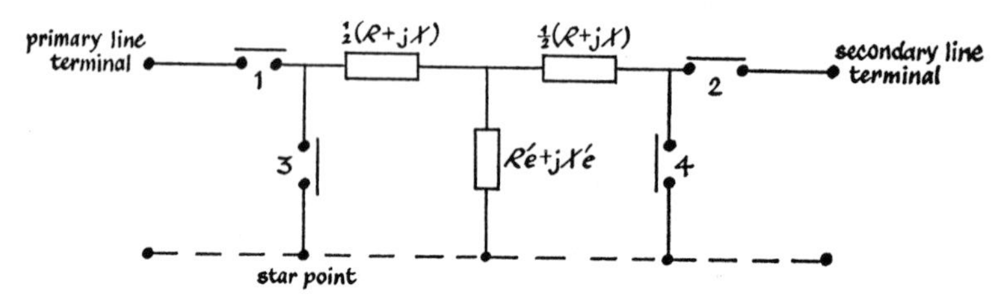

FIG. 6.18. *Zero-sequence equivalent circuit.*

transformers being opposite in sign for negative sequence to that for positive sequence (see section 8.3.5). The impedance to zero-sequence current $\mathbf{Z}_0$ may be about equal to the positive-sequence value or it

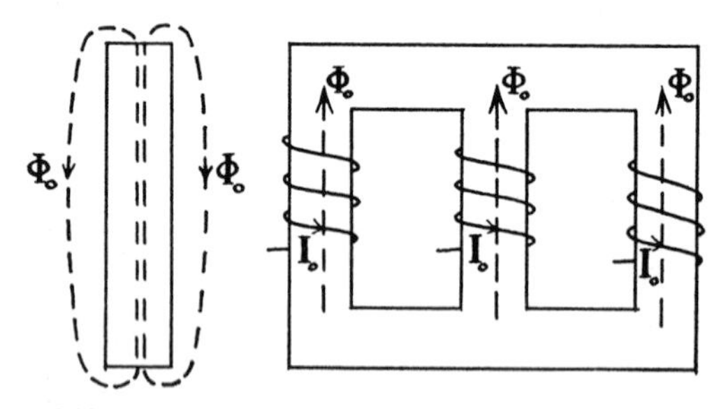

FIG. 6.19. *Zero-sequence flux in three-limb core-type transformer.*

may be infinity. This depends on the winding connections, on whether or not star points are earthed and on the type of magnetic circuit.

Generator	Transformer primary	Transformer secondary	Transformer zero-sequence network	Transformer zero-sequence impedance
			$Z/2$ $Z/2$ Z'_e from supply to load	Infinite on both primary and secondary sides. This applies equally whether the generator star point is earthed or not.
			$Z/2$ $Z/2$ Z'_e from supply to load	Infinite on primary side. $(Z/2+Z'_e)$ on secondary side, where Z'_e is higher for shell, separate single-phase units or 5-limb core-type than for 3-limb core-type. This applies equally whether the generator star point is earthed or not.
			$Z/2$ $Z/2$ Z'_e $\simeq$ Z	$Z_0 = Z_1 = Z_2 = Z$ for shell or separate single-phase units. $Z_0 \simeq 0.85\ Z_1$ for core-type due to mutual coupling. If generator star point is not earthed then equivalent circuit immediately above applies.

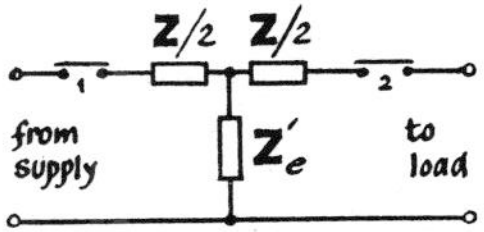

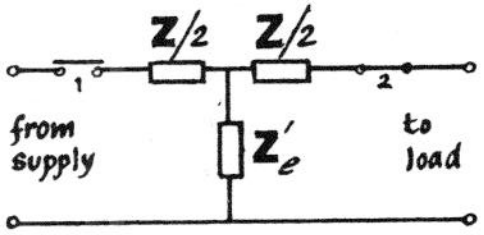

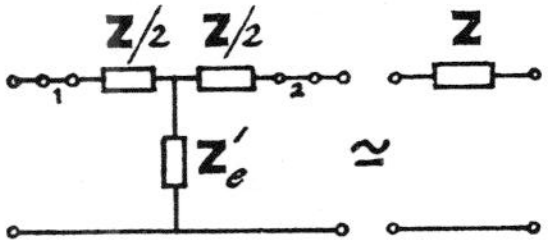

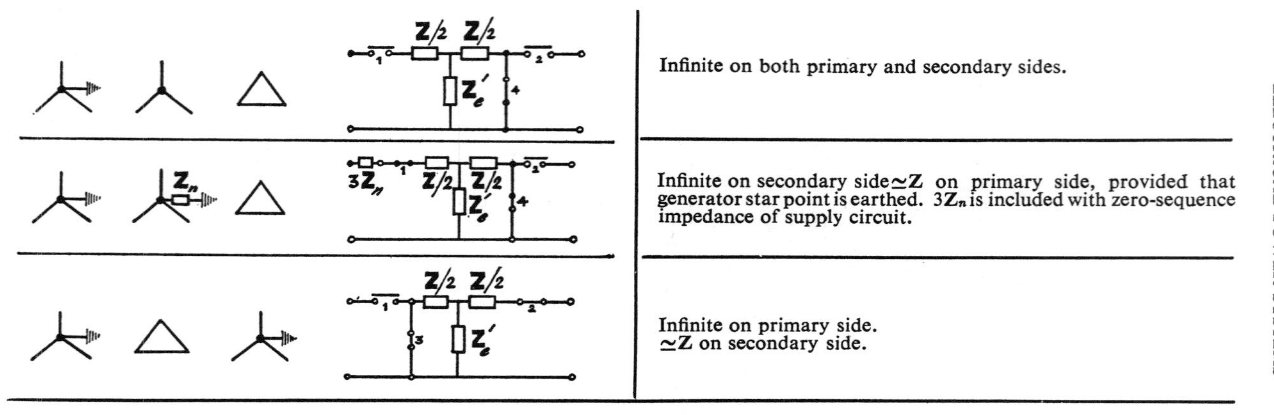

FIG. 6.20. *Transformer zero-sequence equivalent circuits.*

Fig. 6.18 shows a two-winding transformer equivalent circuit in which $R+jX = \mathbf{Z}$ is the total per unit series positive-sequence impedance of one phase of both primary and secondary windings included together, and $R'_e+j X'_e$ is the per unit zero-sequence shunt impedance (often called exciting impedance $\mathbf{Z}'_e$) through which the magnetising and core loss currents flow.

For positive- (or negative-) sequence current, the exciting impedance $\gg R+jX$ (e.g. of the order of 25 per unit compared with 0·1 per unit), so that the exciting impedance may generally be neglected for calculations on a system, and $\mathbf{Z}_1 = \mathbf{Z}_2 = R+jX$. For zero-sequence current, however, the exciting impedance $R_e+j X'_e$ depends upon the core construction. For a shell-type 3-phase transformer or for three single-phase transformers connected to form a 3-phase bank, or for a 3-phase 5-limb core-type transformer, the zero-sequence exciting impedance is of the same order as it is for positive sequence. For a 3-limb core-type unit, however, the zero-sequence fluxes which are in phase with each other in all three limbs as shown in Fig. 6.19, have no return paths in the iron core, and must pass through air, copper, insulation and the tank wall. The zero-sequence flux is therefore limited in value, and the zero-sequence exciting impedance is much smaller than that for positive sequence. It generally lies in the range 0·3 to 4 per unit, being towards the higher value for large power transformers.

It must be noted, of course, that although since both 3rd and other triple frequency harmonic currents are similar to zero-sequence currents in that they are in phase with each other in all three phases of a 3-phase circuit, and the flux paths are also similar because of this, there are important differences. The zero-sequence currents are of fundamental frequency and can therefore be combined in the usual phasor diagrams with the other components of this frequency which are positive- and negative-sequence currents (see Chapter 8). Since a phasor diagram cannot represent components of more than a single frequency, it is not possible to show 3rd or any other harmonic currents in the fundamental frequency phasor diagrams.

Apart from this effect of construction upon the zero-sequence exciting impedance, there is the relatively greater effect of the connection of the windings. Various cases will now be considered. If the star-point on the primary winding is earthed then link 1 in Fig. 6.18 is closed, and if the secondary star-point is earthed link 2

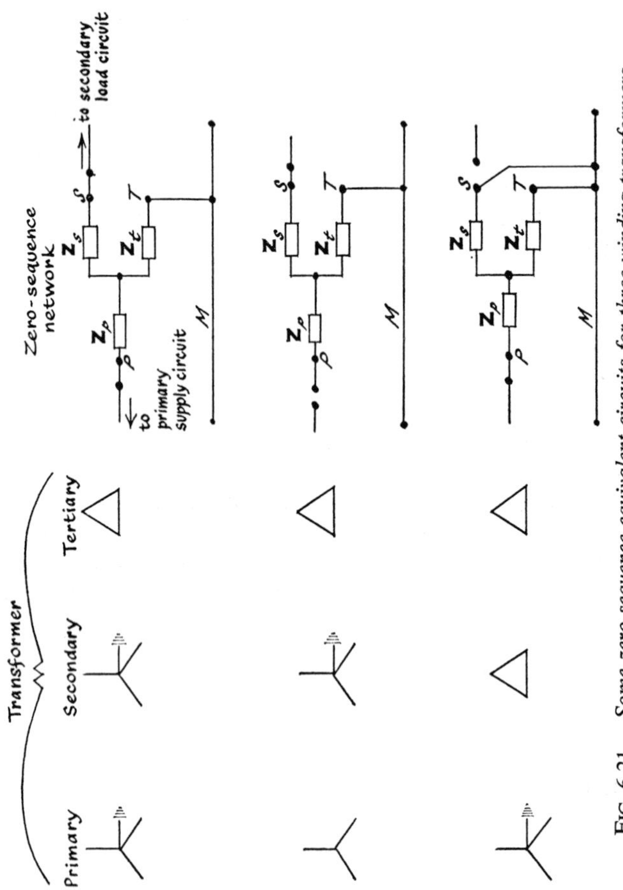

FIG. 6.21. *Some zero-sequence equivalent circuits for three-winding transformers.*

is closed, since the earth path allows flow of zero-sequence current (provided that there is another connection to earth, perhaps at a fault). If the primary winding is delta-connected, link 3 is closed but link 1 is open, while if the secondary winding is in delta, link 4 is closed but 2 is open, since zero-sequence current circulates in a delta-connected winding, but cannot flow in the supply or load circuit connected to it. Fig. 6.20 illustrates the zero-sequence networks due to various connections of 2-winding transformers. It should be noted that zero-sequence current can flow in one winding without automatically having to flow in the other, a case in point being the unbalanced condition which yielded [*6.15*], where zero-sequence current flows in the secondary but not in the primary.

In the case of three-winding transformers, there is the possibility that zero-sequence currents can be caused to flow in the tertiary winding, due to the circuit to which it is connected. In this case a series link (similar to 1 and 2 in Fig. 6.18) is closed in the zero-sequence circuit connecting the tertiary to the external circuit (see Fig. 6.17). If the tertiary winding is delta-connected, then the zero-sequence e.m.f. induced in it can cause current to flow, and the tertiary impedance appears in the zero-sequence network. The zero-sequence current would then be shared between this and the other winding which is not connected to the faulted circuit, according to the respective impedances of their circuits.

The zero-sequence network of a 3-winding transformer can be built up from consideration of Figs. 6.17 and 6.18, according to the way in which the windings are connected. Typical examples are shown in Fig. 6.21 where $\mathbf{Z}_p$, $\mathbf{Z}_s$ and $\mathbf{Z}_t$ are the same impedances as were used in the positive- (or negative-) sequence network of Fig. 6.17.

REFERENCES

HUDSON, A. A. 1966. 'Transformer magnetising inrush current: a resumé of published information', *E.R.A. Report No. 5152.*

MORTLOCK, J. R. & HUMPHREY DAVIES, M. W. 1952. *Power system analysis*, Chapman and Hall, London.

STIGANT, S. A. & LACEY, H. M. 1961. *The J. and P. transformer book*, Johnson and Phillips, London, 9th edition.

B.S. 171, *Power transformers*. British Standards Institution, London.

Examples

1. A single-phase transformer has 480 turns on the primary and 90 turns on the secondary. The mean length of flux path in the iron core is 180 cm and the joints are equivalent to an air gap of 0·1 mm. If the maximum value of flux density is to be 1·1 tesla when 2200 volts at 50-Hz are applied to the primary, calculate:

(*a*) cross-sectional area of core;

(*b*) secondary voltage on no load;

(*c*) primary current and p.f. on no load.

Assume the ampere turns/metre for 1·1 tesla to be 400, the corresponding iron loss to be 1·77 watts/kg at 50 Hz and the density of the iron to be 7·8.

(187 cm^2, 412·5 V, 1·21 A, 0·174)

2. A single-phase transformer is rated at 10 kVA 230/100 volts. When the secondary terminals are open-circuited and the primary winding is supplied at normal voltage (230 V), the current input is 2·6 A at a p.f. of 0·3. When the secondary is short-circuited a voltage of 18 volts on the primary causes full load current in the secondary, the power input to the primary being 240 watts.

Calculate:

(*a*) the efficiency at full load unity p.f.;

(*b*) the load at which maximum efficiency occurs;

(*c*) the value of maximum efficiency.

(95·97%, 8·65 kVA, 96·0%)

3. Two identical 3-phase star/star transformers are connected, one at the generator end and one at the substation end of a 33-kV underground cable. The resistance of the cable per conductor is 7·17 Ω and its inductive reactance is 2 Ω.

Each 6·6/33-kV transformer has an equivalent resistance of 0·04 Ω per phase and leakage reactance of 0·4 Ω per phase referred to the low-voltage side.

If 5000 kW at 6 kV, 0·8 p.f. (lagging), are delivered on the low-voltage side of the substation transformer, calculate the voltage on the low-voltage side of the transformer at the generating station. What assumptions have been made?

(6880 V)

4. Three single-phase 50-kVA 2400/240-volt transformers are connected star/delta in a 3-phase 150-kVA bank to step down the voltage at the load end of a feeder the impedance of which is $0{\cdot}15 + j\,1{\cdot}0\,\Omega$/phase. The line voltage at the sending end of the feeder is 4160 V.

On their secondary sides the transformers supply a balanced 3-phase load through a feeder the impedance of which is $0{\cdot}0005 + j\,0{\cdot}0020\,\Omega$/phase. For each single-phase transformer the short-circuit test readings are 48 V, 20·8 A, and 617 watts measured on the H.V. side.

Find the line voltage at the load when the load draws rated current from the transformers at 0·8 p.f. lagging.

(232·9 V)

5. A 3-phase transformer is connected in star on both primary and secondary sides. It supplies a load through a transmission line. The load has an equivalent resistance to neutral of 400 Ω and reactance of 600 Ω. Each transmission line conductor has resistance of 4 Ω and reactance of 6 Ω. The secondary winding of the transformer has three times as many turns/phase as the primary has.

The transformer has

primary: resistance 0·5 Ω/phase and leakage reactance 2·5 Ω/phase;
secondary: resistance 5·0 Ω/phase and leakage reactance 25 Ω/phase.

If the primary line voltage is 11 kV calculate the secondary line terminal voltage.

(31·2 kV)

6. Two 6600/440-V, star/star connected 3-phase transformers, A of 250 kVA and B of 500 kVA, have the following impedances/phase referred to the secondary side: for A: $R = 0{\cdot}008\,\Omega$, $X = 0{\cdot}035\,\Omega$; for B: $R = 0{\cdot}003\,\Omega$, $X = 0{\cdot}019\,\Omega$.

How will they share a load of 600 kVA at a power factor of 0·8 lagging if the load voltage is 440? The kVA and the p.f. of each transformer are required.

(208·2 kVA, 392·2 kVA, 0·82, 0·785)

7. Two 3-phase star/star transformers are connected in parallel with a primary line voltage of 11 000 V. One has 3 times and the other 3·1 times as many secondary turns as primary turns per phase. Each transformer has a reactance of 100 Ω/phase referred to the

secondary, and the resistance can be neglected. They supply a star-connected load of 1000+j 500 Ω/phase.

Calculate the secondary terminal voltage and its phase angle relative to the secondary e.m.f.s. Indicate how the individual transformer currents can then be calculated.

(University of Leeds) (19,010 V, $\underline{/-2° 15'}$)

8. Two 3-phase transformers *A* and *B*, which have equal transformation ratios and are star-connected on both primary and secondary sides, operate in parallel to supply a load.

Transformer *A* has a primary current of 100 A and its power factor is 0·7 lagging. Calculate the primary current and power factor of transformer *B*.

The primary line voltage is 10 000 and the phase resistances and leakage reactances are: $R_A = 0{\cdot}5\,\Omega$, $X_A = 1{\cdot}5\,\Omega$, $R_B = 0{\cdot}6\,\Omega$ and $X_B = 1{\cdot}8\,\Omega$, all referred to the primary. Both transformers have three times as many turns per phase on the secondary as on the primary.

Calculate the load current and power factor and the equivalent resistance and reactance per phase of the star-connected load.

(University of Leeds) (83·3 A, 0·7 lagging, 61·1 A, 0·71 lagging, 188 Ω, 192 Ω)

9. Two 3-phase transformers which have equal turns ratio, are supplying, in parallel, a load of 4 MW at 0·8 power factor lagging. The data for the transformers are:

	Per unit resistance	*Per unit reactance*
Transformer *A*	0·01	0·05
Transformer *B*	0·015	0·06

Determine the complex power and power factor for each transformer.

(University of London) (2154·6−j 1692·2 kVA
1845·4−j 1307·8 kVA
0·787, 0·818)

10. Two identical, tap-changing, 500-kVA, 11 000/415-V, 3-phase transformers each having an impedance of 0·02+j 0·07 per unit based on rated MVA, are paralleled onto a constant voltage 11 000-V supply and supply a constant impedance load of $0{\cdot}3\underline{/30°}$ Ω/phase which is star-connected on the 415-V side.

If one transformer is on nominal tap and the other is on +5%

(reduced H.V. turns), calculate the H.V. winding current in the former (nominal tap) transformer, and determine its input in watts and vars. Prove the formula used and assume that the percentage impedance of the transformers is independent of tap position.

(L.C.T.) (10·2+j 0·878 A; 194 kW, 16·7 kVAr (leading))

11. Two 3-phase transformers *A* and *B* are star-connected on primary and secondary sides, and both have three times as many secondary as primary turns. They are connected in parallel and supply a delta-connected load of resistance 270 Ω and inductive reactance of 135 Ω/phase. The primary voltage is 11 kV.

Calculate the secondary current of transformer *A* and its phase relationship to the current in one phase of the load. The transformer details are:

Winding	*Resistance* Ω	*Leakage reactance* Ω
primary of *A*	0·4	1·0
primary of *B*	0·4	0·8
secondary of *A*	5·4	7·2
secondary of *B*	3·6	10·8

(University of London) (90 A lagging 25° 54′)

12. Two 3-phase star/star transformers have each a leakage reactance of 0·05 Ω/phase referred to the secondary side. When their primaries are fed in parallel from a supply of normal rated voltage, their corresponding open-circuit secondary phase voltages are 230 V and 232 V respectively. The secondary windings are now connected in parallel to a balanced star-connected load of impedance 0·5 Ω/phase and of power factor 0·9 lagging: calculate

(*a*) the terminal voltage of the load, and

(*b*) the kVA supplied by that transformer whose open-circuit secondary voltage is 230 V.

Neglect magnetising current and all losses.

(University of London) (226 V, 145 kVA)

13. A delta/star 6·6/0·44-kV bank of three identical single-phase transformers supplies a load of 120 kW at unity power factor between line *R* and neutral and a balanced load of 500 kVA at 0·8 lagging power factor.

Determine the current magnitude in each primary winding and each input line, and mark these values on a relevant diagram of connections. Ignore internal drop and no-load current. The phase sequence is *RYB*.

(University of London) (Primary currents 25·3, 25·3, 41·2 A) (Line currents 53, 62, 44 A)

14. The magnetising current in the primary winding of an unloaded single-phase transformer is $i = (0{\cdot}9 \sin \omega t - 0{\cdot}4 \sin 3\omega t)$ amperes and the flux density in the transformer core is $B = [1{\cdot}3 \sin (\omega t - 30°) + 0{\cdot}2 \sin 3\omega t]$ tesla. The net cross-section of the core is 120 cm^2, the fundamental frequency is 50 Hz, and the winding has 60 turns with negligible resistance and leakage reactance.

In series with the primary winding is an air-cored inductor of 54·5 Ω resistance and 0·30 H inductance. Derive an expression for the supply voltage. Calculate the iron loss in the transformer core. (It may be noted that these expressions for B and i are not intended to represent a realistic B/H characteristic.)

(University of London) ($294 \cos (\omega t - 30°) + 49{\cdot}1[\sin \omega t + 1{\cdot}73 \cos \omega t] + 22{\cdot}4 \cos 3\omega t - 21{\cdot}8 \sin 3\omega t$ V: 66 watts)

15. A 3-winding transformer has its windings rated as follows:

primary star-connected	22 kV	200 MVA
secondary star-connected	275 kV	200 MVA
tertiary delta-connected	3·3 kV	

Short-circuit tests gave the following leakage reactances and showed that the resistances may be neglected: with measurements made in the primary circuit with

(*a*) secondary short-circuited and tertiary open the leakage reactance was 0·24 Ω

(*b*) tertiary short-circuited and secondary open the leakage reactance was 0·31 Ω

and with measurements made in the secondary circuit with the tertiary short-circuited and primary open the leakage reactance was 54·3 Ω.

Calculate the per unit impedances of the equivalent circuit (neglecting the no-load current) based on the rated MVA and voltage of the primary winding.

(j 0·042, j 0·057 and j 0·086 per unit)

16. Two 3-phase star/star connected transformers have leakage reactances of 0·005 Ω per phase and 0·004 Ω per phase respectively referred to the secondary side. With their primary windings fed in parallel from a supply of rated voltage and frequency, their corresponding secondary phase voltages on open circuit are $235\underline{/0^\circ}$ and $233\underline{/0^\circ}$ respectively. When the secondary windings supply in parallel a balanced delta-connected load of impedance 0·3 Ω per phase and of power factor 0·8 lagging, calculate (i) the terminal voltage at the load (ii) the kVA supplied to the load by each transformer, (iii) the total kVA supplied to the load.

The magnetising current and all power losses may be neglected in both transformers.

Comment upon the accuracy of the solution.

(University of Leeds) (231 V, 783 kVA, 776 kVA, 1536 kVA)

17. A 20-MVA, 66/22 kV, 3-phase, star/star transformer also carries a 5-MVA, 440-V, delta-connected tertiary winding. Explain briefly the advantages of this arrangement. The transformer supplies balanced 3-phase loads of 12 MVA at 22 kV and 2·5 MVA at 440 V, together with a single-phase load of 2 MVA at 12·7 kV. The load power factors are all unity.

Determine the magnitude of the current in each winding of the transformer neglecting all losses, leakage flux and magnetising current.

(University of London) (162, 136·5, 136·5: 472·5, 315, 315: 2970, 2970, 380 A)

Chapter 7

SYNCHRONOUS MACHINES IN POWER SYSTEMS

7.1 Introduction

The operation of a single synchronous machine is dealt with in textbooks on electrical machines. It is only necessary, therefore, to recall briefly certain principles which are dealt with in detail in such books before proceeding to consider synchronous machines as a part of a power system.

In power system analysis it is often sufficient to represent a synchronous machine by the simple equivalent circuit shown in Fig. 7.1 (see section 1.2). For a 3-phase generator with balanced

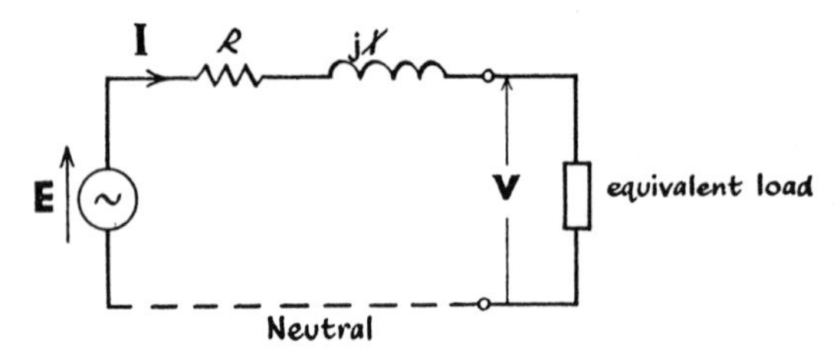

FIG. 7.1 *Approximate equivalent circuit for one phase of a 3-phase cylindrical-rotor alternator under balanced conditions.*

phase e.m.f.s and balanced load, it is sufficient to consider the r.m.s. e.m.f. **E** generated in one phase, and the resistance R and inductive reactance X for one phase. The r.m.s. phase terminal voltage (one line to neutral) **V** is then given by $\mathbf{V} = \mathbf{E} - \mathbf{IZ}$, where **I** is the r.m.s. phase current and $\mathbf{Z} = R + \mathrm{j}\,X$.

This simple equivalent circuit has, however, many restrictions. If a cylindrical-rotor machine were to operate under steady-state conditions without any saturation, then X could be written as X_s the synchronous reactance and it would be a constant. Increasing saturation tends, however, to reduce the effective reactance. Under transient conditions the effective reactance changes, and the rise in its value from a relatively low initial value on the occurrence of short circuit, to X_s after conditions have reached steady state, is discussed in section 7.5.

For a salient-pole machine the simple equivalent circuit of Fig. 7.1 does not always give sufficient accuracy, and it may be necessary to consider the differences between the two main magnetic axes of the machine, and to separate the current **I** and the reactance X into two components, one corresponding to each axis. This is discussed briefly in section 7.4.

7.2 Steady-state Operation of a Single Unsaturated Cylindrical-rotor Alternator

For a cylindrical-rotor alternator, neglect of the harmonics in the spatial distribution of the stator (armature) and rotor (field) ampere-turn distributions, enables the m.m.f.s to be represented vectorially since only the sinusoidal fundamental remains to be considered.

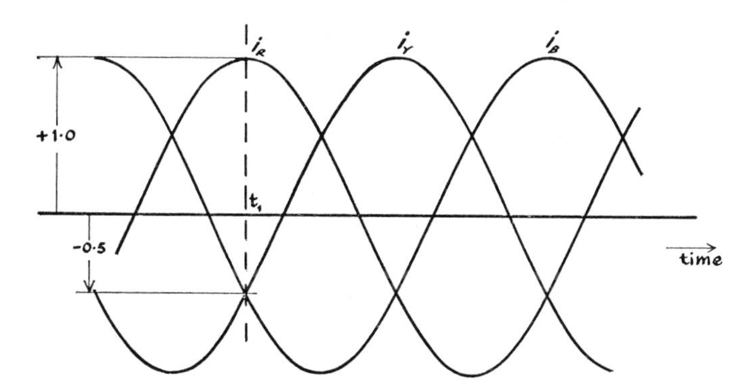

FIG. 7.2 *Time variation of current in a 3-phase stator.*

These m.m.f.s rotate at constant synchronous speed and have constant magnitude for constant field and stator currents.

At the instant t_1 in Fig. 7.2, the stator armature reaction m.m.f. $-$**A** is on the axis of the red phase, since that phase carries maximum current at that instant. This is illustrated in Fig. 7.3, where for simplicity a two-pole machine is shown. The amplitude of the sinusoidal m.m.f. **A** is directly proportional to the r.m.s. stator current **I**. It is more convenient to deal with the amplitudes of the m.m.f.s than with their r.m.s. spatial values, as will be seen when **F** is considered. Bold-face type is used for these m.m.f.s because they have a sinusoidal space distribution, and they are not normally on the same axis. The negative sign for armature reaction at lagging power factor is merely for convenience, so that the component of field m.m.f. which neutralises it is then positive.

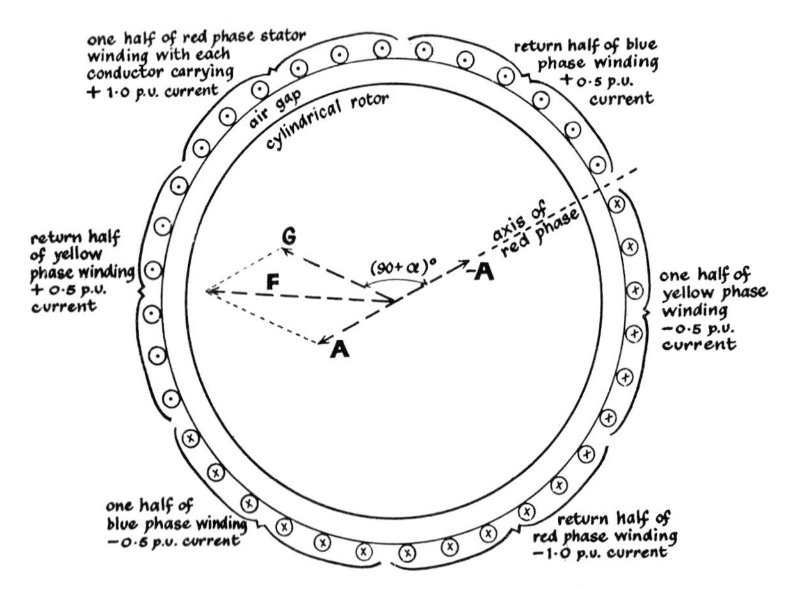

FIG. 7.3. *Distribution of current in a 2-pole 3-phase stator winding and m.m.f. space vectors at the instant t_1 shown in Fig. 7.2.*

When balanced currents flow in the three phases of the alternator, the phasor diagram of Fig. 7.4 shows that an e.m.f. **E** is actually generated in the conductors of one phase. This differs from the terminal voltage **V** due to resistance R of the stator phase conductors, and the leakage reactance X_L set up by that part of the stator flux which does not pass through the main magnetic circuit, and thus does not link with the field winding. The leakage flux links the stator conductors mainly in the end windings and around the slots. Since these leakage flux paths are wholly or partly in air, X_L is a constant for a given machine. For normal lagging power factor load conditions **E** leads the terminal voltage **V** by a small angle, and exceeds it by a small amount, since for 2-pole turbo-alternators X_L is generally between 0·1 and 0·2 per unit (i.e. at full load current $IX_L/V = 0{\cdot}1$ to $0{\cdot}2$).

Fig. 7.4 shows that at the instant when the red phase current **I** has its maximum value, the instantaneous e.m.f. generated in the phase winding is $\sqrt{2}\,\mathbf{E}\cos\alpha$. In the interval since the instant when the peak e.m.f. was induced in the central conductor of the red phase

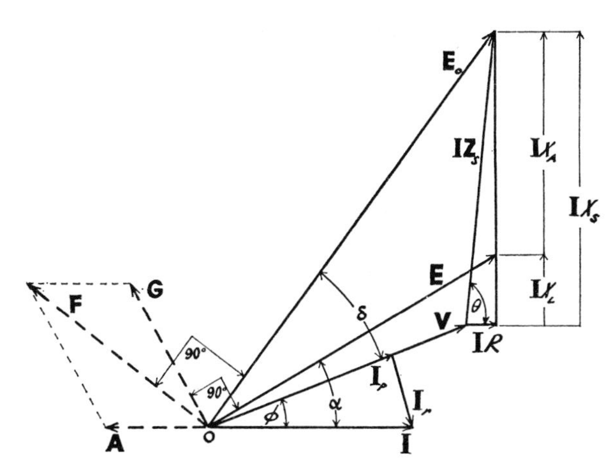

FIG. 7.4. *Voltage and current phasors and m.m.f. space vectors of an unsaturated cylindrical-rotor alternator.*

$\mathbf{V}$ = *terminal phase voltage*
$\mathbf{I}$ = *stator phase current*
$\cos\phi$ = *load power factor*
$\mathbf{I}R$ = *voltage drop due to stator phase resistance*
$\mathbf{I}X_L$ = *voltage drop due to stator phase leakage reactance*
$\mathbf{I}X_S$ = *voltage drop due to stator phase synchronous reactance*
θ = $\tan^{-1} X_S/R$

band, the resultant flux and resultant m.m.f. **G** (which generates **E**) in the air gap must have rotated past this conductor, by a space angle α radians (or a space angle α/p radians for a machine with $2p$ poles). This resultant generating m.m.f. **G** can therefore be shown in Fig. 7.3 leading the space vector $-\mathbf{A}$ by $(\alpha+\pi/2)$ radians. The amplitude of the applied field m.m.f./pole **F** ($= I_f N_f$ where I_f is the d.c. field current and N_f is the number of turns/pole on the rotor) is given by

$$\mathbf{F} = \mathbf{A}+\mathbf{G}.$$

F must lie on the rotor pole centre line (direct axis) and the positions of **A** and **G** vary with load p.f. as discussed below.

Since these space m.m.f.s rotate at synchronous speed under steady-state conditions it is permissible for convenience (even although **F** is due to a d.c. current) to show these space vectors of m.m.f. associated in a diagram with the time phasors of voltage and current. This has been done in Fig. 7.4, choosing arbitrarily to take the armature reaction m.m.f. $-\mathbf{A}$ in line with the current **I** which produces it. This gives an easily-remembered right angle between the resultant generating air-gap m.m.f. **G** and the e.m.f. **E** generated by it, but no physical significance attaches to this angle.

If the saturation of the magnetic circuit is neglected, e.m.f.$\propto$ m.m.f. and there is an e.m.f. $\mathbf{E}_o$ such that then $E_o/F = E/G$. (F denotes the magnitude of **F** as an alternative to writing $|\mathbf{F}|$). $\mathbf{E}_o$ is the open-circuit e.m.f. which would appear in the steady-state after reducing the load current **I** to zero, and maintaining the same value of I_f, i.e. keeping F constant. $\mathbf{E}_o$ also appears 90° behind the corresponding m.m.f. **F**.

The 90° angles between $\mathbf{E}_o$ and **F** and between **E** and **G** are quite arbitrary and have no physical significance at all, since $\mathbf{E}_o$ and **E** are phasors (time varying) but **F** and **G** are space vectors. They must not be confused with the 90° phase angle between an e.m.f. induced in a winding and the flux which is stationary on the axis of the coil, with a magnitude which varies sinusoidally with time, as for example in a transformer.

If saturation and harmonics are neglected in a cylindrical-rotor machine, it may be seen by similar triangles in Fig. 7.4, that the tip of the phasor $\mathbf{E}_o$ must lie on $\mathbf{I}X_L$ produced since the phasor joining **E** to $\mathbf{E}_o$ must be 90° behind **A**. For balanced steady-state operation of an unsaturated cylindrical-rotor machine, the effect of armature reaction within the machine upon the circuit may therefore be represented approximately by using a fictitious armature reactance X_A. The sum of X_A and X_L is called the synchronous reactance X_s, and $X_s \gg X_L$. Values of X_s from 1·5 to 2·0 per unit are common for large 2-pole turbo-alternators. For slow-speed alternators X_s ranges from about 0·5 to 1 per unit. The effective stator impedance/phase under steady-state conditions is then $\mathbf{Z}_s = R + \mathrm{j}\,X_s$, and $\mathbf{Z}_s$ is known as the synchronous impedance.

Since the voltage $\mathbf{I}R$ is only about 0·01 per unit it can be neglected in comparison with $\mathbf{I}X_s$ for some purposes. If Fig. 7.4 is redrawn for $R = 0$ and $\phi = 90°$ lagging, it is seen that **V**, **E** and $\mathbf{E}_o$ are all in phase and lead **I** by 90°. The armature reaction m.m.f. $-\mathbf{A}$ is then directly opposed to the field m.m.f. **F**, so that it is on the direct axis (rotor pole centre line). The field m.m.f. **F** is given by adding G to A algebraically since they are all on the direct axis, and **F** is relatively large. These are the conditions which occur in the zero power factor lagging test which may be carried out on synchronous machines, to give data which with other results enables the values of the components X_A and X_L of X_s to be determined.

Similarly if $R = 0$ and $\phi = 90°$ leading then **V**, **E** and $\mathbf{E}_o$ are again all in phase and lag **I** by 90°. The armature reaction m.m.f. $-\mathbf{A}$ is again on the direct axis but it is now assisting the rotor m.m.f. which is given by $F = G - A$ and **F** is now relatively small for operation at rated terminal voltage. Since the torque and power at which a synchronous machine pulls out of synchronism is almost proportional to $\mathbf{E}_o$ which depends upon **F** (see section 7·7), it follows that system stability may be adversely affected by running machines at relatively low leading power factors, unless special voltage regulators are used (section 7·9).

It can also be seen by redrawing Fig. 7.4 that if **I** were in phase with $\mathbf{E}_o$, the fundamental of the armature reaction m.m.f. wave $-\mathbf{A}$ would be on the interpolar or quadrature axis, so that the air gap flux would be displaced from the direct axis. In general therefore armature reaction at any given power factor and current, causes the air gap m.m.f. **G** to be shifted from the direct axis by an angle which remains constant at all times for that load condition, and alters its magnitude from the air gap m.m.f. **F**.

Worked example 7.1

A 3-phase, cylindrical-rotor, 1000-kVA, 6·6-kV star-connected alternator has the open-circuit and short-circuit characteristics given by the following table:

Field current, A	0	25	50	75	100	150	200
Open-circuit phase voltage, V	0	1200	2400	3450	4000	4500	4700
Short-circuit current, A	0	40	80	—	—	—	—

Assuming that the open-circuit characteristic had coincided with the 'air-gap line' (i.e. neglecting saturation), calculate

(*a*) the synchronous reactance/phase;

(*b*) the open-circuit voltage at the excitation required for full load output at 0·8 power factor lagging.

Using the true open-circuit characteristic, calculate the field current at full load, 0·8 power factor lagging. The stator resistance is negligible and the leakage reactance/phase is 3 Ω.

(University of London)

SOLUTION

Inspection of the table shows that the 'air-gap line' extends up to 50 A excitation as the open-circuit curve is linear in this range, so that $X_s = 1200/40 = 30\ \Omega$/phase.

At full load the stator current is

$$I = \frac{1000 \times 10^3}{\sqrt{3} \times 6600} = 87{\cdot}5 \text{ A}$$

and $V = 6600/\sqrt{3} = 3810$ volts/phase.

In general from Fig. 7.4

$$E_o^2 = (V \cos \phi + IR)^2 + (V \sin \phi + IX_s)^2$$

and $\mathbf{I}R$ is negligible here compared with $\mathbf{I}X_s$

$$E_o^2 = (3810 \times 0{\cdot}8)^2 + (3810 \times 0{\cdot}6 + 87{\cdot}5 \times 30)^2$$

and $E_o = 5780$ volts.

The open-circuit voltage between line terminals would be $\sqrt{3} \times 5780 = 10\,000$ volts.

Alternatively, the solution may be obtained without the phasor diagram by writing,

$$\mathbf{E}_o = \mathbf{V} + \mathbf{I}X_s = 3810\underline{/0^\circ} + 87{\cdot}5\underline{/-36^\circ\ 52'} \times 30\underline{/90^\circ}$$

and this yields the phase angle between $\mathbf{E}_o$ and $\mathbf{V}$ directly.

E, the 'e.m.f. behind leakage reactance' is the e.m.f. actually induced per phase by the resultant field current (or m.m.f. **G**).

From Fig. 7.5

$$|\mathbf{E}| = [(V \cos \phi)^2 + (V \sin \phi + IX_L)^2]^{\frac{1}{2}}$$
$$= [(3810 \times 0{\cdot}8)^2 + (3810 \times 0{\cdot}6 + 87{\cdot}5 \times 3)^2]^{\frac{1}{2}} = 3970 \text{ volts}$$

from the o.c. curve, a phase voltage of 3970 is induced by a field current of 97 A. **G** which can be measured in field amperes if **F** is measured in field amperes instead of ampere turns/pole, is therefore 97 A.

A stator current of 87·5 A sets up armature reaction which requires **A** field amperes to neutralise it. The value of **A** can be found

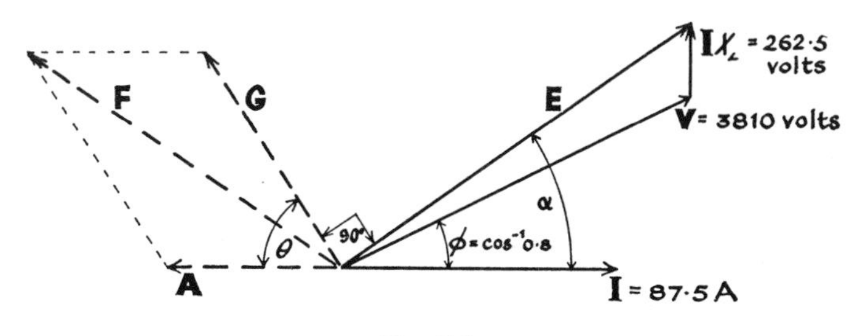

FIG. 7.5.

from consideration of the short-circuit test. Part of the field current $\mathbf{G}_{sc}$ must set up air-gap flux which generates an e.m.f. $\mathbf{I}_{sc}X_L$. For a short-circuit current I_{sc} = say 80 A, $I_{sc}X_L = 240$ volts and from the o.c. curve G_{sc} must be 5 A to generate an e.m.f. of 240 volts. The

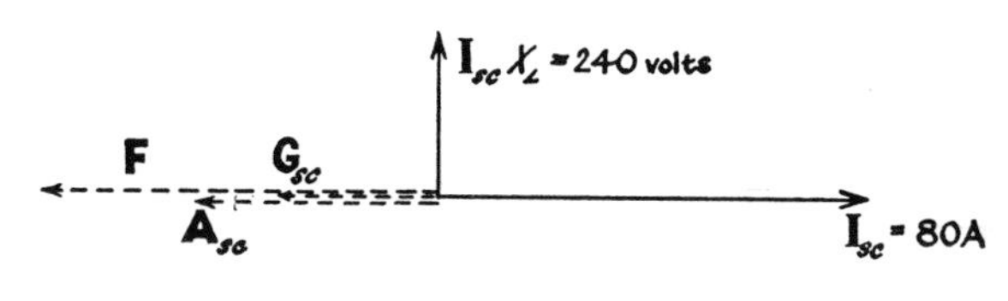

FIG. 7.6.

total field current which is 50 A for a short-circuit stator current of 80 A, must be the sum of the armature-reaction component of the field current $\mathbf{A}_{sc}$ and $\mathbf{G}_{sc}$. A_{sc} for 80 A stator current must therefore be $50-5 = 45$ A since Fig. 7.6 shows that on short-circuit of a stator winding with negligible resistance, all three m.m.f.s act on the same axis and can therefore be added or subtracted algebraically.

Hence for a stator current of 87·5 A, there is a component of field current of $45 \times 87{\cdot}5/80 = 49{\cdot}2$ A which is needed to neutralise the armature reaction.

The total field current **F** can be found by adding **A** and **G** vectorially as shown in Fig. 7.5.

From Fig. 7.5

$$F^2 = G^2 + A^2 + 2GA \cos\theta$$

$$\cos\theta = \sin\alpha = \frac{(3810\times 0{\cdot}6 + 262{\cdot}5)}{3970} = 0{\cdot}6428$$

$$F = [97^2 + 49{\cdot}2^2 + 2\times 97\times 49{\cdot}2\times 0{\cdot}6428]^{\frac{1}{2}}$$

$$= 133{\cdot}9 \text{ A}.$$

A field current of 133·9 A is required to give full-load output at 0·8 power factor lagging.

7.3 Synchronous Impedance and Short-circuit Ratio

If a symmetrical 3-phase short circuit occurs at the terminals of a 3-phase alternator, then after steady-state conditions have been reached, the whole of the e.m.f. $\mathbf{E}_o$ which would be generated in each phase on open-circuit, may be regarded as being absorbed in circulating a short-circuit current $\mathbf{I}_{sc}$ through the synchronous impedance $\mathbf{Z}_s$. Thus for any given field current in Fig. 7.7 the per unit open-circuit e.m.f. may be divided by the corresponding value of per unit short-circuit current to give the per unit synchronous impedance. Alternatively expressing the e.m.f. in volts and the current in amperes

$$Z_s = E_o/I_{sc}\,\Omega.$$

This is sometimes known as a saturated value of synchronous impedance, because although it is constant at low excitation, its value falls as the field current is raised and saturation occurs. If I_{sc} is divided into the e.m.f. given by the straight part of the open-circuit curve which may be produced as shown dotted in Fig. 7.7 and which is known as the air-gap line, then an unsaturated value of synchronous impedance results.

If the machine is run at rated frequency supplying rated current to a load which causes the power factor to be zero lagging, then variation of field current causes the terminal voltage to vary as shown in Fig. 7.7. The value of field current I_{f3} at which the terminal voltage in this zero power factor (z.p.f.) test is zero, will be virtually the same as that required to circulate rated current in the short-circuit test where the power factor is very nearly zero.

Since the short-circuit current is directly proportional to field current, the short-circuit current at the field current I_{f1}, which generates 1·0 per unit voltage on the air-gap line, is $1{\cdot}0 \times I_{f1}/I_{f3}$. The unsaturated per unit synchronous impedance is therefore $1{\cdot}0 \div (I_{f1}/I_{f3}) = I_{f3}/I_{f1}$.

The short-circuit ratio (S.C.R.) is a measure of the relative strengths of rotor (field) and stator ampere-turns. It is the ratio of the field

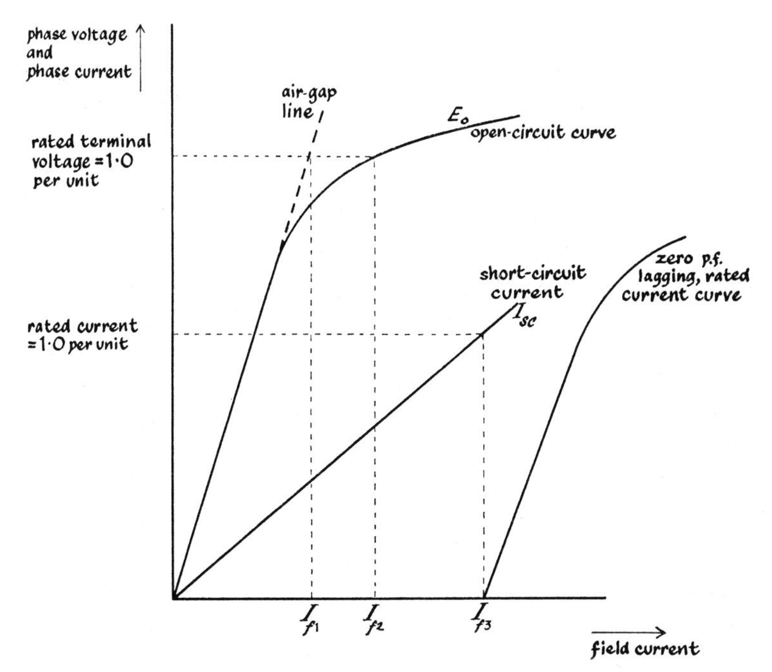

FIG. 7.7. *Open-circuit phase e.m.f., short-circuit phase current and phase terminal voltage with rated current at zero power factor lagging, as a function of field current.*

current which gives rated stator phase voltage on open-circuit to the field current which gives rated stator current on short-circuit.

$$\text{Unsaturated S.C.R.} = I_{f1}/I_{f3}.$$

The unsaturated short-circuit ratio is therefore the reciprocal of the unsaturated per unit synchronous impedance.

The saturated short-circuit ratio $= I_{f2}/I_{f3}$ and this is the value usually quoted as the S.C.R. This is not exactly equal to the reciprocal of the per unit synchronous impedance. The latter is not strictly constant, though for power system calculations a constant effective value is generally assumed. A large 2-pole turbo-alternator may have a synchronous reactance of about 2 per unit and S.C.R. of about 0·5.

7.4 Two-axis Theory of a Salient-pole Alternator

For a cylindrical-rotor machine the permeance of an elemental flux tube crossing the air gap is nearly constant for all angular positions, so that the m.m.f.s shown in Fig. 7.4 are directly proportional to air gap fluxes if saturation of the iron part of the circuit can be neglected.

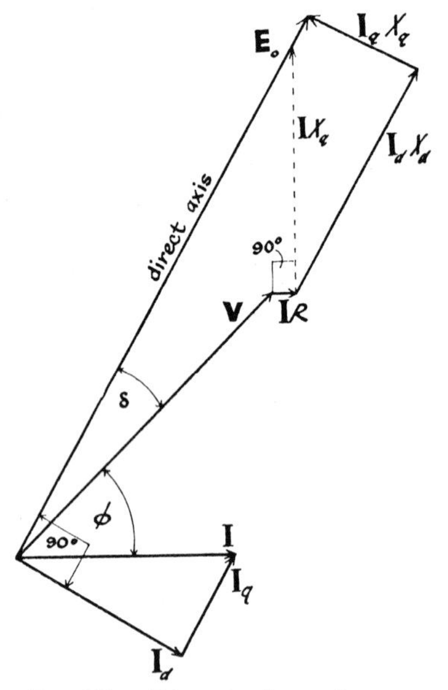

FIG. 7.8. *Two-axis phasor diagram.*

For a salient-pole machine, however, the permeance for a flux tube varies from a maximum $\mathscr{P}_d$ (which would be similar to that for a cylindrical machine) on the direct axis, to a minimum $\mathscr{P}_q$ on the quadrature (interpolar) axis where there is the long air path of the interpolar space. The field m.m.f. **F** acts only on the direct axis, but as indicated in section 7.2, the armature reaction m.m.f. $-$**A**, though rotating synchronously with the rotor m.m.f. **F**, varies in angular position relative to it according to the power factor. In general, since only the fundamental is being considered, the armature reaction m.m.f. can be resolved into two components on two axes separated by 90 electrical degrees. One component is $-\mathbf{A}_d$ on the direct axis and the other is $-\mathbf{A}_q$ on the quadrature axis. These m.m.f.s are deemed to be produced by the components $\mathbf{I}_d$ and $\mathbf{I}_q$ of $\mathbf{I}$ shown in Fig. 7.8, and since $\mathscr{P}_d > \mathscr{P}_q$, the effective quadrature-axis synchronous reactance X_q is less than the corresponding direct-axis value X_d. For salient-pole alternators with X_d of about 0·9 per unit, X_q is likely to be of the order of two-thirds this value, whereas for cylindrical-rotor machines such as 2-pole turbo-alternators with X_d in the range 1·5 to 2·0 per unit, X_q is likely to be only about 5–10% (of 1·5 p.u.) less than this value, due to rotor slots.

It should be noted that $\mathbf{I}_d$ is a reactive lagging current relative to $\mathbf{E}_o$, and creates an armature reaction effect which is included within $\mathbf{I}_d X_d$ leading $\mathbf{I}_d$ by 90°, and hence produces an effect on the direct axis. Both X_d and X_q include the leakage reactance X_L which is assumed to be the same for both. (It is convenient for drawing Fig. 7.8 to know that a phasor $\mathbf{I}X_q$ drawn from the end of the phasor $\mathbf{I}R$ terminates at a point on the phasor $\mathbf{E}_o$, as shown dotted in Fig. 7.8.)

7.5 Transient Conditions

In general, if a short-circuit occurs at or very near the terminals of an alternator, the phase current shown by an oscillogram would be of the form shown in Fig. 7.9. This shows that the current contains

(*a*) a d.c. component which decays approximately exponentially, and

(*b*) an alternating component which decays with time until it reaches a steady-state value.

The d.c. component arises from the fact that if a source of voltage $e = E_m \sin(\omega t + \alpha)$ is applied at $t = 0$, to a circuit containing resistance $R\,\Omega$ and inductance L henry in series, the current which flows is the sum of the steady-state alternating component (particular integral solution) and the transient unidirectional component (complementary function) which dies away exponentially with time constant $T = L/R$ seconds. This latter component can only be zero for circuit closure at a particular point in the voltage wave. The current i is given by

$$i = (E_m/Z)\sin(\omega t + \alpha - \phi) + K\,\mathrm{e}^{-Rt/L}$$

where

$$Z = \sqrt{(R^2+\omega^2L^2)}$$

and

$$\phi = \tan^{-1} \omega L/R.$$

For zero current at $t = 0$

$$K = -(E_m/Z) \sin(\alpha-\phi).$$

The d.c. component will therefore be zero, only if $(\alpha-\phi) = 0$ or π radians. Similarly the d.c. component will have a maximum initial

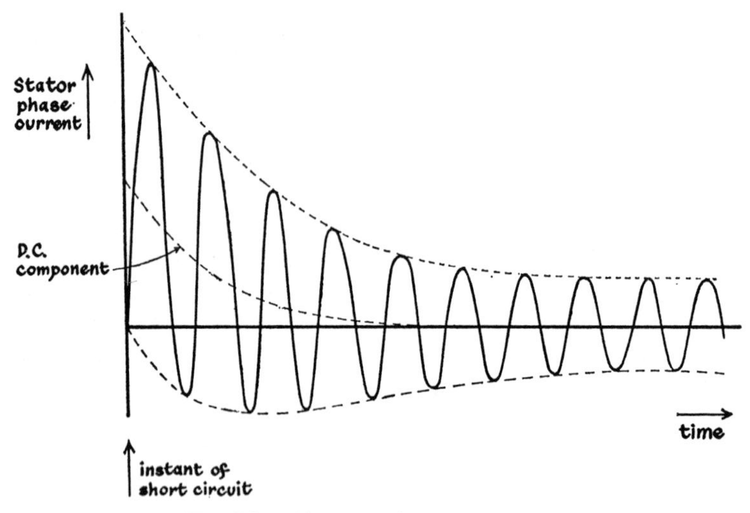

FIG. 7.9. *Alternator short-circuit current.*

value of $-E_m/Z$ which is the peak value of the alternating component, if the circuit is closed when $(\alpha-\phi) = \pm\pi/2$ radians. If $\omega L \gg R$, then $\phi \simeq \pi/2$ radians, so that circuit closure at voltage maximum would give no d.c. component, and closure at voltage zero would cause the maximum d.c. transient current to flow.

This shows why a short-circuit on all three phases of an unloaded alternator gives oscillograms of stator current with a d.c. component of the type shown in Fig. 7.9. Since the three phase voltages are each separated by $2\pi/3$ radians, the amounts of the d.c. components in the three phase currents in any one test or fault will generally all differ, and will vary from one test or fault to another unless the point-on-wave of circuit closure is controlled.

The fact that the alternating component of current falls in

magnitude until the steady-state value is reached as shown in Fig. 7.9, is due to the effective inductive reactance of a synchronous machine increasing from a relatively low initial value at the instant of short-circuit, until it reaches the steady-state value. This steady-state value is generally taken as X_d, the direct-axis synchronous reactance, because on short-circuit the circuit reactance is much greater than the resistance, so that the stator current lags nearly $\pi/2$ radians behind the driving voltage, and the armature reaction m.m.f. established in the steady state is centred almost on the direct axis. Similarly during the period when the reactance of the machine is changing, the values may be taken as direct-axis quantities. It is possible, however, for the fault power factor to be appreciably above zero in some cases, e.g. in resistance-earthed systems during an earth fault, so that reaction in the machines is not necessarily wholly on the direct axis.

It should be noted that in the usual equivalent circuit discussion of transient conditions which is followed at this point, there is the fiction of a constant driving voltage generated in the alternator winding together with a reactance which varies. Only the portion X_L of X_s or X_d is a true reactance, while the rest represents the modification of resultant m.m.f. and generated e.m.f. due to armature reaction. In fact therefore a short-circuit current at a very low lagging power factor reduces the machine e.m.f. more and more as armature reaction builds up.

If the r.m.s. phase e.m.f. on open circuit is E_o and the stator resistance is neglected then the r.m.s. steady-state current $I = E_o/X_d$, so that X_d may be determined from the part of the oscillogram of Fig. 7.9 where the envelope of the current has become constant. In the earlier part of this oscillogram the r.m.s. value of alternating current can be found by dividing the peak-to-peak envelope of current by $2\sqrt{2}$. If this current less the steady-state a.c. value I is now plotted on a log scale as in Fig. 7.10, against a linear time scale, it is found that apart from the first few cycles (of the order of 5), the graph is linear. The slope of this linear part of the curve gives a transient time constant T_d', and producing the straight line back to zero time gives a current $I_d' = E_o/X_d'$. X_d', thus defined, is known as the direct-axis transient reactance. For a 2-pole turbo-alternator X_d' may lie between 0·1 and 0·25 per unit, and for water-wheel generators the range is similar.

If the r.m.s. current, over the first few cycles, in excess of the values shown by the dotted line in Fig. 7.10, is now plotted on a log scale against a linear time scale, a straight line is given with a slope which yields the sub-transient time constant T''_d. This new curve (or that of Fig. 7.10) intercepts the axis at zero time at a current $I''_d = E_o/X''_d$. X''_d, defined by this equation, is known as the direct-axis sub-transient reactance.

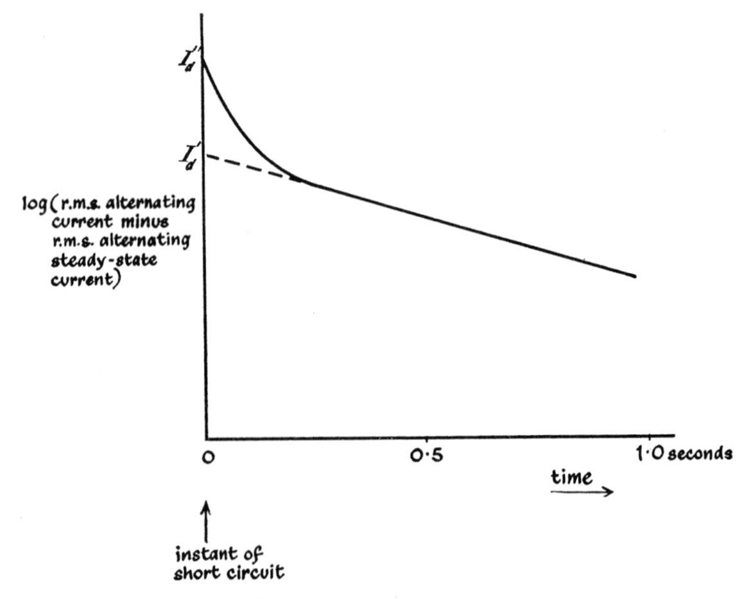

FIG. 7.10. *Log plot of subtransient and transient components of short-circuit current on linear time scale.*

In order to discuss the reasons for the variation in reactance, the changes in flux due to armature reaction must be considered. This will be done only briefly here as it is dealt with in more detail in many books on machines. At the instant prior to short-circuit there is some flux on the direct axis linking both stator and rotor, due only to rotor m.m.f. if the machine is on open-circuit, or due to the resultant of rotor and stator m.m.f. if some stator current is flowing. When there is a sudden increase of stator current on short-circuit, the flux linking stator and rotor cannot change instantaneously due to eddy currents flowing in various rotor circuits, which oppose this change. Since therefore the stator m.m.f. is unable at first to establish any armature reaction, that part of the synchronous reactance corresponding to this, X_A, does not exist initially. The initial reactance, the sub-transient value X''_d, is therefore similar in value to the leakage reactance X_L which exists under steady-state conditions. For 2-pole turbo-alternators X''_d may be between 0·07 and 0·18 per unit, while for water-wheel alternators the range may be 0·1 to 0·3 per unit, though it may exceed this if there are no damper windings.

The eddy currents which cause the initial sub-transient time constant T''_d, flow in the steel tooth tips on either side of the air gap in a cylindrical-rotor alternator. In the case of slow-speed alternators driven by water-turbine or diesel engine, these eddy currents flow in damper windings embedded in the pole faces. Since the time constant of decay of the eddy currents in the stator or rotor steel surface paths or the damper winding T''_d is generally about one or two cycles, the direct-axis sub-transient reactance X''_d is only used in calculations if the effect of the initial current is important, as for example when determining the current flowing in a circuit breaker when it closes on to a short-circuit.

If the machine had steel laminations without damper windings, the sub-transient component of the short-circuit current would virtually vanish, and the current would fall exponentially with the transient time constant T'_d. This fall occurs as the eddy currents induced into the field winding decay, since the field winding closed through the relatively low resistance of the exciter forms a closed loop. Since on short-circuit the d.c. current in the field rises to X_d/X'_d times its original value, the transient time constant $T'_d \simeq (X'_d/X_d)T'_{do}$ if the fault is at the alternator terminals. T'_{do} the transient open-circuit time constant is the ratio of field circuit inductance to resistance, and X'_d/X_d may be about 1/10 or 1/5. If there is an external circuit impedance $R_{ext}+\text{j}\, X_{ext}\,\Omega$ between the machine and the fault with $R_{ext} \ll X_{ext}$, then

$$T'_d \simeq \frac{X'_d+X_{ext}}{X_d+X_{ext}} T'_{do}.$$

The transient time constant T'_d is usually of the order of one second. The reactance used in determining the time constant T_a of the d.c. component of the stator current may be taken approximately as the negative-sequence reactance X_2 of a stator phase (see section

7·6). This is because most of the decay of the d.c. component occurs during the sub-transient period, and as discussed in section 7·6, X_2 is approximately the average of X''_d and X''_q, the latter being the quadrature-axis sub-transient reactance.

If a 3-phase short circuit occurs at the terminals of an unloaded alternator with r.m.s. phase voltage $V_m/\sqrt{2}$, at an instant when the d.c. component of current in one phase has its maximum value, then that phase carries a d.c. component of

$$I_{dc} = \frac{V_m}{X''_d} e^{-t/T_a} \qquad [7.1]$$

where $T_a \simeq X_2/2\pi fR$ seconds and R is the stator resistance/phase.

The alternating component of current in each phase has an r.m.s. value of

$$I_{ac} = \frac{V_m}{\sqrt{2}}\left[\left(\frac{1}{X''_d}-\frac{1}{X'_d}\right)e^{-t/T''_d}+\left(\frac{1}{X'_d}-\frac{1}{X_d}\right)e^{-t/T'_d}+\frac{1}{X_d}\right] \qquad [7.2]$$

If the fault is not at the terminals, any circuit impedance is included together with any fault impedance in [7.1] and [7.2], and reduces the magnitudes of the various components of current and modifies the time constants T_a and T'_d. For faults some distance from the point of generation the resistance may be sufficiently large to reduce the time constants T_a and T'_d to very low values. Furthermore the constant impedances of lines and transformers between the fault and the alternators tend to swamp the variation of alternator reactance. The r.m.s. a.c. component of the fault current is then almost constant.

It is more usual for a fault to occur on a machine when it is delivering a pre-fault load current **I** at a lagging power factor cos ϕ than when it is unloaded. The phasor diagram of Fig. 7.8, with stator resistance neglected is re-drawn in Fig. 7.11, showing the pre-fault load current **I** and allowing the stator reactances on the two axes to take up the appropriate values corresponding to the sub-transient, transient and steady-state conditions. This gives the three fictitious driving voltages $\mathbf{E}_o$, $\mathbf{E}'$ and $\mathbf{E}''$ which may be considered to be effective during the various periods of time after the fault. It is usual to assume for many power system studies that during whichever of these three fault periods is of interest, sub-transient, transient or steady-state, the r.m.s. current can be considered to remain constant at a level appropriate to that period. Where variation of current must be considered within one of these periods, it is generally more convenient to keep the reactance constant and vary the source voltage behind it, rather than to consider a varying reactance. The voltage $\mathbf{E}_o$, which may be called a voltage behind synchronous reactance, would be equal to the actual phase voltage on open-circuit $\mathbf{E}_{oc}$, if all the assumptions of the simple analysis of section 7.2 were in fact true, e.g. if there were no saturation and all harmonics were zero.

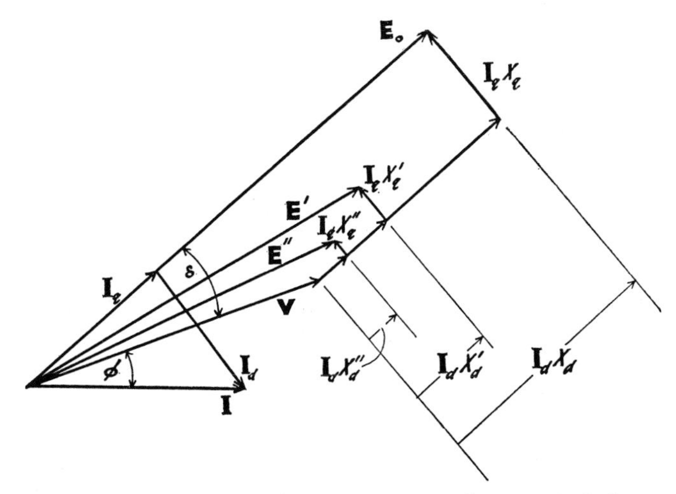

FIG. 7.11. *Two-axis phasor diagram showing pre-fault current and the voltages behind subtransient and transient reactances.*

When the fault current in a circuit with high ratio of inductive reactance to resistance, has fallen to the steady-state value, lagging almost $\pi/2$ radians behind the voltages, it is generally satisfactory for power system analysis to represent the machine by a source with an e.m.f. $\mathbf{E}_o$ and a reactance X_d. Similarly, if the machine has solid rotor or damper windings, and conditions in the first one or two cycles are critical, it may be represented approximately during that period by an e.m.f. $\mathbf{E}''$ (sometimes called the voltage behind sub-transient reactance) and a constant reactance X''_d. For some purposes, e.g. in stability studies, this initial period is less critical, and it is possible to use the simple approximation of an e.m.f. $\mathbf{E}'$, the voltage behind transient reactance, and a reactance X'_d. This simple equivalent circuit generally gives sufficient accuracy for calculations of the machine angle in the first swing, but more detailed representation is needed for some transient stability studies, e.g. where pole-slipping occurs (Day and Parton, 1965). It should be noted that if either of these voltages **E**′ or **E**″ is used, it does not lead the terminal voltage **V** by the load angle δ. These approximations are more justifiable if, in addition to the previous assumptions of no harmonics, no saturation, zero stator resistance and no change in speed during the fault, the quadrature- and direct-axis permeances and hence

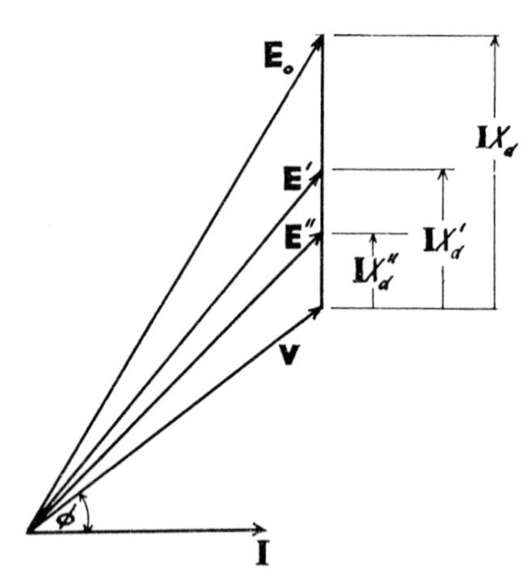

FIG. 7.12. *Approximate phasor diagram for cylindrical-rotor alternator with pre-fault current and voltages behind subtransient and transient reactances.*

reactances are assumed to be equal, as is almost true for a turbo-alternator with cylindrical rotor. The ends of the phasors **V**, **E**″, **E**′ and $\mathbf{E}_0$ in Fig. 7.11 are then co-linear and Fig. 7.11 reduces to Fig. 7.12.

It is therefore possible approximately to represent the loaded generator which is suddenly subjected to an increase in current due to a fault, by a source of e.m.f. **E**″ in series with reactance X_d'' or, an e.m.f. **E**′ and a reactance X_d' where

$$\left.\begin{aligned} \mathbf{E}'' &= \mathbf{V} + \mathrm{j}\,\mathbf{I}X_d'' \\ \text{and} \qquad \mathbf{E}' &= \mathbf{V} + \mathrm{j}\,\mathbf{I}X_d' \end{aligned}\right\}. \qquad [7.3]$$

These e.m.f.s may again be termed 'voltages behind subtransient and transient reactance respectively'. It is usual to assume that both e.m.f. and reactance remain constant during whichever of the two non-steady-state periods is being considered, if the fault is not immediately adjacent to the machine.

Fig. 7.13 shows one phase of an alternator with a pre-fault load current **I** which is supplied to a balanced load of impedance $\mathbf{Z}_L$/phase, through a system in which the impedances of transformers

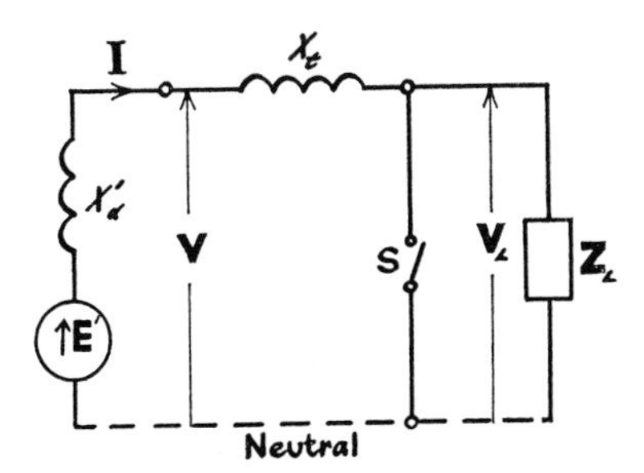

FIG. 7.13. *Equivalent circuit for a fault on a loaded alternator during the transient period.*

and lines total X_t. The equivalent-circuit alternator voltage behind transient reactance, which is regarded as circulating the current during the transient period after the 3-phase fault occurs at the load, is $\mathbf{E}'$ and this is related to the actual pre-fault terminal voltages at the load and generator, $\mathbf{V}_L$ and $\mathbf{V}$ respectively, by

$$\mathbf{E}' = \mathbf{V}_L + \mathrm{j}\,\mathbf{I}(X'_d + X_t) = \mathbf{V} + \mathrm{j}\,\mathbf{I}X'_d.$$

If a 3-phase fault of zero impedance occurs at the load, this is equivalent to closing switch S in Fig. 7.13. By Thevenin's theorem (which can be used since the reactances are assumed to remain constant during the period of interest), the current flowing through the short-circuit during the transient period is

$$\mathbf{I}' = \frac{\mathbf{V}_L(\mathbf{Z}_L + X_t + X'_d)}{\mathbf{Z}_L(X_t + X'_d)}. \qquad [7.4]$$

The current flowing in the generator during the transient period is $\mathbf{I}' + \mathbf{I}$.

In the case of a synchronous motor which is receiving power from the 3-phase supply, a fault which causes the supply voltage to fall, may cause the machine to supply power from its stored energy into the system and the fault, until the fault is cleared or the speed drop reduces the internal e.m.f. sufficiently relative to the supply voltage. The changes in reactance of the synchronous motor will be similar to those in the alternator, and [*7.3*] with a reversed sign for the reactance voltage drop, will give the effective machine voltages behind sub-transient and transient reactances.

Worked example 7.2

A 22-kV 100-MVA alternator with 0·25 per unit transient reactance is supplying a load through a transmission line of reactance 0·05 per unit at 100 MVA. The load at a particular time is equivalent to a 50-MVA synchronous motor with 0·20 per unit transient reactance, which is taking 40 MW at 0·8 power factor leading, with a terminal voltage of 21·9 kV.

If a 3-phase short circuit occurs at the alternator terminals calculate the currents in each of the two machines and in the fault during the transient period.

SOLUTION

For 100 MVA and 22 kV the base current is

$$100\times 10^6/(\sqrt{3}\times 22\times 10^3) = 2625 \text{ A}.$$

The magnitude of the load current is

$$40\times 10^6/(\sqrt{3}\times 21{\cdot}9\times 10^3\times 0{\cdot}8) = 1320 \text{ A} = 1320/2625 = 0{\cdot}503 \text{ per unit}.$$

If the motor terminal voltage is the reference phasor, then the load current is $0{\cdot}503(0{\cdot}8+\text{j}\,0{\cdot}6) = 0{\cdot}402+\text{j}\,0{\cdot}302$ per unit.

The pre-fault generator and motor terminal voltages are therefore $\mathbf{V}_G = (21{\cdot}9/22{\cdot}0+\text{j}\,0)+\text{j}\,0{\cdot}05(0{\cdot}402+\text{j}\,0{\cdot}302) = 0{\cdot}9799+\text{j}\,0{\cdot}0201$ per unit and $\mathbf{V}_M = 21{\cdot}9/22{\cdot}0+\text{j}\,0 = 0{\cdot}995+\text{j}\,0$ per unit.

The generator 'voltage behind transient reactance' is

$$\begin{aligned}\mathbf{E}'_G &= 0{\cdot}9799+\text{j}\,0{\cdot}0201+\text{j}\,0{\cdot}25(0{\cdot}402+\text{j}\,0{\cdot}302)\\ &= 0{\cdot}9044+\text{j}\,0{\cdot}1206 \text{ per unit}\end{aligned}$$

and the motor 'voltage behind transient reactance' is

$$\begin{aligned}\mathbf{E}'_M &= 0{\cdot}995+\text{j}\,0-\text{j}\,0{\cdot}40(0{\cdot}402+\text{j}\,0{\cdot}302)\\ &= 1{\cdot}1158-\text{j}\,0{\cdot}1608 \text{ per unit}.\end{aligned}$$

The total fault current during the transient period, may now be found by superposition of the two components $\mathbf{I}'_G$ and $\mathbf{I}'_M$, which flow through switch S (Fig. 7.14) when it closes to represent the fault.

Since the fault is regarded as a short-circuit having zero impedance, there can be no current fed by machine G into machine M, and

$$\mathbf{I}'_G = \mathbf{E}'_G/\mathrm{j}\,0{\cdot}25$$

$$\mathbf{I}'_G = (0{\cdot}9044+\mathrm{j}\,0{\cdot}1206)/\mathrm{j}\,0{\cdot}25 = 0{\cdot}4824-\mathrm{j}\,3{\cdot}6176 \text{ per unit}$$

$$\mathbf{I}'_G = 1265-\mathrm{j}\,9490 \text{ A}$$

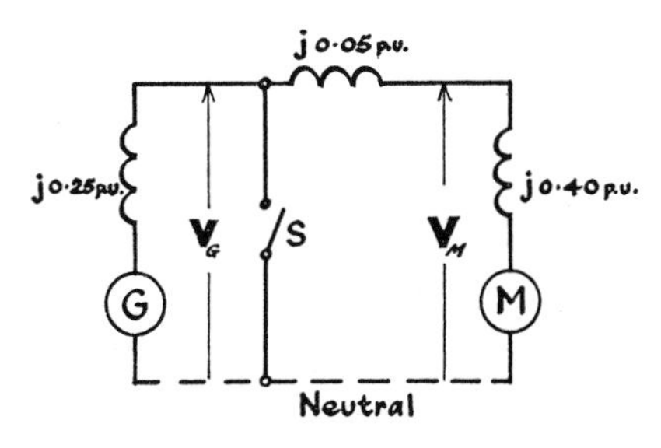

FIG. 7.14.

and similarly

$$\mathbf{I}'_M = \mathbf{E}'_M/\mathrm{j}\,0{\cdot}45 = (1{\cdot}1158-\mathrm{j}\,0{\cdot}1608)/\mathrm{j}\,0{\cdot}45$$

$$\mathbf{I}'_M = -0{\cdot}3573-\mathrm{j}\,2{\cdot}4795 \text{ per unit}$$

$$\mathbf{I}'_M = -938-\mathrm{j}\,6500 \text{ A}$$

The total current in each phase of the short-circuit

$$= \mathbf{I}'_M+\mathbf{I}'_G = 327-\mathrm{j}\,15\,990 \text{ A}.$$

Worked example 7.3

Two 22-kV 3-phase 50-Hz generators, one of 60 MVA and 0·18 per unit transient reactance, and the other of 100 MVA and 0·2 per unit transient reactance, are connected to a busbar from which a 150-MVA 22/132-kV transformer with 0·12 per unit leakage reactance, feeds a transmission line of inductance 27·8 mH. At the other end of the line, a second 150-MVA 132/11-kV transformer with 0·09 per unit leakage reactance, supplies a busbar at which the voltage is 10·55 kV when the following two loads are connected to it: (*a*) a balanced 3-phase load of 40 MW at 0·8 power factor lagging and (*b*) a 20-MVA synchronous motor with 0·16 per unit transient reactance, taking 20 MVA at 0·9 power factor leading. The two generators share the reactive volt-amperes of the system equally but share the power in proportion to their ratings.

If a 3-phase fault with an effective arc resistance of 0·0484 Ω between each line and earth, occurs at the motor terminals, when the above loads are being supplied, calculate the current in each of the generators and that in the machine which was previously motoring, during the transient period after the fault occurs.

SOLUTION

If 100 MVA is taken as the base, then at 132 kV base current = $10^8/(\sqrt{3}\times 132\,000) = 437{\cdot}37$ A.

The per unit transmission line reactance at 100 MVA and 132 kV

$$= \mathrm{j}\,\frac{27{\cdot}8\times 2\pi\times 50}{1000}\times\frac{437\times\sqrt{3}}{132\,000} = \mathrm{j}\,0{\cdot}05 \text{ p.u.}$$

Generator A, rated at 60 MVA, has $\mathrm{j}\,X_d' = \mathrm{j}\,0{\cdot}18\times 100/60 =$ $\mathrm{j}\,0{\cdot}3$ per unit at 100 MVA.

The synchronous motor M has $\mathrm{j}\,X_d' = \mathrm{j}\,0{\cdot}16\times 100/20 = \mathrm{j}\,0{\cdot}8$ per unit.

The 22/132-kV transformer has leakage reactance of $\mathrm{j}\,0{\cdot}12\times 100/150 = \mathrm{j}\,0{\cdot}08$ per unit and the 132/11-kV transformer has reactance of $\mathrm{j}\,0{\cdot}06$ per unit.

The fault resistance between each line and earth is $437\times 12\times 0{\cdot}0484\times\sqrt{3}/11\,000 = 0{\cdot}04$ per unit, since the base current on the 11-kV system is twelve times that on the 132-kV system, i.e. 5248 A.

The 40-MW load is carrying a pre-fault current of

$$\frac{40\times 10^6}{\sqrt{3}\times 10\,550\times 0{\cdot}8}(0{\cdot}8-\mathrm{j}\,0{\cdot}6)\text{ A} = 2736(0{\cdot}8-\mathrm{j}\,0{\cdot}6)\text{ A}$$

taking the pre-fault voltage at the motor busbar as the reference phasor throughout the calculations.

The impedance of this load is $10\,550/[\sqrt{3}\times 2736(0{\cdot}8-\mathrm{j}\,0{\cdot}6)] =$ $2{\cdot}23(0{\cdot}8+\mathrm{j}\,0{\cdot}6)$ Ω and this is

$$\frac{5248\times 2{\cdot}23\times\sqrt{3}}{11\,000}(0{\cdot}8+\mathrm{j}\,0{\cdot}6) = 1{\cdot}4716+\mathrm{j}\,1{\cdot}1037 \text{ per unit.}$$

The 40-MW load current of $2736(0{\cdot}8-\mathrm{j}\,0{\cdot}6) = 2736(0{\cdot}8-\mathrm{j}\,0{\cdot}6)/$ $5248 = 0{\cdot}41707-\mathrm{j}\,0{\cdot}3128$ per unit.

Before the fault, the motor current $= 20\times10^6(0{\cdot}9+\mathrm{j}\,0{\cdot}436)/(\sqrt{3}\times 10\,550) = 985+\mathrm{j}\,477$ A $= (985+\mathrm{j}\,477)/5248 = 0{\cdot}1877+\mathrm{j}\,0{\cdot}0909$ per unit.

The total pre-fault current supplied from the two generators via the two transformers and the line is $0{\cdot}41707-\mathrm{j}\,0{\cdot}3128+0{\cdot}1877+\mathrm{j}\,0{\cdot}0909 = 0{\cdot}60477-\mathrm{j}\,0{\cdot}2219$ per unit.

These pre-fault currents and the voltage of $10\,550/\sqrt{3} = 10\,550/11\,000 = 0{\cdot}96$ per unit, are shown with the system per unit impedances in Fig. 7.15. The synchronous reactances of the machines X_A, X_B and X_M and their internal e.m.f.s $\mathbf{E}_A$, $\mathbf{E}_B$ and $\mathbf{E}_M$ are not known, but they are not needed for the present calculation.

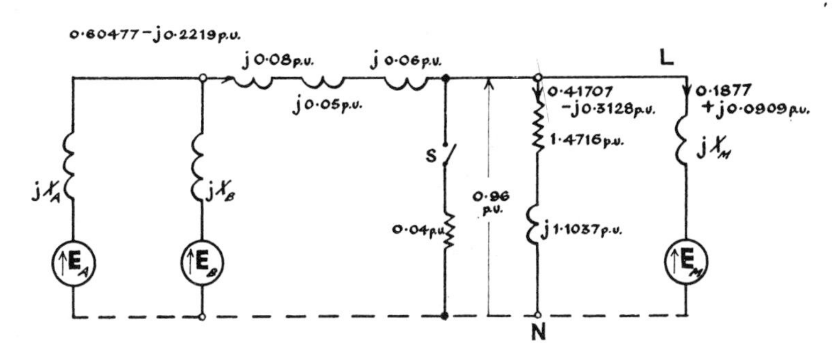

FIG. 7.15. *Conditions before the fault.*

When the fault comes on the system, i.e. when switch S is closed, it is assumed that X_A, X_B and X_M immediately take up the values j 0·3, j 0·2 and j 0·8 per unit, and remain constant at these values during the period of interest, and that the machine e.m.f.s immediately take up the values of the voltages behind transient reactance and remain at these constant r.m.s. values.

The current flowing in the fault can be found at once by Thevenin's theorem. The impedance at terminals LN due to that part of the system on the transmission line side with machines A and B not generating any e.m.f. $= \mathrm{j}\,0{\cdot}19+\mathrm{j}(0{\cdot}3\times0{\cdot}2/0{\cdot}5) = \mathrm{j}\,0{\cdot}31$ per unit from the moment when the fault comes on.

On the load side of terminals LN from the moment of fault initiation, there is an impedance of $\mathrm{j}\,0{\cdot}8(1{\cdot}4716+\mathrm{j}\,1{\cdot}1037)/(1{\cdot}4716+\mathrm{j}\,1{\cdot}9037) = 0{\cdot}1632+\mathrm{j}\,0{\cdot}5894$ per unit.

The impedance of the whole circuit at terminals LN is therefore j 0·31(0·1632+j 0·5894)/(0·1632+j 0·8994) = 0·01877+j 0·20656 per unit.

The current flowing in the fault arc is therefore 0·96+j 0/(0·04+0·01877+j 0·20656) = 1·2233−j 4·2995 per unit.

The phase (line to neutral) voltage at the point of fault $\mathbf{V}_{LN}$ = 0·04(1·2233−j 4·2995) = 0·048932−j 0·17198 per unit and the current flowing in the load which previously took 40 MW is 0·048932 −j 0·17198/(1·4716+j 1·1037) = −0·034815−j 0·090754.

The three machines together supply a current of 1·2233−j 4·2995 −0·03482−j 0·09075 = 1·18848−j 4·3903 per unit.

The division of this current between the three machines may be found by establishing their effective voltages behind transient reactance using [*7.3*].

In the pre-fault condition machines *A* and *B* together supply 0·60477−j 0·2219 per unit current, and their individual currents $\mathbf{I}_A$ and $\mathbf{I}_B$ must now be found.

The terminal voltage at the generators before the fault, can be seen from Fig. 7.15 to be 0·96+j 0+j 0·19(0·60477−j 0·2219) = 1·00216+j 0·11491 per unit.

The complex power output per phase from the generators into the system is therefore

$$\mathbf{VI}^* = (1{\cdot}00216+\mathrm{j}\,0{\cdot}11491)(0{\cdot}60477+\mathrm{j}\,0{\cdot}22188)$$

$$P+\mathrm{j}\,Q = 0{\cdot}58058+\mathrm{j}\,0{\cdot}29184 \text{ per unit.}$$

Since the machines share the power in proportion to their ratings

A supplies 3/8 × 0·58058 = 0·21772 per unit power

B supplies 5/8 × 0·58058 = 0·36286 per unit power

and both *A* and *B* supply +j 0·14592 per unit reactive volt-amperes.

If $\mathbf{I}_A = a-\mathrm{j}\,b$ per unit current then

$$0{\cdot}21772 = \text{real part of } (1{\cdot}00216+\mathrm{j}\,0{\cdot}11491)(a+\mathrm{j}\,b) = 1{\cdot}00216a-0{\cdot}11491b$$

and

$$0{\cdot}14592 = \text{'imaginary' part of } (1{\cdot}00216+\mathrm{j}\,0{\cdot}11491)(a+\mathrm{j}\,b) = 0{\cdot}11491a+1{\cdot}00216b.$$

Solving these two equations a = 0·23091 and b = 0·11913 and if $\mathbf{I}_B = c-\mathrm{j}\,d$ per unit current then

$$a+c = 0{\cdot}60477$$

$$c = 0{\cdot}37386$$

and as

$$b+d = 0{\cdot}22188$$

$$d = 0{\cdot}10275.$$

The pre-fault currents in the two generators are therefore

$$\mathbf{I}_A = 0{\cdot}23091 - \mathrm{j}\,0{\cdot}11913$$
$$\mathbf{I}_B = 0{\cdot}37386 - \mathrm{j}\,0{\cdot}10275.$$

From [7.3] the voltages behind transient reactance are respectively

$$\mathbf{E}'_A = 1{\cdot}00216 + \mathrm{j}\,0{\cdot}11491 + \mathrm{j}\,0{\cdot}3(0{\cdot}23091 - \mathrm{j}\,0{\cdot}11913)$$
$$\mathbf{E}'_B = 1{\cdot}00216 + \mathrm{j}\,0{\cdot}11491 + \mathrm{j}\,0{\cdot}2(0{\cdot}37386 - \mathrm{j}\,0{\cdot}10275)$$
$$\mathbf{E}'_A = 1{\cdot}03790 + \mathrm{j}\,0{\cdot}18418$$
$$\mathbf{E}'_B = 1{\cdot}02271 + \mathrm{j}\,0{\cdot}18968.$$

The voltage behind transient reactance $\mathbf{E}'_M$ in the machine which was previously motoring is given by

$$\mathbf{E}'_M = 0{\cdot}96 + \mathrm{j}\,0 - \mathrm{j}\,0{\cdot}8(0{\cdot}1877 + \mathrm{j}\,0{\cdot}0909)$$
$$= 1{\cdot}03272 - \mathrm{j}\,0{\cdot}15016$$

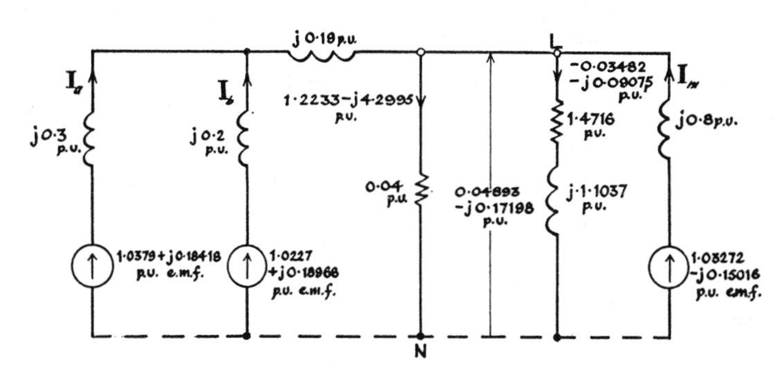

FIG. 7.16. *During the transient period of fault.*

Fig. 7.16 shows the three per unit e.m.f.s and the terminal voltage at the fault, the per unit impedances for each branch and the per unit currents in the two line to neutral branches. The currents $\mathbf{I}_a$, $\mathbf{I}_b$ and $\mathbf{I}_m$ in the three machines can now be found.

From the right-hand mesh

$$0{\cdot}048932 - \mathrm{j}\,0{\cdot}17198 = 1{\cdot}03272 - \mathrm{j}\,0{\cdot}15016 - \mathrm{j}\,0{\cdot}8\,\mathbf{I}_m.$$

so that $\mathbf{I}_m = 0{\cdot}027275 - \mathrm{j}\,1{\cdot}2297$ per unit and since base current in the 11-kV system is 5248 A, the machine which was previously motoring, feeds a current of 143 − j 6453 A into the system during the transient period.

The total current supplied by machines A and B = 1·2233 −

$j\ 4{\cdot}2995 - 0{\cdot}034815 - j\ 0{\cdot}090754 - 0{\cdot}027275 + j\ 1{\cdot}2297 = 1{\cdot}1612 - j\ 3{\cdot}1606$.

The terminal voltage of these two machines

$$= 0{\cdot}04893 - j\,0{\cdot}17198 + j\,0{\cdot}19(1{\cdot}1612 - j\,3{\cdot}1606)$$
$$= 0{\cdot}64944 + j\,0{\cdot}04865 \text{ per unit.}$$

For machine A

$$1{\cdot}0379 + j\,0{\cdot}18418 - j\,0{\cdot}3\,\mathbf{I}_a = 0{\cdot}64944 + j\,0{\cdot}04865$$

so that $\mathbf{I}_a = 0{\cdot}4518 - j\,1{\cdot}295$ per unit.

The base current at 22 kV is 2624 A, and $\mathbf{I}_a = 1186 - j\,3398$ A, and for machine B

$$1{\cdot}02271 + j\,0{\cdot}18968 - j\,0{\cdot}2\,\mathbf{I}_b = 0{\cdot}64944 + j\,0{\cdot}04865$$

so that $\mathbf{I}_b = 0{\cdot}7052 - j\,1{\cdot}8663$ per unit. As a check, $\mathbf{I}_a + \mathbf{I}_b = 1{\cdot}1570 - j\,3{\cdot}1613$ per unit.

The error in this total compared with the value given above is less than 0·4% in the in-phase component and 0·02% in the quadrature component. This illustrates that in a solution obtained manually by log tables or by hand-calculating machine as here, even working to a number of decimal places, can still leave noticeable errors which are aggravated by the subtraction of quantities of similar size. The problem also illustrates the long and tedious calculations which even a very simple power system involves, and which clearly in a real system with its large numbers of machines, transformers, transmission lines, etc., are too lengthy to calculate by hand. Both of these problems are overcome by the use of digital computers, and much of the material of Volume 2 is concerned with writing the equations involved in power system analysis, in a form which is suitable for handling on a digital computer.

7.6 Unbalanced Loading or Faults: Sequence Reactances

Little consideration is given here to unbalanced conditions, but in dealing with unbalanced faults (see Chapter 8), the negative- and zero-sequence impedances of synchronous machines are involved. The reasons for these being different from the positive-sequence impedance are therefore outlined here. (It is suggested that the part of Chapter 8 dealing with symmetrical components may be usefully studied, before reading this section.)

The m.m.f. wave due to positive-sequence currents in the stator winding rotates synchronously with the rotor and its m.m.f. (i.e. at the same speed and in the same direction as the m.m.f. wave due to the direct current in the rotor field winding). The m.m.f. due to negative-sequence stator currents, however, moves at twice synchronous speed with respect to the rotor, since it rotates at synchronous speed in the opposite direction. It therefore moves alternately past the direct and quadrature axes, and since it sets up a varying armature reaction effect, the negative-sequence reactance X_2 is a subtransient reactance which is taken as the average between the positive-sequence values corresponding to the direct and quadrature axes. For this reason $X_2 \simeq \frac{1}{2}(X_d'' + X_q'')$, i.e. it is the mean of the direct- and quadrature-axis subtransient reactances. It is not usually necessary to distinguish any time variation of X_2 [or X_o] during transient conditions, since there is no normal constant armature reaction to be affected. X_2 is generally in the range 0·07 to 0·2 per unit for 2-pole turbo-alternators. For salient pole machines with damper windings X_2 is generally within the range 0·1 to 0·3 per unit, but may exceed this range if there are no damper windings. There will be a negative-sequence resistance which takes account of power dissipated in the rotor poles or damper winding by the double supply frequency induced currents. The duration and amount of unbalanced faults or loading must be limited by negative-sequence relaying, so that these losses do not cause excessive heating of the pole faces or damper bars.

Zero-sequence currents in the stator windings do not produce a rotating magnetic field, since the phase m.m.f.s are all in time phase. If each phase winding produced a sinusoidal space m.m.f., then with the rotor removed, the flux at a point on the axis of the stator due to to zero-sequence current would be zero at every instant. When the flux in the air gap or the leakage flux around slots or end connections is considered, no point in these regions is equidistant from all three phase windings of the stator. Furthermore, the m.m.f. produced by a phase winding departs from a sine wave, by amounts which depend upon the arrangement of the winding. The actual zero-sequence reactance therefore depends on the chording and breadth factors of the stator phase winding. It varies considerably according to the winding details, but for 2-pole turbo-alternators may lie between 0·02 and 0·15 per unit, and for hydro-electric generators may lie between 0·04 and 0·2 per unit.

7.7 Operation of Synchronous Machines Connected to Infinite Busbars

If a cylindrical-rotor alternator, for which $X_d \simeq X_q$, operates under balanced steady-state conditions while connected to the grid system, then the system frequency and busbar voltage are virtually unaffected by changes in the loading or excitation of this machine. The machine is then said to be connected to infinite busbars.

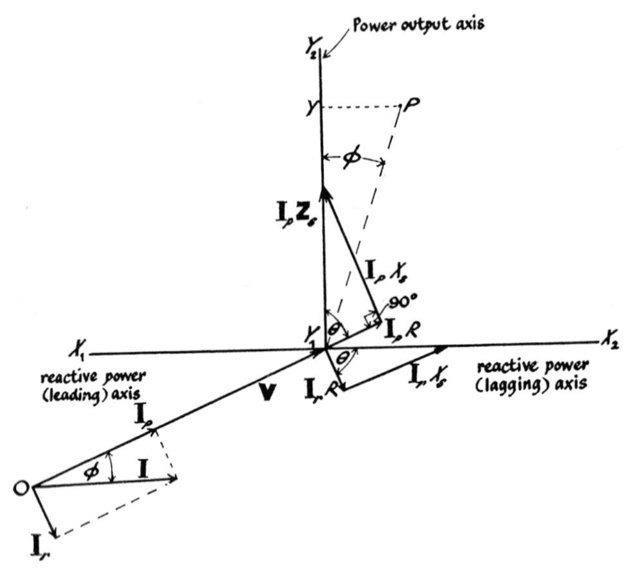

FIG. 7.17. *Phasor diagram for steady-state operation of a cylindrical-rotor alternator connected to infinite busbars.*

The terminal voltage **V** in Fig. 7.4 is therefore constant, but variation of prime mover input power can vary the in-phase component of current I_p, and variation of excitation can vary the e.m.f.s $\mathbf{E}_o$ and **E** and the reactive or quadrature component of current I_r. Variation of power output, i.e. variation of I_p, with I_r zero would cause the e.m.f. $\mathbf{E}_o$ behind synchronous impedance, to move along a locus which is a line making an angle $\theta = \tan^{-1} X_s/R$ with the terminal voltage phasor **V**. Thus a scale may be marked in power output along the power axis $Y_1 Y_2$ as shown in Fig. 7.17 (see section 1.2.2). Similarly a scale may be marked in reactive volt-amperes or MVAr, along an axis $X_1 X_2$ which is at right angles to $Y_1 Y_2$ and which is the locus of $\mathbf{E}_o$ if $I_p = 0$ and I_r is varied.

For a phase current $\mathbf{I} = I_p - \mathrm{j}\, I_r = \mathbf{I}\underline{/-\phi}$ where the reference phasor is the phase terminal voltage **V**, the e.m.f. behind synchronous impedance is found by joining O to a point P on a line radiating from Y_1 making an angle ϕ with the power axis, and with an intercept on that axis $Y_1 Y$. This distance $Y_1 Y$ on the power scale represents a power output $3VI_p$. Only one scale, either in voltage or in MVA, is

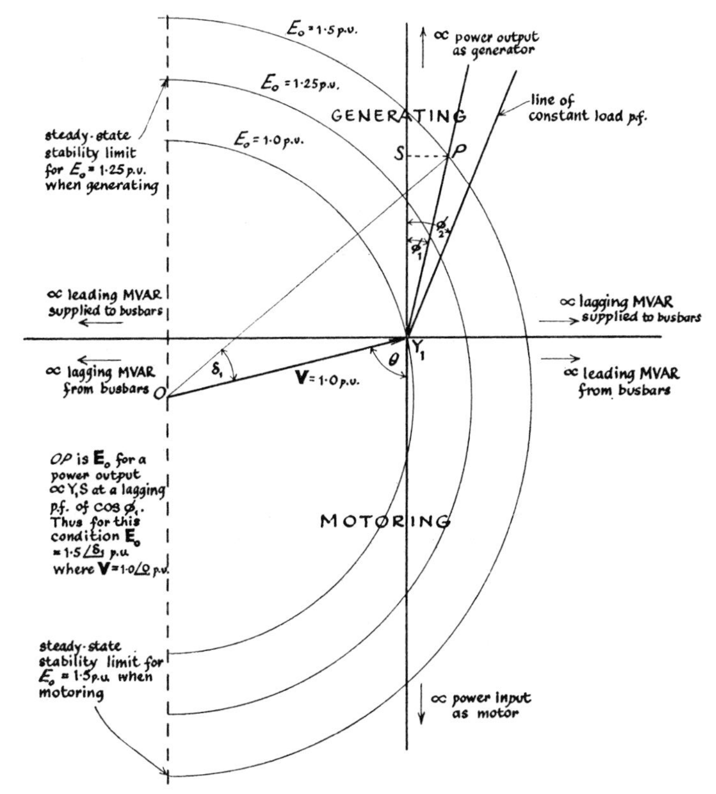

FIG. 7.18. *Cylindrical-rotor machine operating in steady state on infinite busbars.*

therefore arbitrary, and fixing one scale determines the other. Fig. 7.17 may be converted from a voltage diagram to an MVA diagram by multiplying every phasor by $3V/Z_s$, and this will in fact be done in Fig. 7.21.

For each circle for a given value of E_o shown in Fig. 7.18 there is a given constant field current I_f. If there were no saturation, E_o would be proportional to I_f and it would then have been possible to mark the curve for $E_o = 1{\cdot}0$ p.u. as $I_f = 1{\cdot}0$ p.u., that for $E_o = 0{\cdot}75$ p.u. as $I_f = 0{\cdot}75$ p.u., etc. Curves of constant field current may be taken approximately as circles with centre O and radii can be obtained from the open-circuit curve for the machine. Since $X_s \gg R$, $\theta \to \pi/2$ radians and point O in Fig. 7.18 may be taken to lie on the MVAr axis. On the MVA scale, the radius OY_1 which is that of an excitation circle giving rated voltage on open-circuit is approximately the MVA rating × short-circuit ratio. This may be seen as follows:

If rated current (1·0 per unit) flows at rated terminal voltage (1·0 per unit), then the per unit IX_d voltage drop ≃ rated current/S.C.R. = 1·0/S.C.R. Also $IX_d = IX_s$ for a cylindrical-rotor machine. From Fig. 7.17 this voltage drop is

(*a*) along the power axis Y_1Y_2 if the p.f. = 1·0, i.e. it represents 1·0 per unit power,

(*b*) along the MVAr axis X_1X_2 if the p.f. = 0, i.e. it represents 1·0 per unit MVAr,

(*c*) along Y_1P if the p.f. = $\cos\phi$ as in the general case, i.e. it represents 1·0 per unit MVA.

A distance along the voltage scale of 1/S.C.R. per unit therefore represents rated MVA, or conversely a voltage of 1·0 per unit is represented by a distance on the MVA scale of rated MVA × short-circuit ratio.

If for the generating mode, the load angle δ between $\mathbf{V}$ and $\mathbf{E}_o$ (see Fig. 7.4) is gradually increased by increasing the load by infinitesimally small increments until $\delta = \theta$, then any further increase in power input which increased δ beyond θ would cause the power output to fall. The machine rotor would then tend to accelerate out of synchronism due to the difference power. Similarly if in the motoring mode the angle between $\mathbf{V}$ and $\mathbf{E}_o$ increased to $-(\pi-\theta)$, the limit of stability would again be reached. Thus the dotted line shown parallel to the power axis in Fig. 7.18 is the steady-state stability limit.

The complex input or generated power/phase from the prime-mover to the alternator under steady-state conditions is (see section 1.2.3 and Appendix 4).

$$\mathbf{S}_1 = P_1 + \mathrm{j}Q_1 = \mathbf{E}_o\mathbf{I}^* = \mathbf{E}_o\left(\frac{\mathbf{E}_o - \mathbf{V}}{\mathbf{Z}_s}\right)^*.$$

The power input/phase P_1 is therefore

$$P_1 = \frac{E_o}{Z_s}(E_o \cos\theta - V\cos(\theta+\delta)) \qquad [7.5]$$

since if $\mathbf{E}_o$ is the reference phasor, $\mathbf{V} = V\underline{/-\delta}$ and $\mathbf{Z}_s = Z_s\underline{/\theta}$. Similarly the complex output power/phase is

$$\mathbf{S}_2 = P_2 + \mathrm{j}Q_2 = \mathbf{VI}^*$$

$$P_2 = \frac{V}{Z_s}(E_o\cos(\theta-\delta) - V\cos\theta). \qquad [7.6]$$

Equations [7.5] and [7.6] which differ by the stator copper loss/phase I^2R, show that as δ varies, maximum input or generated power occurs when $\delta = (\pi-\theta)$ radians, but maximum output power occurs when $\delta = \theta$.

If the stator resistance R can be neglected in comparison with the synchronous reactance X_s, as is true for many purposes, then [7.5] and [7.6] both reduce to

$$P = \frac{VE_o}{X_s}\sin\delta. \qquad [7.7]$$

The steady-state power variation with load angle for both generating and motoring with constant excitation, is therefore approximately sinusoidal as shown by the continuous curve in Fig. 7.19.

If a synchronous motor with negligible stator resistance is connected to infinite busbars of phase voltage $\mathbf{V}$, and is operating on load with a leading power factor, the phasor diagram of Fig. 7.20 may be drawn. For simplicity, this is shown for the case of an unsaturated cylindrical-rotor machine, though synchronous motors are commonly salient pole machines. If the machine were operated on no load and its losses were negligible, then $\mathbf{E}$, the phase e.m.f. generated by the d.c.-excited rotor winding, would lie along OY, and if the excitation were such that $E > V$ the current would lead $\mathbf{V}$ by $\pi/2$ radians and if $E < V$ the current would lag $\mathbf{V}$ by $\pi/2$ radians.

When the mechanical load is increased on the machine to bring it to the condition shown in Fig. 7.20, it can be seen that the rotor pole axis has fallen back behind its original position by the load angle δ from the no-load position. The motor power is $VI\cos\phi$ and, since $IX_s\cos\phi$ may be seen from Fig. 7.20 to equal $E\sin\delta$, it may be rewritten as

$$\frac{VIX_s \cos \phi}{X_s} = \frac{VE \sin \delta}{X_s}.$$

The variation of power with load angle is therefore the same for motoring or generating modes (equation [7.7]), and this is illustrated in Fig. 7.19, where negative values of δ, i.e. **E** lagging **V**, apply to the motoring condition.

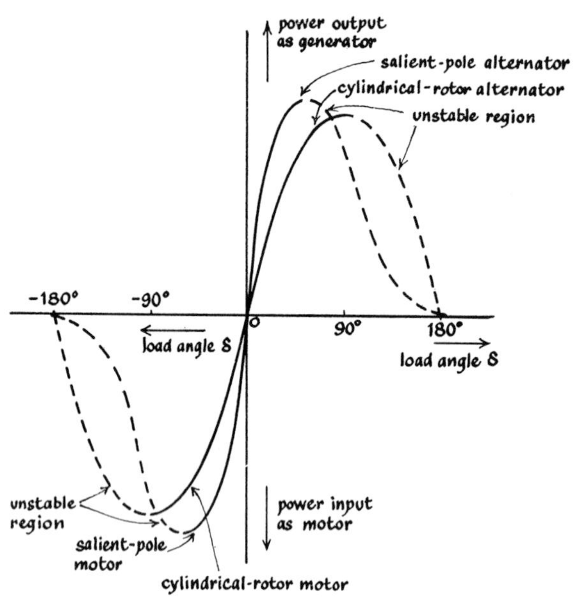

FIG. 7.19. *Steady-state variation of power with load angle for generating and motoring modes of cylindrical-rotor and salient-pole machines.*

If an alternator supplies power P to a synchronous motor via an external circuit, then

$$P = \frac{E_o E \sin \delta}{X}$$

where $\mathbf{E}_o$ and **E** are again the phase e.m.f.s of generator and motor respectively, δ is the angle between $\mathbf{E}_o$ and **E**, and X is the sum of the machine synchronous reactances and the external circuit reactance (the resistance of which is assumed to be negligible in comparison with its reactance).

For a salient-pole alternator in which the stator resistance may be neglected in comparison with its synchronous reactance, the power output P may be deduced, either in a similar way to that used above for the cylindrical-rotor machine, or from the phasor diagram of Fig. 7.8 (with $R = 0$). Thus from Fig. 7.8 since

$$I \cos \phi = I_q \cos \delta + I_d \sin \delta,$$
$$V \cos \delta = E_o - I_d X_d \quad \text{and} \quad V \sin \delta = I_q X_q$$

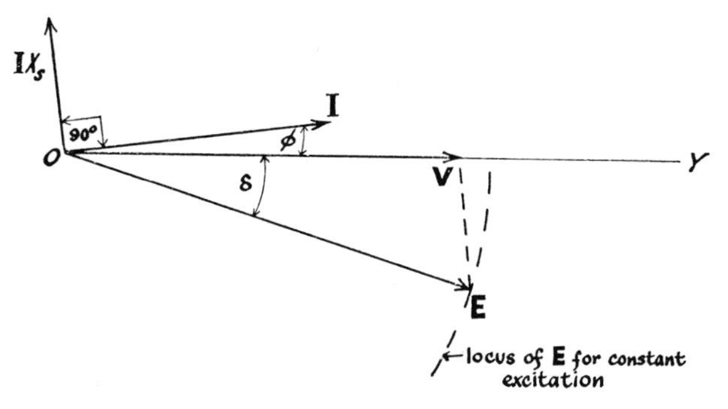

FIG. 7.20. *Phasor diagram of cylindrical-rotor synchronous motor operating with leading power factor.*

the expression for power may be written

$$P = V(I_q \cos \delta + I_d \sin \delta)$$
$$= V\left(\frac{V \sin \delta}{X_q} \cos \delta + \frac{(E_o - V \cos \delta)}{X_d} \sin \delta\right)$$
$$= \frac{VE_o \sin \delta}{X_d} + V^2\left(\frac{\sin \delta \cos \delta}{X_q} - \frac{\sin \delta \cos \delta}{X_d}\right)$$
$$= \frac{VE_o \sin \delta}{X_d} + \frac{V^2(X_d - X_q) \sin 2\delta}{2X_d X_q}. \qquad [7.8]$$

Comparison with [7.7] shows an additional power term due to reluctance torque arising from the saliency of the rotor. As Fig. 7.19 shows, the maximum power now occurs at an angle below $\pi/2$ radians, and the salient-pole machine is more 'stiff' and more stable than a cylindrical-rotor machine.

It is not in fact possible to operate a cylindrical-rotor machine at, or very close to, the steady-state stability limit of $\delta \simeq \pm\pi/2$, since it could only apply to an ideal system, in which no changes of connection or faults occurred, and in which all load changes took place very slowly. Sudden changes in load, switching in or out of parts of the system and faults, could all cause loss of stability, unless margins are fixed sufficiently far from the limit.

These sudden changes, particularly faults, can cause large differences to occur between the input power from a prime mover and the output power of the alternator into the system, during the interval before governor action occurs. This power difference is usually an accelerating one since the alternator generally feeds less power (though more current) into the faulted system than it was previously supplying, and this causes an initial increase with subsequent oscillation of the alternator rotor or load angle δ. This problem of transient stability for given conditions of loading and given load change or fault and duration of fault, is dealt with in Volume 2.

The regions of the chart of Fig. 7.18 in which a machine may operate safely under steady-state conditions are dealt with in the next section. In the section after that one, the reasons for the extension of this safe region at leading power factors due to the action of fast-acting voltage regulators, are considered.

7.8 Stability Margin and Operational Limits for a Cylindrical-rotor Alternator

The limits to the regions of the power chart shown in Fig. 7.18, within which it is possible to operate a cylindrical-rotor alternator safely under steady-state conditions, are set by

(*a*) the MW rating of the prime mover,

(*b*) the MVA rating of the alternator,

(*c*) the maximum excitation current of alternator field and exciter (usually limited by rotor heating),

(*d*) instability in the leading power factor region.

Fig. 7.21 has been converted from the voltage diagram of Fig. 7.18 to a power diagram, by multiplying all phasors of Fig. 7.18 by $3V/X_s$, for the case where stator resistance is neglected. It shows the chart for a machine rated at 0·8 p.f. lagging (i.e. full-load power is $0{\cdot}8 \times$ rated MVA) and it is drawn for operation at rated terminal voltage. Unit rotor current is shown arbitrarily as that value which gives rated voltage on open-circuit, and the maximum field current then lies in the neighbourhood of 2·5 per unit. This latter value may

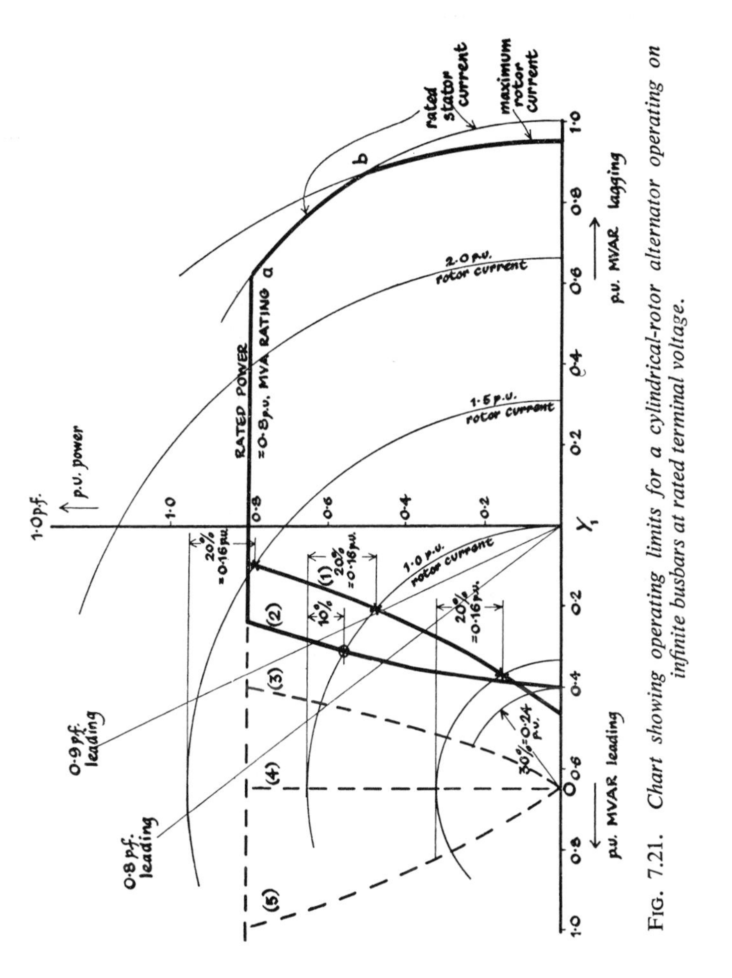

FIG. 7.21. *Chart showing operating limits for a cylindrical-rotor alternator operating on infinite busbars at rated terminal voltage.*

be designed to be that which gives rated power at rated MVA and rated power factor. In this case, the curve of maximum rotor current would intersect the rated MW line at the same point at which it is intersected by the rated stator current curve (which has Y_1 as its centre), so that the operating region limit *ab* due to the MVA limit would shrink to zero. The centre of the excitation circles *O*, as explained in section 7.7 may be taken to be displaced along the MVAr axis from Y_1 by approximately the MVA rating × short-circuit ratio, so that in Fig. 7.21 a short-circuit ratio of about 0·65 has been used.

The safe operating region is that within the heavy lines bounded by rated power at the top, and maximum rotor current (and possibly rated stator current) on the right. The left-hand limit depends upon the method of voltage regulation.

The boundary marked (1) is obtained as follows. Draw excitation circles centre *O*: in Fig. 7.21 they are drawn for 0·5, 1·0, 1·5 and 2·0 per unit. Draw lines parallel to the MVAr base tangential to each circle at the maximum power point. Allow a 20% of rated power margin from these maximum power values, i.e. as rated power is 0·8 per unit here, draw another set of lines parallel to the first and 0·16 per unit power below them. When the points of intersection of each of these new lines with its appropriate excitation circle are joined, the limit marked (1) is given. In fact when hand control of excitation is employed a margin of 20% is allowed at full load, but this is increased to 35% at light load, due to the possibility of the load increments at small loads exceeding those at greater loads. This type of margin is illustrated in curve (2) for the different values of the margins which are discussed next.

Boundary (2) applies if a normally inactive automatic voltage regulator is used, with a 10% margin of rated power at full load and 30% margin at light load. Thus the change from hand control to this regulator has permitted operation in the region bounded by curves (1) and (2) and the rated MW line.

It should be noted that Fig. 7.21 applies to operation at rated terminal voltage. If the terminal voltage is increased above this value the margin boundaries (1) and (2) move to the left, and if it is reduced they move to the right.

Since for a salient-pole alternator, the second term in [*7.8*] represents torque which is independent of excitation when operating on infinite busbars, such a machine can deliver up to 25% of its rating with no d.c. excitation, and the area of leading power factor operation in Fig. 7.21, in which the machine can run safely, is greater than for a cylindrical-rotor machine.

Actual tests on alternators connected to the grid system (Mason, Aylett and Birch, 1959) have shown that machines can in fact operate up to the lines (3) (see Fig. 7.21) with hand control, and up to (4) with normally inactive voltage regulators, so that the margins in use are very safe.

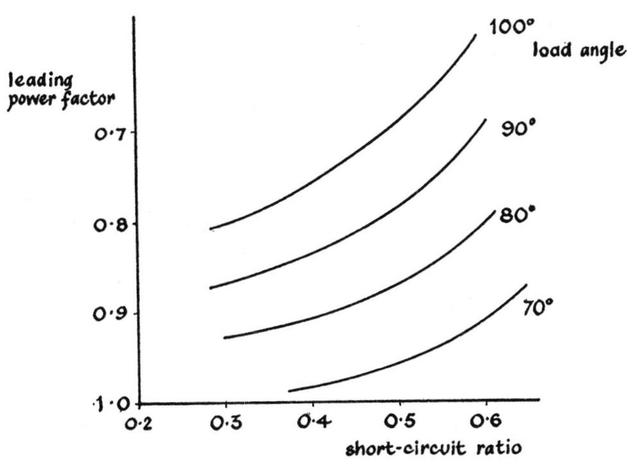

FIG. 7.22. *The effect of operation at leading power factor with a large load angle upon the short-circuit ratio.*

Tests with a continuously-acting voltage regulator showed that the machine could operate up to line (5) in Fig. 7.21 without loss of synchronism. It therefore appears to be possible to operate alternators at load angles up to at least 90° and still maintain a reasonable stability margin, if continuously-acting voltage regulators are used, together with leading-reactive-power limiters. The latter prevent excessive reduction of excitation in the event of a sudden rise in system voltage.

As Fig. 7.22 illustrates, operation at larger load angles for a given leading power factor limit, enables a machine to be designed with a smaller short-circuit ratio. The advantages of this, as well as the reasons why a continuously-acting regulator makes this possible, are discussed in the next section.

7.9 Operation at Leading Power Factor with Fast Continuously-acting Voltage Regulators

If continuously-acting automatic voltage regulators are used, it is possible to operate alternators safely at fairly low leading power factors. This has two valuable results.

(*a*) The generator short-circuit ratio can be reduced, which makes the machine cheaper, since it reduces the weight per megawatt of output. More particularly, since larger output machines are more efficient, and transport difficulties restrict the maximum transportable weight, a saving in weight by reduction in short-circuit ratio, leads to a larger and more efficient machine. The largest part of the stator of a 1350 MVA, 4-pole alternator exported to the USA in 1976 weighed 310 t. In some countries there are railway vehicles capable of moving 420 to 550 t and an ultimate capacity of up to 600 t is considered possible. The use of fast-acting voltage regulators with negligible zone of insensitivity (or dead-band), have enabled short-circuit ratios to be reduced to about 0·5 for generators of power factors of 0·9 or 0·85 lagging connected to the 275-kV system. This, together with such major advances as cooling of stator and rotor by hydrogen, internal water-cooling of stator and rotor conductors, and higher-quality steels, makes it possible to design 2-pole machines of 1300 to 2000 MW operating at load angles up to about 80°. Recent developments in designs for slotless generators or superconducting generators, appear to offer the possibility of machines in the range 2000 to 10 000 MW.

(*b*) It helps the operation of some power stations at leading p.f. for some load conditions which arise now that 275 kV or 400 kV is used for transmission, e.g. where relatively long 275-kV cables require appreciable charging kVA, as in the London area, or during the night when lines are lightly loaded.

The effect of the fast-acting voltage regulator on releasing the machine from the normal limits of stability, may be seen from the following approximate and simplified analysis. A single round-rotor machine connected to infinite busbars through a line, is equivalent to two round-rotor machines connected together, provided that rotor current effects can be neglected, i.e. there is equivalence in the steady state. $\mathbf{E}_1$ and X_1 are the e.m.f. and synchronous reactance of the exporting machine, and $\mathbf{E}_2$ is either the infinite bus-voltage or the e.m.f. of the motoring machine, and X_2 is the line reactance or the motoring machine synchronous reactance.

The phase power flow P (from Fig. 7.23) is $IE_1 \cos(\delta/2)$ and

$$I = \frac{2\,E_2 \sin(\delta/2)}{(X_1+X_2)}$$

$$P = \frac{E_1 E_2}{(X_1+X_2)} 2 \sin(\delta/2) \cos(\delta/2)$$

$$= \frac{E_1 E_2 \sin \delta}{(X_1+X_2)}. \qquad [7.9]$$

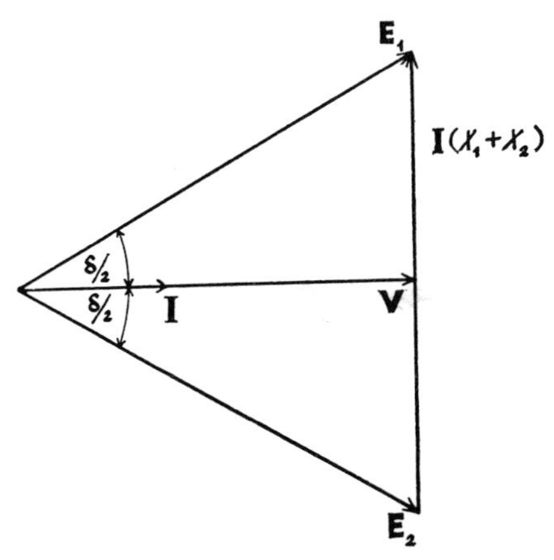

FIG. 7.23. *Phasor diagram for a single machine connected to infinite busbars through a line, or two machines connected together.*

If for simplicity it is assumed that $X_1 = X_2 = 1{\cdot}0$ p.u. and that the excitation is set constant at the value which gives rated current, i.e. $I_f = 1{\cdot}0$ p.u. at unity power factor and rated terminal voltage $V = 1{\cdot}0$ p.u., then it follows from Fig. 7.23 that

$$E_1 = E_2 = 1/\cos 45° = \sqrt{2}$$

so that from [*7.9*]

$$P = \sin \delta \qquad [7.10]$$

which gives the steady-state power limit of $P = 1{\cdot}0$ at $\delta = 90°$, and the sine curve of the cylindrical-rotor machine as shown in Fig. 7.19.

If, however, the excitation is constantly and instantaneously adjusted by a very fast-acting voltage regulator, which maintains the terminal voltage V at 1·0 and the power factor at 1·0, then from Fig. 7·23, $E_1 = E_2 = 1/\cos(\delta/2)$ and from [*7.9*]

$$P = \frac{\sin \delta}{2 \cos^2 (\delta/2)} = \tan (\delta/2) \qquad [7.11]$$

Thus the power transfer now increases continuously as δ increases from 0 to 180°, instead of decreasing when δ exceeds 90°.

If the load increase were made in very small increments, at the end of each of which the excitation was adjusted to return V to 1·0 p.u., then examining the system for steady-state stability in the usual way (i.e. by checking whether the power transfer is increased or decreased when there is a small increase in δ at constant excitation), shows that [*7.10*] again applies instead of [*7.11*]. Ideally, therefore, the voltage regulator should have an infinite amplification factor, with no time lags in the regulator, exciter or main machine field. For this purpose it is usual to have a static network as the voltage sensing element, feeding a signal to a static amplifier. On the main shaft there may be three exciters: a permanent magnet generator with its rectified output feeding the field of an a.c. pilot exciter, which in turn has its output rectified to supply the field of the main a.c. exciter. In this latter circuit the signal from the voltage-sensing circuit may be introduced via a magnetic amplifier. Silicon diodes or thyristors are generally used for rectification, including that between the main a.c. exciter and the alternator field, where the power may exceed 2 MW for a 500-MW machine. In some cases rotating armature a.c. exciters are used with rectifiers mounted within the hollow shaft, so that brush maintenance is avoided.

7.10 Governors and Frequency Control

In a power system, it is necessary to control the flow of power and of reactive power (vars), and the magnitudes of the voltages. The power input to the network depends on the prime-mover input power, and is controlled by governor settings. The most efficient power stations (base load stations) run 24 hours a day with loads at or near their ratings, while slightly less efficient stations are running with spare generating capacity (spinning reserve), and others are brought on to the bars as the load builds up, in the order of their economic efficiency. Where the transmission distances and losses are large, it is possible to analyse the optimum use of stations and routing of power by digital computer studies, though other considerations such as stability margins on line outages, may cause the optimum economical conditions to be impracticable. The var flow in the system may be controlled by varying the excitation of alternators and by tap change on transformers.

When a single generator supplies a load, the frequency is controlled by the prime-mover governor, which keeps the speed almost constant by varying the mechanical power input, and which must have a

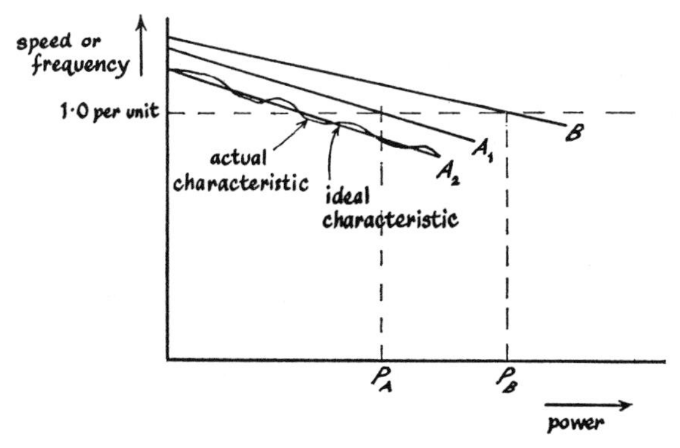

FIG. 7.24. *Variation of frequency with output power.*

slightly drooping characteristic of speed against power output for stability. The speed and frequency can be changed for a given load by raising or lowering this characteristic by means of the governor speed changer. This is illustrated in Fig. 7.24 where A_1 and A_2 are different speed changer settings of machine A. If the excitation is increased, without increase in prime-mover input power, the rise in voltage and power output, cause the governor to reduce the frequency and thus the voltage. Conditions settle down to operation with the original terminal voltage but reduced frequency. The vars supplied only change in so far as the frequency changes, so that altering the excitation has little effect on the power factor of the machine. This is very different from the effect of excitation changes on a machine operating in parallel with other synchronous machines.

When several machines are operating in parallel they must have a common frequency, and Fig. 7.24 illustrates the case of two machines A and B in parallel. If their governor settings are A_1 and B, then they operate at 1·0 per unit frequency (50 Hz in Great Britain) and deliver powers of P_A and P_B respectively. The two machines A and B connected to a busbar with a phase voltage V, supply a load which is constant while the prime-mover power inputs are kept constant. The components of current $I_{Ap} = I_A \cos \phi_A$ and $I_{Bp} = I_B \cos \phi_B$ in phase with the busbar voltage must remain constant while their prime-mover powers are constant, so long as the terminal voltage V remains constant.

If the machines have synchronous reactances/phase of X_A and X_B, and negligible stator resistance, then for power factors $\cos \phi_A$ and $\cos \phi_B$, their internal e.m.f.s (or voltages behind synchronous reactance) are $\mathbf{E}_A$ and $\mathbf{E}_B$ as shown in Fig. 7.25. Since, for constant

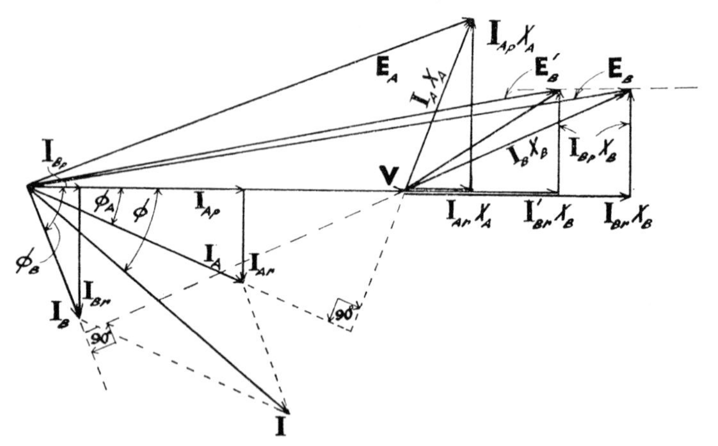

FIG. 7.25. *Phasor diagram for two cylindrical-rotor alternators in parallel.*

prime-mover power, $I_{Ap}X_A$ and $I_{Bp}X_B$ must remain constant, a change in excitation of either machine must cause its e.m.f. to move along a locus parallel to the terminal voltage phasor **V**. If the excitation of B is decreased so that its e.m.f. becomes $\mathbf{E}'_B$, B now supplies $V(I_{Br} - I'_{Br})$ less vars/phase.

If there were only these two machines connected to a constant load impedance, A would tend to supply the vars dropped by B. If the excitation of A is kept constant, an increase in vars, i.e. in $I_{Ar}X_A$ would cause $\mathbf{E}_A$ to move on an arc of a circle so that I_{Ap} would tend to decrease. There would therefore be a tendency for the load power and the terminal voltage to fall, bringing some rise in frequency of an amount depending upon the governor characteristics. It would be possible to restore the voltage and power to their original levels by increasing the excitation of *A*, so that there would be a transfer of vars from *B* to *A*.

Normally, however, an alternator is effectively connected to 'infinite busbars', so that the terminal voltage *V* and system frequency are virtually unaffected by changes in excitation or power supply to a given machine. Machine *A* would then be unaffected by the reduction in excitation of *B*. Excitation therefore controls only the var output of the generator concerned, while the governor speed setting controls the power output by altering the steam valve positions. The total var demand of the system is met by all the stations connected to it. If in a given power station there are a number of sets with the same power factor rating, they should share the vars in proportion to their ratings whilst maintaining a given busbar voltage, and this is done automatically by means of voltage regulators.

Compound regulators, which are dependent upon current as well as on voltage are used, because otherwise whichever regulator acted first upon a change of busbar voltage, would cause its set to drop or take up a disproportionate part of the change in vars. The more reactance there is between the groups of machines in different power stations, the less the effect of machine excitation on the var distribution, and tap-changing on interconnector transformers gives control over the sharing of the vars.

Worked example 7.4

Two 3-phase alternators working in parallel have excitations such that on open-circuit their terminal phase voltages are $\mathbf{E}_1$ and $\mathbf{E}_2$. They have synchronous impedances $\mathbf{Z}_1$ and $\mathbf{Z}_2$. If they deliver a

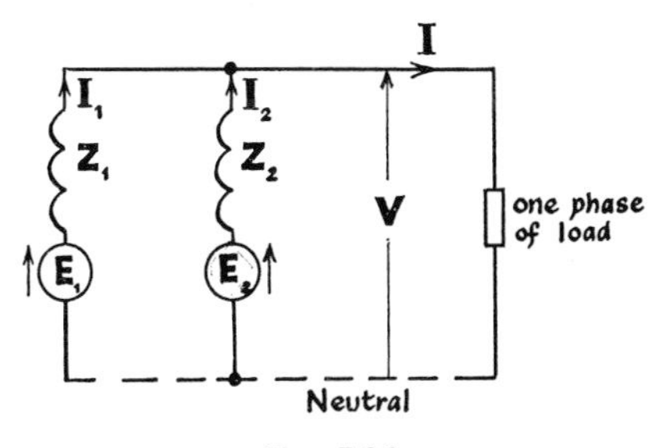

FIG. 7.26.

current **I** to each phase of a load, obtain expressions for the currents $\mathbf{I}_1$ and $\mathbf{I}_2$ in the two alternators.

Two identical 5000-kVA, 6·6-kV, 3-phase turbo-alternators are in parallel equally excited and sharing equally a load of 8000 kW at 6·6 kV and 0·8 power factor lagging. Calculate the excitation, in terms of no-load excitation at rated voltage, of each alternator assuming a constant synchronous reactance per phase of 2·0 per unit and neglecting stator resistance.

If the excitation voltage of one alternator is to be reduced by 15%, determine by how much the excitation voltage of the other alternator must be increased, in order to avoid a change of terminal voltage and of steam supply to either set.

(University of London)

SOLUTION

V is the terminal voltage across one phase of the load. $\mathbf{I}_1 = (\mathbf{E}_1 - \mathbf{V})/\mathbf{Z}_1$ and $\mathbf{I}_2 = (\mathbf{E}_2 - \mathbf{V})/\mathbf{Z}_2$.

$$\mathbf{I} = \mathbf{I}_1 + \mathbf{I}_2$$

$$\mathbf{I} = \frac{\mathbf{E}_1}{\mathbf{Z}_1} + \frac{\mathbf{E}_2}{\mathbf{Z}_2} - \mathbf{V}\left[\frac{1}{\mathbf{Z}_1} + \frac{1}{\mathbf{Z}_2}\right]$$

$$\mathbf{V} = \frac{\mathbf{Z}_1\mathbf{Z}_2}{(\mathbf{Z}_1 + \mathbf{Z}_2)}\left[\frac{\mathbf{E}_1}{\mathbf{Z}_1} + \frac{\mathbf{E}_2}{\mathbf{Z}_2} - \mathbf{I}\right]$$

$$\mathbf{I}_1 = \frac{\mathbf{E}_1}{\mathbf{Z}_1} - \frac{\mathbf{Z}_2}{(\mathbf{Z}_1 + \mathbf{Z}_2)}\left[\frac{\mathbf{E}_1}{\mathbf{Z}_1} + \frac{\mathbf{E}_2}{\mathbf{Z}_2} - \mathbf{I}\right] = \frac{\mathbf{E}_1 - \mathbf{E}_2}{\mathbf{Z}_1 + \mathbf{Z}_2} + \mathbf{I}\frac{\mathbf{Z}_2}{\mathbf{Z}_1 + \mathbf{Z}_2}.$$

Similarly

$$\mathbf{I}_2 = \frac{\mathbf{E}_2 - \mathbf{E}_1}{\mathbf{Z}_1 + \mathbf{Z}_2} + \mathbf{I}\frac{\mathbf{Z}_1}{\mathbf{Z}_1 + \mathbf{Z}_2}.$$

The first terms of these two equations give the current $\mathbf{I}_c$ which circulates round the local circuit of the two stators, and the last terms are the share of the total load current taken by each machine, if each has an equal e.m.f. and there is no circulating current.

In general, the load sharing between two alternators could be found from these equations, but in this case they share the load equally so that $\mathbf{I}_1 = \mathbf{I}_2 = \mathbf{I}/2$, and are equally excited so that $\mathbf{E}_1 = \mathbf{E}_2$.

$$\mathbf{E}_1 = \mathbf{E}_2 = [(1\cdot0 \times 0\cdot8)^2 + (1\cdot0 \times 0\cdot6 + 2\cdot0)^2]^{\frac{1}{2}}$$
$$= (0\cdot8^2 + 2\cdot6^2)^{\frac{1}{2}} = 2\cdot72,$$

The excitation for this load is therefore 2·72 times that which would give rated voltage of 1·0 p.u. (i.e. $6600/\sqrt{3}$) per phase on open-circuit. This assumes of course that saturation can be neglected,

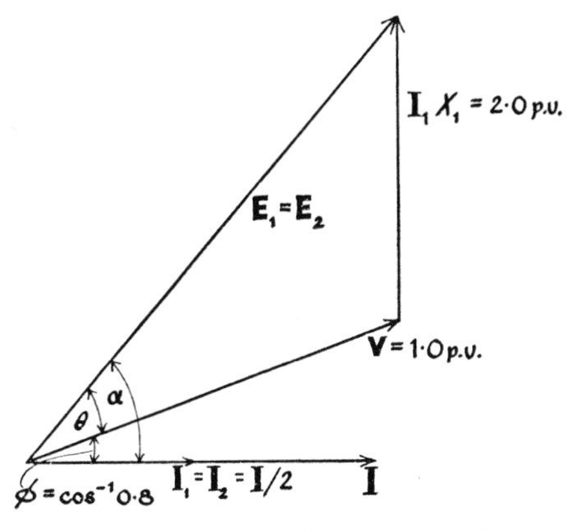

FIG. 7.27.

which must be so if the synchronous reactance is taken as constant as this question specifies.

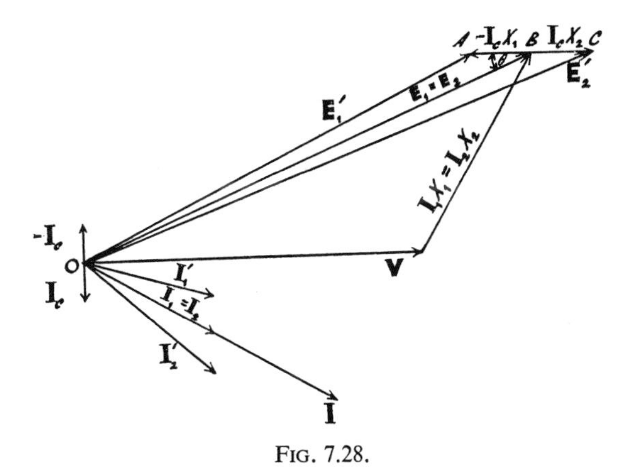

FIG. 7.28.

When the excitation of machine 1 is reduced so that $E_1' < E_2'$, a circulating current $\mathbf{I}_c = (\mathbf{E}_2' - \mathbf{E}_1')/2X$, (where $X = X_1 = X_2$), flows out of machine 2 and $-\mathbf{I}_c$ flows out of machine 1, i.e.

$$\mathbf{I}_1' = \mathbf{I}/2 - \mathbf{I}_c \quad \text{and} \quad \mathbf{I}_2' = \mathbf{I}/2 + \mathbf{I}_c.$$

Since the steam supply to neither prime-mover is changed, this current $\mathbf{I}_c$ must be in quadrature with the terminal voltage $\mathbf{V}$. Fig. 7.28 shows the effect of reducing E_1' to $0{\cdot}85\,E_1$. From Fig. 7.27, $\theta = \alpha - 36°\ 52'$ and $\alpha = \sin^{-1}(2{\cdot}0+0{\cdot}6)/2{\cdot}72 = 72°\ 57'$

$$\theta = 36°\ 5'.$$

In Fig. 7.28,

$$OB = 2{\cdot}72 \text{ p.u.}$$
$$OA = 0{\cdot}85 \times 2{\cdot}72 = 2{\cdot}31 \text{ p.u.}$$

and

$$AB = BC$$
$$\underline{/OBA} = \theta = 36°\ 5'$$
$$2{\cdot}31^2 = 2{\cdot}72^2 + AB^2 - 2\times 2{\cdot}72\ AB \cos 36°\ 5'$$

hence

$$AB = I_cX = 0{\cdot}54 \text{ p.u.}$$
$$OC^2 = (E_2')^2 = 2{\cdot}72^2 + 0{\cdot}54^2 - 2\times 2{\cdot}72 \times 0{\cdot}54 \cos 143°\ 55'$$
$$= 10{\cdot}03$$

and

$$E_2' = 3{\cdot}17 \text{ p.u.}$$

The % increase in the e.m.f. of the second machine is

$$100 \times \frac{3{\cdot}17 - 2{\cdot}72}{2{\cdot}72} = 16{\cdot}6\%.$$

It may be noted that data is given by which the actual magnitudes of currents and voltages can be found but that they are not strictly required by the question. Although a phasor diagram such as Fig. 7.28 helps to show the conditions it is of course not essential to a solution, e.g. the last part of the problem could be solved as follows:

Taking $\mathbf{V} = 1{\cdot}0\underline{/0°}$ as reference phasor so that $\mathbf{I}_c = 0 - \mathrm{j}\,I_c$

$$\mathbf{E}_1' = \mathbf{E}_1 - (-\mathrm{j}\,\mathbf{I}_c)(\mathrm{j}\,2) = \mathbf{E}_1 - 2\mathbf{I}_c$$
$$\mathbf{E}_1 = 2{\cdot}72\underline{/\theta} = 2{\cdot}72\underline{/36°\ 5'}$$
$$\mathbf{E}_1' = (2{\cdot}72 \cos 36°\ 5' - 2I_c) + \mathrm{j}\,2{\cdot}72 \sin 36°\ 5'$$
$$E_1' = 0{\cdot}85E_1 = 0{\cdot}85 \times 2{\cdot}72 = 2{\cdot}31 = [(2{\cdot}193 - 2I_c)^2 + 1{\cdot}6^2]^{\frac{1}{2}}$$

hence

$$I_c = 0{\cdot}265 \text{ per unit.}$$
$$\mathbf{E}_2' = \mathbf{E}_1 + 2\mathbf{I}_c$$

$$E'_2 = [(2{\cdot}193+2I_c)^2+1{\cdot}6^2]^{\frac{1}{2}}$$
$$= 3{\cdot}17 \text{ per unit.}$$

Due to the drooping characteristics which each governor must have for stability, the power/frequency characteristic of a power system can generally be represented approximately as

$$\Delta P+K\Delta f = 0 \qquad [7.12]$$

where Δf is the small change in frequency associated with a small power change ΔP in load or generation. The constant K depends upon the governor and load characteristics. For the British Grid System during daytime operation, K is about 5000 MW/Hz.

If two systems A and B are interconnected, a sudden change ΔP on A will affect the common frequency, and since the disturbance is shared between the two systems according to their power/frequency characteristics, there will be a change in power transfer between the two systems and this is equal to the change in power in system B. Thus

$$\left.\begin{aligned}\Delta P_A+K_A\Delta f &= 0\\ \Delta P_B+K_B\Delta f &= 0\end{aligned}\right\} \qquad [7.13]$$

and

$$\Delta P_A+\Delta P_B = \Delta P. \qquad [7.14]$$

From [7.13] and [7.14]

$$\Delta P_A = \frac{K_A}{K_A+K_B}\Delta P \qquad [7.15]$$

$$-\Delta P_B = \frac{-K_B}{K_A+K_B}\Delta P = \Delta P_i \qquad [7.16]$$

where ΔP_i is the change in power transfer in the interconnector, with a positive sign denoting power flow from A to B.

If these two systems were disconnected, the same disturbance ΔP in one system which would only affect that system, would cause a change in frequency in that system, which was also the change in the difference between the two system frequencies, viz. for a disturbance ΔP in A,

$$\Delta f_A = \Delta(f_A-f_B) = -\frac{\Delta P}{K_A}. \qquad [7.17]$$

From [7.16] and [7.17]

$$\Delta P_i = -\frac{K_B}{K_A+K_B} \cdot \Delta P = \frac{K_A K_B}{K_A+K_B} \cdot \Delta(f_A - f_B). \qquad [7{\cdot}18]$$

Equation [7.18] gives the power transfer in an a.c. interconnecter between two systems which were previously operating independently at frequencies f_A and f_B. Since after interconnection they must operate at the same frequency as long as synchronism in maintained, then $\Delta(f_A - f_B) = f_A - f_B$.

In this country at present there is no automatic frequency-control scheme, although considerable development work has been done and such control will probably be a part of the progress towards automation of power generation and transmission. At present, an increase in load demand causes a small drop in frequency, as the extra power is supplied from the kinetic energy of the generating sets. This small speed change causes increased power input to the turbines through governor action, and the extra power comes initially from the thermal stored energy of the boilers until the new firing rate takes effect. This process can be automatic or manual, and at present is the latter. When steady-state is again reached after this load increase, there will have been a drop in frequency proportional to the increase (equation [7.12]). If the effect of a number of load changes over a period causes a drop in frequency which requires correction, one or more stations will be ordered to increase output. This will cause the frequency to rise so that the other generators will shed some load by governor action.

Interconnection between two systems by a d.c. link as in the cross-Channel link between Britain and France, avoids the automatic frequency control equipment, which would be necessary for the interconnector to carry only a very small fraction of the generation in the two systems. The asynchronous link also avoids stability limitations, and there are no cable or line leading vars to be compensated and to be carried in the conductors, and the cable insulation can be more efficiently stressed. Difficulties in supplying the additional load required in large cities in Britain, e.g. the increasing difficulty of obtaining wayleaves for E.H.V. lines coming into the cities, and the charging current of high-voltage a.c. cables approaching their rating in a length of relatively few km, might eventually cause the adoption of high-voltage d.c. links between load centres and the transmission lines terminating in the city outskirts.

REFERENCES

DAVIES, M., MORGAN, F. & BIRD, J. I. 1959, 'Power frequency characteristics of the British Grid System', *Proc. I.E.E.,* **106**, Part A, No. 26, pp. 154–167.

FENWICK, D. R. & WRIGHT, W. F., 'Review of trends in excitation systems and future developments', 1976, *Proc. I.E.E.,* **123**, pp. 413–420.

HORSLEY, W. D. 1963. 'Turbo-type generators—a review of progress', *Proc. I.E.E.,* **110**, No. 4, pp. 695–702.

KIRTLEY, J. L. & FURUYAMA, M. 1975, 'A design concept for large superconducting alternators', I.E.E.E., Trans, PAS 94, pp 1264–1269.

LORCH, H. O., 'Feasibility of turbogenerator with superconducting rotor and conventional stator', 1973, *Proc. I.E.E.*, **120**, pp 221–227 & 1256–1259.

MASON, T. H., AYLETT, P. D. & BIRCH, F. H. 1959, 'Turbo-generator performance under exceptional operating conditions', *Proc. I.E.E.,* **106,** Part A, No. 29, pp. 357–380.

MILLER, T. J. E. & HUGHES, A. 1977, 'Comparative design and performance analysis of air-cored and iron-cored synchronous machines', *Proc. I.E.E.,* **124**, pp 127–132.

SPOONER, E., 'Fully slotless turbogenerators', 1973, *Proc. I.E.E.,* **120**, pp 1507–1518 & **122**, pp 75–79.

VICKERS, V. J., 'Recent trends in turboalternators', 1974, *Proc. I.E.E.,* **121** (IIR), pp 1273–1306.

Examples

1. A salient-pole alternator has direct- and quadrature-axis synchronous reactances of 0·7 and 0·4 per unit respectively. If the machine is operating at normal voltage and full load at 0·8 power factor lagging, to what value will the terminal voltage rise if the load is disconnected? Neglect stator resistance and saturation.

(University of London) (1·517 per unit)

2. The 3-phase busbars of a power station are connected to alternators of large capacity. Another alternator is now connected to the busbars, its stator winding having synchronous reactance of 4 Ω/phase and negligible resistance. This alternator delivers 100 A at 0·8 power factor lagging. Calculate its open-circuit phase e.m.f. and its angle of displacement δ from the terminal voltage.

The steam supply to the alternator is increased and δ is thus increased by 10°. Calculate the new load current delivered by the alternator and its power factor.

The excitation is then changed until the power factor of the alternator is again 0·8 lagging. Calculate the load current now delivered by the alternator.

The busbar voltage remains unchanged at 11 kV.

(University of London) (6600 V, 2° 15′, 370 A, 0·98 p.f. 455 A)

3. A 3-phase, star-connected, 6·6-kV, 50-Hz synchronous generator with a cylindrical rotor has the open-circuit characteristic at rated speed given by the following data:

Line voltage (kV)	3·0	4·0	5·0	6·0	7·0	8·0
Field current (A)	38	51	64	80	105	172

The armature resistance is negligible. The field current is 180 A for full-load armature current at 6·6 kV and zero power factor lagging, and 72 A for full-load current in the stator windings on steady short-circuit.

Determine, in each case for full-load stator current:

(*a*) the stator leakage-reactance voltage per phase,
(*b*) the armature-reaction m.m.f. in terms of field current,
(*c*) the field current required for operation at rated terminal voltage and a load power factor of 0·8 lagging.

(University of London) (415 V, 63 A, 154 A)

4. A 12-pole,50-Hz 3-phase star-connected synchronous motor gave the following open-circuit characteristic at 500 r.p.m.:

Field current A	0	20	40	60	80
Line voltage	0	1250	2150	2650	3050

The stator resistance and synchronous reactance/phase are 0·5 and 4·0 Ω respectively. Determine the input current and the field current when the motor is operating at unity power factor from a 2200-volt, 50-Hz, 3-phase supply.

The load is such that the motor converts 450 kW into mechanical power.

(H.N.C.) (124 A, 44 A)

5. A 1500-kVA, 3-phase, 3·3-kV, star-connected alternator has the following open-circuit characteristic:

Generated line e.m.f. (volts)	2080	2800	3300	3700	4000
Field current (A)	50	75	100	125	150

The machine is delivering two-thirds of full-load current at rated voltage and 0·8 power factor lagging. The stator phase resistance is 0·04 Ω and armature reaction on this load can be considered to be equivalent to a field current of 50 A. If the synchronous reactance at this condition is 2·5Ω, calculate the stator leakage reactance/phase. (Comment upon the accuracy of the solution.)

(University of London) (0·31 Ω)

6. A 3-phase, star-connected alternator has a maximum continuous rating of 75 MVA, 60 MW, a synchronous reactance/phase of 1·5 per unit and negligible stator resistance. Draw its performance chart for constant terminal voltage on a per unit basis, taking 1 p.u. as 75 MVA, and insert on the chart the stability-limit line allowing a safety margin of 10% of rated MW. What is the minimum excitation (expressed as a percentage of rated terminal voltage) which would give an output of 60 MW within this stability-limit line?

(L.C.T.) (132%)

7. Two 3-phase, star-connected alternators are working in parallel. Their ratings are 10 000 kVA and 5000 kVA at 6·6 kV. Each alternator is delivering its full-load current to a load at 0·8 power factor lagging. Each generated phase voltage is $7000/\sqrt{3}$ leading the terminal phase voltage by 10°. The stator windings have negligible resistance.

Calculate (*a*) the terminal voltage, (*b*) the synchronous reactance of each alternator and (*c*) the equivalent impedance of the load/phase.

The excitation and steam supply are so adjusted that the 5000-kVA alternator increases its generated voltage by 10% but the generated voltages of both machines maintain their displacement of 10° from the terminal voltage. Calculate the terminal voltage under these conditions.

(University of London) (6·0 kV, 0·98 Ω, 1·96 Ω, 2·12+j 1·58 Ω, 6·25 kV)

8. A 3-phase generator has a direct-axis synchronous reactance of 1·0 p.u. and a quadrature-axis synchronous reactance of 0·65 p.u. per phase.

Draw a phasor diagram for the machine when operating at full-load at a power factor of 0·8 lagging and estimate therefrom (*a*) the load angle and (*b*) the per-unit no-load e.m.f. Neglect armature resistance and the effect of saturation.

(University of London) (20·3°, 1·77 per unit)

9. Two identical star-connected alternators are running in parallel. The generated phase voltages are $\mathbf{E}_1 = 6700$ volts and $\mathbf{E}_2 = 6500$ volts with $\mathbf{E}_2$ leading $\mathbf{E}_1$ by 10°. Calculate the power factor of the load if the total load current is 500 A lagging $\mathbf{E}_1$ by 37°. The synchronous reactance of each alternator is 10 Ω/phase and resistance is negligible.

(University of London) (0·93)

10. A 3-phase star-connected, 5000-kVA, 6·6-kV alternator has a stator winding synchronous reactance/phase of 3·6 Ω and negligible resistance.

Calculate the open-circuit voltage at the excitation which is necessary in order to deliver full load at 0·8 lagging power factor. The alternator is connected to 'infinite busbars' at 6·6 kV.

If the power supply to the alternator is kept constant, calculate (*a*) the current and power factor at which the alternator will operate when the excitation is increased to a value which would give a 10% greater open-circuit voltage; (*b*) the current and power factor obtained when the excitation is reduced to the minimum possible operating value.

(University of London) (8·5 kV, 532 A, 0·66, 1120 A, 0·31)

11. Two 3-phase generators in parallel supply a total load of 6 MW at 0·8 power factor lagging. The frequency of one by itself falls from 51 to 49·75 Hz when loaded by 10 MW, and the other from 51 to 49·5 Hz when loaded by 2 MW. Determine the power supplied by each when in parallel, and the power factor of the first if that of the second is 0·71 lagging.

(H.N.C.) (5143 kW, 857 kW, 0·815)

12. Two identical 3-phase, star-connected alternators are connected in parallel to supply a star-connected load having a resistance of 4 Ω and an inductive reactance of 3 Ω/phase. The excitation phase voltage of each alternator is 5 kV and these voltages are in phase with each other. The stator winding of each alternator has a synchronous reactance of 3 Ω/phase and negligible resistance. Calculate the terminal voltages of the alternators.

An identical alternator is connected to 'infinite busbars' the voltage of which is 6600 V. Steam supply and excitation are adjusted until it delivers 500 A at 0·8 power factor lagging. If the excitation voltage is increased by 10%, calculate the power factor at which it will operate and the current it delivers.

(University of London) (7210 V, 0·65, 610 A)

13. A 3-phase, star-connected synchronous generator is running on 11-kV infinite busbars. Its open-circuit characteristic is:

Field current A	50	110	140	180
o.c. voltage (line) kV	7	12·5	13·75	14·8

Calculate the generated and output powers corresponding to an excitation of 160 A and a power angle of 50°. The stator resistance and synchronous reactance/phase are 0·5 Ω and 9 Ω respectively.

(L.C.T.) (14·1 MW, 13·35 MW)

14. A 3·3-kV, 500-kVA, 3-phase, 50-Hz, 300 r.p.m. synchronous motor has its excitation adjusted to 1·5 per unit. If the per unit values of the direct-axis and quadrature-axis synchronous reactances of the motor are 0·8 and 0·65 respectively, determine the power angle at which it will fall out of step and the torque corresponding to this angle. The machine may be assumed to be unsaturated.

(L.C.T.) (81·7°, 30 600 Nm)

15. An alternator is rated at 60 MW, 11 kV, 0·8 power factor lagging and has a synchronous reactance of 2·0 per unit and negligible resistance. Choosing a suitable 3-phase MVA scale, draw the performance chart for this machine and show on it the stability limit line for under-excitation allowing a safety margin of 10%.

Deduce a formula relating a voltage scale to a 3-phase MVA scale, and hence calculate the per-phase star voltage scale to which the diagram is drawn.

(L.C.T.) (For 1 cm = 20 MVA, phase voltage = 1·88 cm)

16. A power system A has a power frequency characteristic such that a drop in frequency of 0·1 Hz is caused by an additional load of 250 MW, and the corresponding load change on an adjacent network B for the same frequency deviation is 400 MW. At a certain time, A is running at 49·85 Hz and B at 50·00 Hz. What would

be the power transfer if the two systems were to be paralleled via a short a.c. interconnector?

(L.C.T.) (231 MW, B to A)

17. A 100-MVA, 3-phase, 50-Hz generator has the following per unit reactances and time constants:

$$X_d = 1{\cdot}2, \quad X_d' = 0{\cdot}3, \quad X_d'' = 0{\cdot}2, \quad T_a = 0{\cdot}25 \text{ seconds.}$$
$$T_d' = 1{\cdot}0 \text{ seconds}, \quad T_d'' = 0{\cdot}04 \text{ seconds.}$$

If, when generating rated voltage at rated frequency, a symmetrical 3-phase short-circuit occurs, determine the asymmetrical current in phase a, 3 cycles later. Assume that the fault occurs when $e_a = 0$.

(L.C.T.) (6·6 per unit)

Chapter 8

POWER SYSTEM FAULT CALCULATIONS

8.1 Introduction

A fault is said to occur on a power system when abnormally high currents flow due to the partial or complete failure of the insulation at one or more points; the complete failure of insulation is called a short-circuit. Fault calculations involve finding the voltage and current distribution throughout the system during the fault. Abnormally high current may flow as a consequence of abnormally high voltages on the system due to lightning or switching surges (see Volume 2), which puncture through, or cause flashover across the surface of otherwise healthy insulation, and the resulting damage (or ionisation of the surrounding insulation) causes a follow-through power arc.

The phrase 'abnormally high current' must be interpreted relative to the current which normally exists in the circuit. Usually there is little or no current leakage to earth so that an earth-fault current could mean any current greater than zero. Also it is possible for a fault current at times of light load—Sunday morning in summer—to be less than peak-load current in winter. Light load infers few generators far apart on the system so giving a high source impedance which limits fault current.

Symmetrical three-phase faults will be dealt with first; then asymmetrical faults such as earth-faults or line-to-line faults, and for these the method of symmetrical components will be used.

When MVA is mentioned in connection with faults and circuit breakers, it must be clearly understood that the voltage by which the current is multiplied to give the volt-amperes, is the rated voltage of that part of the system and not the actual voltage at the fault or at the circuit breaker. Since the rated voltage is always known, MVA (in this connection) is an indirect way of giving the fault current. When fault current passes through a two-winding transformer, it is changed in magnitude, but fault MVA is unchanged.

When a fault occurs at a point in a power system, the corresponding fault MVA is referred to as the fault level at that point, and, unless otherwise stated, will be taken to refer to a 3-phase symmetrical short-circuit.

The per-unit (p.u.) system of units discussed in Appendix 1 will be used throughout this chapter.

8.2 Symmetrical 3-phase Short-circuit

8.2.1 BASIC CALCULATIONS

This type of fault is defined as a simultaneous short-circuit across all three phases. It occurs infrequently, as for example, when a line, which has been made safe for maintenance by clamping all three phases together to earth, is accidentally made alive; or when, due to slow fault clearance, an earth-fault spreads across to the other two phases; or when a mechanical excavator cuts quickly through a whole cable. It is an important type of fault in that it results in an easy calculation and a pessimistic answer being almost the worst case with regard to fault-level. (An exception to this statement is dealt with in section 8.4.9.)

The calculation is easy because, being a completely symmetrical circuit, it can be treated per-phase-star, i.e. equipment is star-connected (or is replaced by its star-equivalent if delta-connected, as shown in Appendix 2) so that calculations can take place on one phase of the circuit. Calculations in this single-phase circuit using phase (line-to-neutral) e.m.f.s. and phase impedances yield a phase current, which defines the current in the other two phases, since they are equal to it in magnitude and displaced from it by 120° and 240°. When the fault is asymmetrical, this simple theory no longer applies and the more involved theory of symmetrical components is needed, as shown in sections 8.3 and 8.4.

The estimation of the MVA rating required of a circuit breaker is usually made on the assumption that it must clear a 3-phase fault because, as that is generally the worst case, it is reasonable to assume that the circuit breaker can clear any other fault. The circuit breaker rated breaking capacity in MVA must therefore be equal to, or greater than, the 3-phase fault level MVA.

Since circuit breakers are manufactured in preferred standard sizes, e.g. 250, 500, 750 MVA, high precision is not necessary when calculating the 3-phase fault level at a point in a power system. It is usual therefore to make the following assumptions:

(*a*) Immediately prior to the fault, the system is on no-load at rated frequency and all currents are neglected except the fault current. Unless otherwise stated, it is usual to assume that all generators are running at their rated voltages, e.g. 11 and 33 kV, and are interconnected by transformers operating at the appropriate ratio, e.g. 11/33 kV.

(*b*) All the generators have their e.m.f.s in phase. This means that the system is not swinging, that is, that synchronising currents have pulled all machines into step (or synchronism) prior to the fault.

The combined effect of (*a*) and (*b*) is that all generators can be replaced by a single generator since all e.m.f.s are equal and in phase. The word equal must be interpreted as meaning equal when expressed in per unit terms or in volts on the same voltage level. This procedure is sometimes called paralleling the sources on to a common e.m.f. busbar (shown dotted in the diagrams because it does not physically exist). The impedances of the alternators are then treated as part of the power system impedance network.

(*c*) System resistance is neglected and only the inductive reactance of the system is allowed for. This gives minimum system impedance and maximum fault current and a pessimistic answer; the great advantage gained from this assumption is that all reactances, currents and voltages can be treated arithmetically. The student should check that the difference between Z and X when $X = 5R$ is only about 2%, and this is acceptable when it is remembered that the system impedances are probably not known to an accuracy better than about 5% of their own values. The X/R ratio for alternators is about 20/1, for transformers about 10/1 and for high voltage lines about 8/1 to 3/1. The neglect of resistance is not so acceptable for light overhead lines of 11 kV and less, or for cables where X and R can often be nearly equal, but in many such cases the system is a radial (series) circuit and $\mathbf{Z} = R + \mathrm{j}\,X$ can be used without great difficulty. The procedure is to reduce the system to a single reactance, then to calculate the current at the fault and to retrace the steps of the calculation to find the current and voltage distribution throughout the system.

All the above assumptions are merely convenient when obtaining an estimate of 3-phase fault level at a point in a power system. For more accurate work, the load currents can be allowed for by using network theory, e.g. Thevenin (Helmholtz) or superposition; and resistance can be included in the system impedances. By using digital computers, the accurate methods of calculation are becoming economic propositions.

Assuming a symmetrical 3-phase short-circuit on a system where the rated voltage is V_r, let E volts be the e.m.f./phase (of the common e.m.f. busbar) behind a reactance of $X\,\Omega$/phase representing the whole system reduced to a single reactance. Then the phase fault current is

$$I_f = E/X \text{ amperes.} \qquad [8.1]$$

E and X must relate to the same voltage level in systems involving transformers. E is often taken to be the rated phase voltage V_r (see assumption (*a*)), although it will in general exceed it by an amount which is not large if the fault current is being calculated for the first cycle or two after the short-circuit (see section 7.5), as is commonly required. The total 3-phase fault-level VA is therefore

$$(\text{VA})_f = 3\, V_r\,.\, E/X \text{ volt-amperes.} \qquad [8.2]$$

Changing to the per-unit notation (see Appendix 1) and using the subscript b for the base value of any quantity, then

$$I_{f\cdot pu} = I_f/I_b, \quad E_{pu} = E/V_b \quad \text{and} \quad X_{pu} = X/X_b.$$

Appendix 1 shows that [8.1] may be written as

$$I_{f\cdot pu} = E_{pu}/X_{pu}. \qquad [8.3]$$

The base volt-amperes/phase $= V_b I_b$ so that the per-unit fault volt-amperes are given by

$$(\text{VA})_{f.pu} = \frac{V_r\,.\,E/X}{V_b I_b} = \frac{E}{V_b}\cdot\frac{V_r}{I_b X}.$$

It may be noted that the 3 in [8.2] has cancelled out in the per-unit volt-amperes.

If the base voltage is equal to the rated voltage, i.e. $V_b = V_r$

$$(\text{VA})_{f\cdot pu} = E_{pu}/X_{pu}. \qquad [8.4]$$

As discussed above E is often taken as equal to V_r and is thus equal to V_b, i.e. $E = 1{\cdot}0$ p.u.

Equations [8.3] and [8.4] then reduce to

$$I_{f \cdot pu} = 1/X_{pu} \qquad [8.5]$$

and

$$(\text{VA})_{f \cdot pu} = 1/X_{pu}. \qquad [8.6]$$

Worked example 8.1

Fig. 8.1(a) shows four identical alternators in parallel each rated at 11 kV, 25 MVA, and each having a (subtransient) reactance of 16%

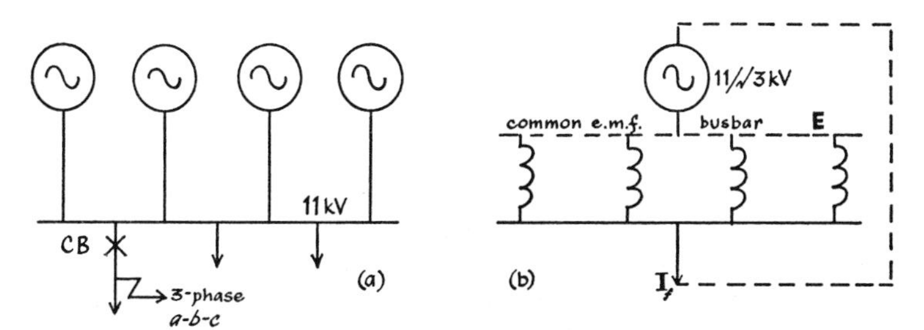

FIG. 8.1. *Three-phase fault.*
(*a*) *Actual system* (*single line*). (*b*) *Equivalent circuit.*

on its (continuous thermal) rating. Find the 3-phase fault level at one of the outgoing feeders.

SOLUTION

Base 11 kV, 25 MVA. The fault MVA from one alternator is

$$(\text{VA})_{f \cdot pu} = 1/X_{pu} = 1/0{\cdot}16 = 6{\cdot}25 \text{ p.u.}$$
$$(\text{VA})_f = 6{\cdot}25 \times 25 = 156 \text{ MVA.}$$

The total fault MVA = $6{\cdot}25 \times 25 \times 4 = 625$ MVA. Alternatively, Fig. 8.1(b) shows the four alternators paralleled on to a common e.m.f. busbar and the alternator reactances treated as part of the system reactance, and the total reactance is $0{\cdot}16/4 = 0{\cdot}04$ p.u. and the fault MVA is

$$(\text{VA})_{f \cdot pu} = 1/0{\cdot}04 = 25 \text{ p.u.}$$
$$(\text{VA})_f = 25 \times 25 = 625 \text{ MVA.}$$

8.2.2 CURRENT-LIMITING REACTORS

In example 8.1, the fault level is 625 MVA, whereas the nearest circuit-breaker standard ratings are 500 and 750 MVA. If, for economic reasons, it is necessary to use 500 MVA circuit breakers, then the fault current must be reduced by increasing the system reactance. If the system had been in the design stage, the student should check that specifying alternators with 20% reactance (sub-transient, see section 7.5) would have been a solution. If however the system exists then it is necessary to add separate reactances either in series with each alternator, or in series with each feeder, or,

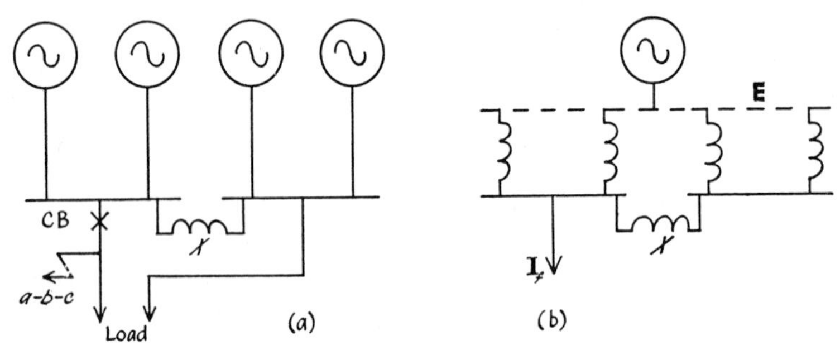

FIG.8.2. *Sectionalised busbar.*
(*a*) *Actual system.* (*b*) *Equivalent circuit.*

as is more usual, to sectionalise the busbar and connect sections via a busbar reactor as shown in Fig. 8.2.

Worked example 8.2

Using the data of example 8.1 but connected as in Fig. 8.2, calculate the reactance of the current limiting reactor X to limit the fault level on a feeder to 500 MVA.

SOLUTION

Base 11 kV, 25 MVA.

Two alternators will supply $625/2 = 312{\cdot}5$ MVA directly to the fault, so the other section must be limited to $500-312{\cdot}5 = 187{\cdot}5$ MVA $= 187{\cdot}5/25$ p.u. $= 1/X_t$ where X_t is the per unit reactance of the total right hand section. Thus $X_t = 25/187{\cdot}5 = 0{\cdot}133$ p.u. $= 13{\cdot}3\%$. Since two alternators in parallel have a combined reactance of $16/2 = 8\%$, the busbar reactor must provide the remaining reactance, i.e. $13{\cdot}3-8 = 5{\cdot}3\%$.

The ohmic value of the reactance can be calculated as follows:

$$\text{base voltage} = 1\ \text{p.u.} = 11/\sqrt{3} = 6{\cdot}36\ \text{kV}$$

$$\text{base current} = 1\ \text{p.u.} = 25/(\sqrt{3}\times 11) = 1{\cdot}31\ \text{kA}$$

$$\text{base reactance} = 1\ \text{p.u.} = 6{\cdot}36/1{\cdot}31 = 4{\cdot}86\ \Omega$$

and

$$0{\cdot}053\ \text{p.u. reactance} = 4{\cdot}86\times 0{\cdot}053 = 0{\cdot}258\ \Omega.$$

The specification of current-limiting reactors is given in B.S.S. 171 and some salient points are now given. Rated current is the current the reactor can carry continuously without exceeding a specified temperature rise dependent on the type of insulation used. In the example 8.2, the reactor might be rated to carry a current corresponding to 25 MVA at 11 kV, i.e. 1·31 kA, so that, should one alternator break down, up to 25 MVA could be taken from the other section. Notice that the voltage used is the rated system voltage and not the voltage between the terminals of one phase of the reactor.

$$\text{Over-current factor} = \frac{\text{r.m.s. symmetrical through-fault current}}{\text{rated current}}$$

$$= 187{\cdot}5/25 = 7{\cdot}5$$

Over-current time is the time in seconds that the reactor can carry the above fault current without suffering damage. A typical time would be 3 seconds. On the basis that during a fault lasting at most only a few seconds, the transformer will absorb all the I^2Rt heating effect (and dissipate none), the relation between the fault current and the permissible fault time is given approximately by I^2t = constant. This indicates, for example, that the reactor could carry half the above fault current for about 12 seconds. The reactor must be designed to withstand the electromagnetic forces corresponding to

$$2{\cdot}55\times \text{r.m.s. rated current}\times \text{over-current factor}$$

$$= 2{\cdot}55\times \text{r.m.s. symmetrical rated through-fault current.}$$

The reason for the factor 2·55 is given in the discussion of circuit breaking in Volume 2.

Current-limiting reactors should have a reactance high enough to limit the fault current, but not so high as to cause excessive voltage drop due to load current.

The use of busbar section reactors as shown in Fig. 8.2 is common at all voltage levels. Reactors are occasionally inserted into sections of the transmission network either to limit fault level, or to control the sharing of load current between lines in parallel forming a closed ring between two stations.

If X is the reactance of the reactor in Ω/phase, I the thermal rated (load) current in amperes, and V the rated line-to-neutral voltage of the system, then at rated current the voltage between terminals of one phase $= IX$ volts

$$= IX/V \text{ p.u.} = X_{pu}$$

total 3-phase reactive volt-amperes $= 3(IX)I = 3I^2X$ VAr

$$= 3I^2X/(3VI) = X_{pu}.$$

Thus in example 8.2, the reactor would have a through-fault rating of 187·5 MVA (9·85 kA at 11 kV line), a through-load rating of 25 MVA (1·31 kA at 11 kV line), a voltage between terminals of one phase of 5·3% of $11/\sqrt{3}$ kV = 337 V = 1310 × 0·258 V, and an MVAr rating of 5·3% of 25 MVA = 1·323 MVAr = 3 × 1·31 × 0·337 MVAr.

8.2.3 FAULT IN-FEEDS: SOURCE IMPEDANCE

Fig. 8.2 illustrates the layout of a power station prior to the mid-1930s. Each station supplied its own local load independently of other power stations. In the mid-1930s the 132-kV Grid was established to interconnect the larger power stations. Fig. 8.3 shows a typical arrangement. Each of the 11-kV busbar sections is connected to the Grid via a 50-MVA, 12·5% reactance transformer. The remainder of the Grid system is shown (dotted) as an equivalent alternator. If a 3-phase fault occurs on the 132-kV busbar, the MVA fed to the fault by this equivalent alternator (but excluding the power-station shown) is called the Grid in-feed and is an indirect way of giving the source impedance of this equivalent alternator. Thus if the Grid in-feed were given as 1500 MVA, then on a base of 100 MVA, the Grid in-feed MVA = 1500/100 = 15 p.u. and the Grid source reactance = 1/15 = 0·0667 p.u. Fig. 8.4 shows the system reduced to a common e.m.f. busbar and with all reactances expressed in per unit on 100 MVA base. It is shown in Appendix 1 that machine reactances are proportional to base MVA. The reduction of the network to a single reactance will involve the use of the star-delta transformation—see Appendix 2.

Worked example 8.3

Calculate the fault level for a 3-phase symmetrical short circuit on an 11-kV feeder in the system shown in Figs. 8.3 and 8.4.

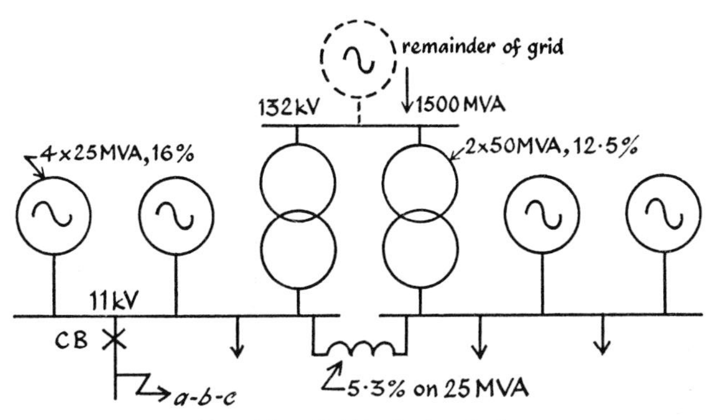

FIG. 8.3. *Grid in-feed to fault.*

SOLUTION

Base 100 MVA.

Replacing the mesh-connected p.u. reactances connected to nodes 2, 3 and 4 by a star-connected set with a star-point node numbered 5, then the latter are

$$X_{25} = (0{\cdot}25 \times 0{\cdot}25)/0{\cdot}712 = 0{\cdot}0878 \text{ p.u.}$$

$$X_{35} = X_{45} = (0{\cdot}25 \times 0{\cdot}212)/0{\cdot}712 = 0{\cdot}0743 \text{ p.u.}$$

Then from Fig. 8.4(a) and (b), between nodes 1 and 5 there are two reactances in parallel of $(0{\cdot}0667+0{\cdot}0878)$ and $(0{\cdot}32+0{\cdot}0743)$ i.e. $(0{\cdot}1545 \times 0{\cdot}3943)/(0{\cdot}0667+0{\cdot}0878+0{\cdot}32+0{\cdot}0743) = 0{\cdot}1105$ p.u.

The reactance between nodes 1 and 3 is then 0·32 in parallel with $(0{\cdot}1105+0{\cdot}0743)$ i.e. $(0{\cdot}32 \times 0{\cdot}1848)/0{\cdot}5048 = 0{\cdot}117$ p.u.

The fault MVA level at the 11-kV feeder near the circuit breaker (i.e. node 3) is thus $100/0{\cdot}117 = 855$ MVA.

This shows that the circuit breakers at the supply end of the 11-kV feeders would need to be up-rated to the next standard size, viz. 1000 MVA.

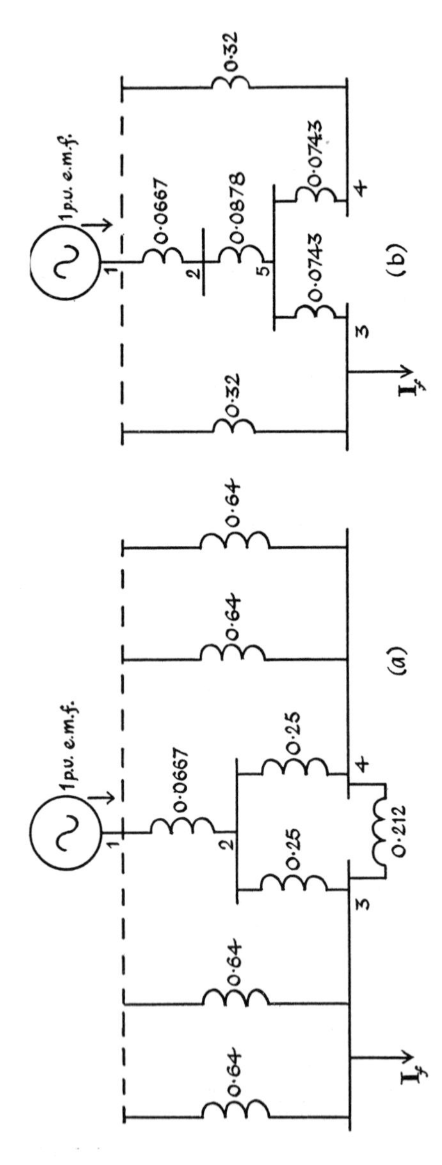

FIG. 8.4. (*a*) *Equivalent circuit for Fig.* 8.3 *on* 100-*MVA base.* (*b*) *Simplified equivalent circuit.*

During the period when the demand for electricity was doubling about every 8 to 10 years, new power stations and transmission lines were being added to the Grid system. Thus the internal impedance of the Grid system was falling and its fault level rising. Also, in any year, the fault level is highest when the Grid is running under maximum (peak) load conditions, usually in December or January. Under these conditions, the fault level at all major busbars on the Grid system is calculated, and referred to as the maximum plant fault level. Although minimum Grid impedance is associated with peak loading conditions, the same fault levels would obtain if the same machines were all running but on no-load.

8.3 Symmetrical Components

8.3.1 GENERAL PRINCIPLES

The majority of faults in a power system are not symmetrical 3-phase short-circuits, but occur between one line and earth or, less commonly, occur between lines where they may also involve earth. Also there is usually a finite fault impedance rather than a short-circuit. Thus instead of a completely balanced 3-phase circuit which can be solved by calculating the current flowing in one of the three phases, as in section 8.2, an unbalanced 3-phase circuit must now be solved. This can most conveniently be done by using the symmetrical components of an unbalanced system of currents or voltages, because this yields three (fictitious) single-phase networks only one of which contains a driving voltage (e.m.f.). The impedances in these three networks do include some which are different from those used in the analysis of completely balanced 3-phase networks (which strictly should all be prefaced by the words 'positive-sequence' when used or defined). For example, in Chapter 7 it is shown that a synchronous machine under balanced conditions may be regarded as having effective reactances called synchronous, transient and sub-transient at different times, and these are all positive-sequence reactances, but that when the current is unbalanced other reactances also are needed (section 7.6). For other equipment these other reactances are discussed in the respective chapter dealing with that equipment (sections 2.6, 3.7 and 6.10), and the use of them is discussed in this chapter (sections 8.3.2, 8.3.3 and 8.3.4).

The method of symmetrical components can be applied to any polyphase system containing any number of phases, but the 3-phase system is the only one of interest here. It can be applied to a set of unbalanced 3-phase currents or voltages and will be used for both in this chapter (section 8.4), though the equations relating the original (real or physical) quantities to their symmetrical components will be stated in terms of currents. The symmetrical components of a line current may be said to be fictitious, since metering the line current directly gives the real current which may be unbalanced, i.e. unequal in magnitude to the other two line currents and/or not at 120° with respect to each of them. On the other hand it is possible to obtain a measure of any one of the symmetrical components of an unbalanced current (or voltage), by connecting a metering circuit in which the other components do not flow (and this has to be done for some circuits used in power systems, see section 8.5).

The theory of symmetrical components states that, by the Superposition Principle in a linear network, any set of asymmetrical 3-phase phasors of sequence *abc*, e.g. currents $\mathbf{I}_a$, $\mathbf{I}_b$ and $\mathbf{I}_c$, can be replaced by the sum of three sets of currents:

(i) a positive-sequence set of three symmetrical 3-phase currents (i.e. all numerically equal and all displaced from each other by 120°), having the same sequence *abc* as the original set, and denoted by $\mathbf{I}_{a1}$, $\mathbf{I}_{b1}$ and $\mathbf{I}_{c1}$.

(ii) a negative-sequence set of three symmetrical 3-phase currents having the sequence opposite to that of the original set, viz. *acb* and denoted by $\mathbf{I}_{a2}$, $\mathbf{I}_{b2}$ and $\mathbf{I}_{c2}$.

(iii) a zero-sequence set of three currents all equal both in magnitude and in phase, i.e. having no sequence, and denoted by $\mathbf{I}_{a0}$, $\mathbf{I}_{b0}$ and $\mathbf{I}_{c0}$, and these all equal $\mathbf{I}_0$.

Fig. 8.5(a) illustrates these three sets of symmetrical components. It will be shown later that the relative values of the three components of current in any one line (I_{a1}, I_{a2} and I_{a0}), and the relative phase angles between them, will vary from one asymmetrical fault condition to another. Fig. 8.5(a) can therefore illustrate only one particular set of conditions. It must be noted that although the sequence *acb* of negative-sequence is opposite to that of positive-sequence, the set of negative-sequence phasors must still be regarded as rotating in the same direction (anticlockwise) as the positive- (or zero-) sequence phasors.

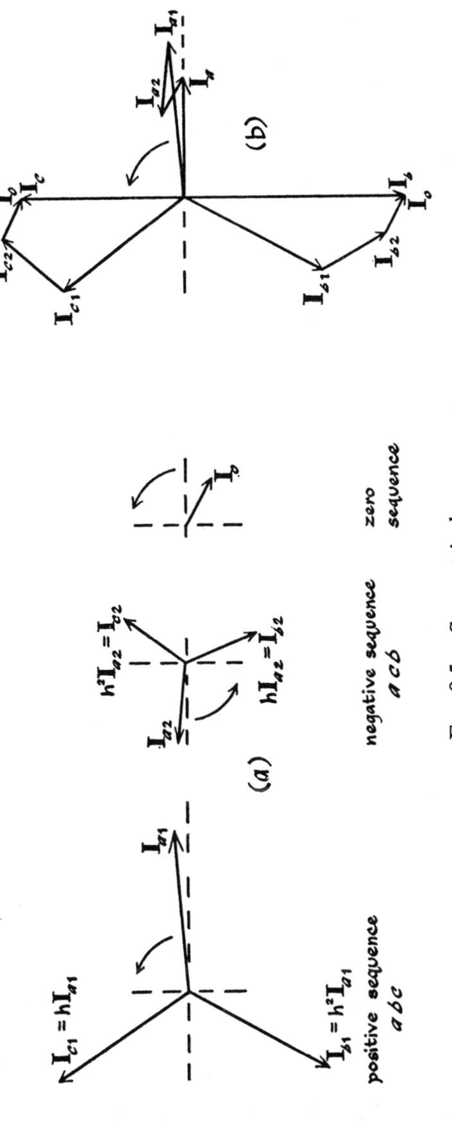

FIG. 8.5. *Symmetrical components.*
(*a*) *Sequence components* (*b*) *Addition of sequence components*

The negative-sequence currents when added to the positive-sequence currents, will give a resultant which is a set of three asymmetrical 3-phase currents, but the phasor sum of these three currents will be zero since the two components sets are both symmetrical 3-phase sets, i.e. if

$$\mathbf{I}_a = \mathbf{I}_{a1}+\mathbf{I}_{a2}$$
$$\mathbf{I}_b = \mathbf{I}_{b1}+\mathbf{I}_{b2}$$
$$\mathbf{I}_c = \mathbf{I}_{c1}+\mathbf{I}_{c2}.$$

Addition gives

$$\mathbf{I}_a+\mathbf{I}_b+\mathbf{I}_c = 0+0.$$

Under unbalanced conditions, however, there will generally be a resultant total current which is the sum of the three line currents, e.g. when there is an earth-fault. The current flowing in the earth path which is the sum of the three line currents must be the sum of the three zero-sequence currents flowing in the three lines, since the positive- and negative-sequence components sum to zero. Thus earth-path current

$$= \mathbf{I}_{a0}+\mathbf{I}_{b0}+\mathbf{I}_{c0}$$
$$= 3\mathbf{I}_{a0} = 3\mathbf{I}_0. \qquad [8.7]$$

The three sets of sequence currents, each being symmetrical in themselves, can be calculated on a single-phase (or per-phase-star) basis, with the actual current flowing in any one of the phases given by the sum of its three sequence components, i.e.

$$\begin{aligned}\mathbf{I}_a &= \mathbf{I}_{a1}+\mathbf{I}_{a2}+\mathbf{I}_{a0}\\ \mathbf{I}_b &= \mathbf{I}_{b1}+\mathbf{I}_{b2}+\mathbf{I}_{b0}\\ \mathbf{I}_c &= \mathbf{I}_{c1}+\mathbf{I}_{c2}+\mathbf{I}_{c0}.\end{aligned} \qquad [8.8]$$

Fig. 8.5(b) shows the particular unbalanced currents given by adding the symmetrical components shown in Fig. 8.5(a).

By using the h operator discussed in Appendix 3, [8.8] is reduced to

$$\begin{aligned}\mathbf{I}_a &= \mathbf{I}_{a1}+\mathbf{I}_{a2}+\mathbf{I}_{a0}\\ \mathbf{I}_b &= \mathrm{h}^2\mathbf{I}_{a1}+\mathrm{h}\mathbf{I}_{a2}+\mathbf{I}_{a0}\\ \mathbf{I}_c &= \mathrm{h}\mathbf{I}_{a1}+\mathrm{h}^2\mathbf{I}_{a2}+\mathbf{I}_{a0}.\end{aligned} \qquad [8.9]$$

In the symmetrical component method the equations are simpler if only faults which are symmetrical with respect to one phase are considered (this will be seen to be true of all faults considered in this chapter) and this phase is taken as the reference phase and is called the *a* phase (whether it is *R*, *Y* or *B* depends on the actual system).

The three equations in [*8.9*] can now be solved for the three component currents in terms of the three actual system currents, provided that the determinant

$$\begin{vmatrix} 1 & 1 & 1 \\ h^2 & h & 1 \\ h & h^2 & 1 \end{vmatrix} \neq 0.$$

The student should satisfy himself that the determinant has the value $j\,3\sqrt{3}$ and is thus not zero. To solve for $\mathbf{I}_{a1}$, the second equation in [*8.9*] may be multiplied by h, and the third by h^2, and these two equations may now be added to the first. The result is

$$\begin{aligned} \mathbf{I}_{a1} &= (1/3)(\mathbf{I}_a + h\mathbf{I}_b + h^2\mathbf{I}_c) \\ \mathbf{I}_{a2} &= (1/3)(\mathbf{I}_a + h^2\mathbf{I}_b + h\mathbf{I}_c) \\ \mathbf{I}_{a0} &= (1/3)(\mathbf{I}_a + \mathbf{I}_b + \mathbf{I}_c). \end{aligned} \qquad [8.10]$$

Using [*8.10*], the student should show that if the actual currents form a symmetrical 3-phase set, they contain only positive-sequence components, and that there are no zero-sequence components of current in a 3-phase, 3-wire system having no leakage or charging currents to earth. (Students familiar with matrices should write [*8.9*] in matrix form and by matrix inversion obtain [*8.10*].)

The above three components of current in phase *a* can be considered to flow in three fictitious single-phase circuits, which are coupled together at the point of fault in a manner which depends upon the nature of the fault. If there were any mutual coupling between these three networks, the analysis would be more complicated. Before considering these networks therefore, we shall consider the conditions under which no mutual coupling exists between them, and show that these conditions are met in a power system. It is sufficient to consider the components of current $\mathbf{I}_{a1}$, $\mathbf{I}_{a2}$ and $\mathbf{I}_{a0}$, since the currents in phases *b* and *c* can be determined from [*8.9*].

8.3.2 CONDITIONS FOR NO MUTUAL COUPLING BETWEEN THE THREE SEQUENCE NETWORKS

Fig. 8.6 shows a 3-phase system having balanced self-inductive reactance X_L/phase and mutual-inductive reactance X_M between pairs of phases. The line currents are $\mathbf{I}_a$, $\mathbf{I}_b$ and $\mathbf{I}_c$ and they may be unbalanced.

$$\mathbf{V}_{aa'} = \mathbf{I}_a\mathbf{X}_L+\mathbf{I}_b\mathbf{X}_M+\mathbf{I}_c\mathbf{X}_M$$
$$\mathbf{V}_{bb'} = \mathbf{I}_a\mathbf{X}_M+\mathbf{I}_b\mathbf{X}_L+\mathbf{I}_c\mathbf{X}_M$$
$$\mathbf{V}_{cc'} = \mathbf{I}_a\mathbf{X}_M+\mathbf{I}_b\mathbf{X}_M+\mathbf{I}_c\mathbf{X}_L$$

where $\mathbf{X} = \mathrm{j}\,X$.

The positive-sequence component of the series voltage drop, with $\mathbf{V}_{aa'}$ as the reference phasor, is (from [*8.9*])

$$\begin{aligned}\mathbf{V}_{aa'1} &= \tfrac{1}{3}(\mathbf{V}_{aa'}+\mathrm{h}\mathbf{V}_{bb'}+\mathrm{h}^2\mathbf{V}_{cc'})\\ &= \tfrac{1}{3}[\mathbf{X}_L(\mathbf{I}_a+\mathrm{h}\mathbf{I}_b+\mathrm{h}^2\mathbf{I}_c)-\mathbf{X}_M(\mathbf{I}_a+\mathrm{h}\mathbf{I}_b+\mathrm{h}^2\mathbf{I}_c)]\\ &= \mathbf{I}_{a1}(\mathbf{X}_L-\mathbf{X}_M).\end{aligned}$$

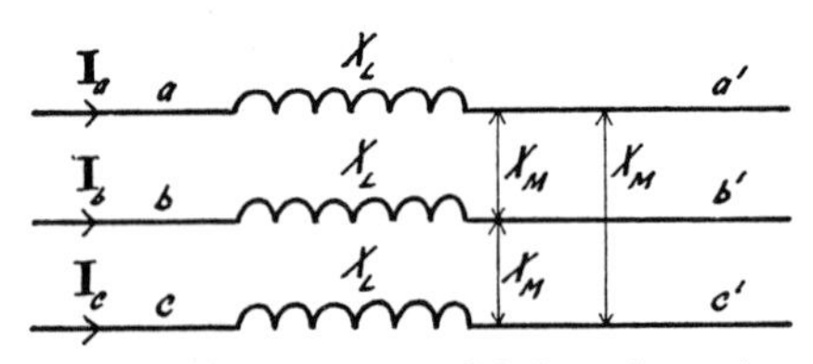

FIG. 8.6. *3-phase system with balanced impedances.*

The positive-sequence reactance/phase can be defined as

$$\mathbf{X}_1 = \mathbf{V}_{aa'1}/\mathbf{I}_{a1} = \mathbf{X}_L-\mathbf{X}_M.$$

Similarly, the negative-sequence reactance/phase =

$$\mathbf{X}_2 = \mathbf{X}_L-\mathbf{X}_M = \mathbf{X}_1 \text{ (for static equipment)}$$

and the zero-sequence reactance/phase =

$$\mathbf{X}_0 = \mathbf{X}_L+2\mathbf{X}_M$$

The equation for $\mathbf{V}_{aa'}$ can be rewritten in terms of the sequence components of current as

$$\begin{aligned}\mathbf{V}_{aa'} &= \mathbf{X}_L(\mathbf{I}_{a1}+\mathbf{I}_{a2}+\mathbf{I}_{a0})+\mathbf{X}_M(-\mathbf{I}_{a1}-\mathbf{I}_{a2}+2\mathbf{I}_{a0})\\ &= \mathbf{I}_{a1}(\mathbf{X}_L-\mathbf{X}_M)+\mathbf{I}_{a2}(\mathbf{X}_L-\mathbf{X}_M)+\mathbf{I}_{a0}(\mathbf{X}_L+2\mathbf{X}_M)\\ &= \mathbf{I}_{a1}\mathbf{X}_1+\mathbf{I}_{a2}\mathbf{X}_2+\mathbf{I}_{a0}\mathbf{X}_0. \qquad [8.11]\end{aligned}$$

This result of [*8.11*] shows that, providing the system reactances are balanced from the points of generation right up to the fault, as they are in a power system, each sequence current causes a voltage drop of its own sequence only. There is therefore no mutual coupling between the three sequence networks. The same result may be shown more rigorously by matrix methods with impedance in the neutral and mutual coupling between the three lines and neutral included.

This has been left to Volume 2 where other transformations of coordinate systems will be dealt with by matrix methods.

8.3.3 THE THREE SEQUENCE NETWORKS

In section 8.3.2 it has been shown that positive-sequence currents do not give rise to voltages of negative- or zero-sequence; that negative-sequence currents give voltages of negative-sequence only, and that zero-sequence currents give voltages of zero-sequence only. We can therefore regard each current as flowing within its own network through impedances of its own sequence only, such as the reactances $\mathbf{X}_1$, $\mathbf{X}_2$ and $\mathbf{X}_0$ referred to in section 8.3.2.

Consideration of a synchronous machine's windings shows that a normal machine without any fault can only generate balanced e.m.f.s of normal sequence, i.e. positive-sequence *abc*. Thus an e.m.f. $\mathbf{E}_{an1}$ only is generated in the positive-sequence circuit of the *a* phase, and $\mathbf{E}_{an2}$ and $\mathbf{E}_{an0}$ are zero. $\mathbf{E}_{an1}$ is the line-to-neutral e.m.f. of the generator. $\mathbf{Z}_1$ in Fig. 8.7 is the single positive-sequence impedance which appears between the generation and the point of fault, when all the impedances of the system have been reduced as discussed in section 8.2 (since the system is balanced $\mathbf{Z}_{a1} = \mathbf{Z}_{b1} = \mathbf{Z}_{c1} = \mathbf{Z}_1$, and similarly for the other two sequences). The positive-sequence voltage of phase *a* to earth at the point of fault $\mathbf{V}_{ae1}$ is thus given by

$$\mathbf{V}_{ae1} = \mathbf{E}_{an1} - \mathbf{I}_{a1}\mathbf{Z}_1. \qquad [8.12]$$

For the negative-sequence, since $\mathbf{E}_{an2} = 0$, the negative-sequence component of the voltage of phase *a* to earth at the point of fault is

$$\mathbf{V}_{ae2} = -\mathbf{I}_{a2}\mathbf{Z}_2. \qquad [8.13]$$

For the zero-sequence, if we consider the source to be a single alternator with its star-point earthed through an impedance $\mathbf{Z}_n$, then [*8.7*] shows that $\mathbf{Z}_n$ carries a current of $3\mathbf{I}_{a0}$ causing a voltage drop of $3\mathbf{I}_{a0}\mathbf{Z}_n$. To give the same voltage drop the zero-sequence network of Fig. 8.7 must therefore contain an impedance $3\mathbf{Z}_n$ as shown. Thus

$$\mathbf{V}_{ae0} = -\mathbf{I}_{a0}(\mathbf{Z}_0 + 3\mathbf{Z}_n) = -\mathbf{I}_{a0}\mathbf{Z}_0' \qquad [8.14]$$

and the p.d. between phase *a* and earth at the point of fault is $\mathbf{V}_{ae}$ given by

$$\mathbf{V}_{ae} = \mathbf{V}_{ae1} + \mathbf{V}_{ae2} + \mathbf{V}_{ae0} = \mathbf{E}_{an1} - \mathbf{I}_{a1}\mathbf{Z}_1 - \mathbf{I}_{a2}\mathbf{Z}_2 - \mathbf{I}_{a0}\mathbf{Z}_0'. \qquad [8.15]$$

Similarly $\mathbf{V}_{be}$ and $\mathbf{V}_{ce}$ may be determined from [8.9]

$$\mathbf{V}_{be} = \mathrm{h}^2\mathbf{V}_{ae1} + \mathrm{h}\mathbf{V}_{ae2} + \mathbf{V}_{ae0} = \mathrm{h}^2(\mathbf{E}_{an1} - \mathbf{I}_{a1}\mathbf{Z}_1) - \mathrm{h}\mathbf{I}_{a2}\mathbf{Z}_2 - \mathbf{I}_{a0}\mathbf{Z}'_0 \quad [8.16]$$

and

$$\mathbf{V}_{ce} = \mathrm{h}\mathbf{V}_{ae1} + \mathrm{h}^2\mathbf{V}_{ae2} + \mathbf{V}_{ae0} = \mathrm{h}(\mathbf{E}_{an1} - \mathbf{I}_{a1}\mathbf{Z}_1) - \mathrm{h}^2\mathbf{I}_{a2}\mathbf{Z}_2 - \mathbf{I}_{a0}\mathbf{Z}'_0. \quad [8.17]$$

Already then one point of difference has emerged between the zero- and negative-sequence networks. Other differences will emerge when we consider briefly in the next section how the impedances $\mathbf{Z}_1$, $\mathbf{Z}_2$ and $\mathbf{Z}_0$ may be obtained for a power system with many items of interconnected equipment.

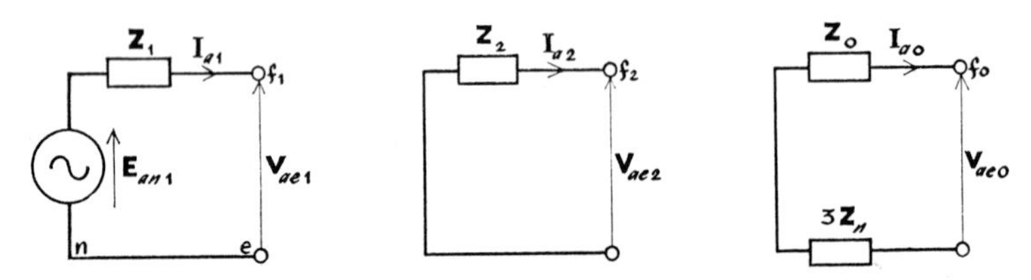

FIG. 8.7. *The three sequence networks.*

Clearly no current will flow in the three sequence networks when they are isolated as shown in Fig. 8.7, but in section 8.4 the interconnection of these networks will be derived to represent various unbalanced fault conditions.

8.3.4 $\mathbf{Z}_1$, $\mathbf{Z}_2$ AND $\mathbf{Z}_0$ FOR AN INTERCONNECTED POWER SYSTEM

The process of reducing the positive-sequence impedances/phase of an interconnected network to a single value between generation and fault, is exactly that dealt with in section 8.2, where though it was not said then, all impedances were in fact the positive-sequence impedances of the equipment.

All static equipment has the same impedance to negative-sequence current as to positive-sequence current (see sections 2.6, 3.7 and 6.10). For synchronous machines, the positive-sequence reactance varies with time (section 7.6), and the negative-sequence reactance is not necessarily equal to it, although the difference may be small during the first few cycles after a fault. The lay-out of the negative-sequence circuit is identical to that of the positive-sequence network since both are balanced 3-phase systems, and they will only differ in the values of reactances for synchronous machines and the lack of negative-sequence e.m.f. (see Fig. 8.8(b) and (c)). $\mathbf{Z}_2$ is thus found by the same process of reduction to a single impedance by combining elements in series or parallel, and using delta-star transformations.

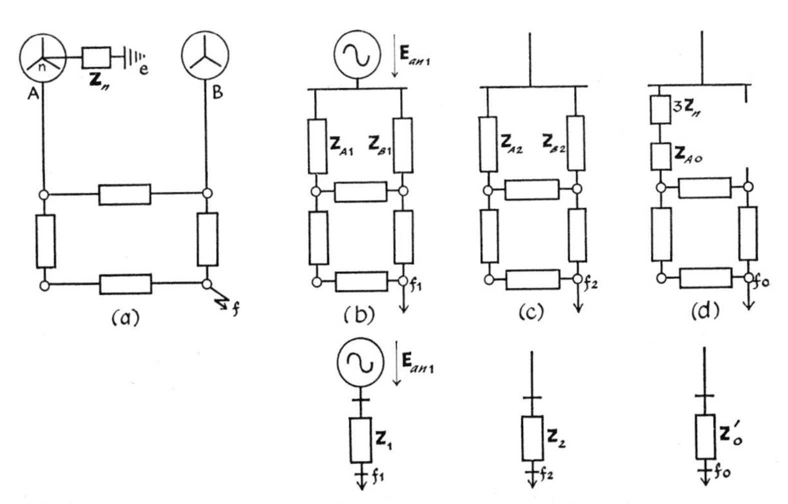

FIG. 8.8. *Unbalanced fault in inter-connected system.*
(*a*) *System* (*b*) *Positive-sequence network*
(*c*) *Negative-sequence network* (*d*) *Zero-sequence network*

In the zero-sequence network, however, an unearthed star-point of an alternator or transformer winding, causes $\mathbf{Z}_0$ to be infinite for that part of the circuit since if there is no path to earth $\mathbf{I}_0 = 0$ (see [*8.7*] and section 6.10). Thus in the zero-sequence network of Fig. 8.8(d), the zero-sequence impedance of alternator B does not appear because it is in series with an infinite impedance. The zero-sequence network generally has different values of impedance, for every item of equipment, from the positive-sequence network (see sections 2.6, 3.7, 6.10 and 7.6), though in some transformers the impedances may be taken as equal (see section 6.10).

If a transformer has a delta-connected winding, then again an open-circuit is introduced into the zero-sequence network since zero-sequence currents circulate around a delta-connected winding without flowing in the 3-phase circuit connected to it. This is discussed in section 6.10.

The alternator is the only equipment with a time-varying reactance to positive-sequence current. The sub-transient reactance is used in the positive-sequence network if for example the maximum mechanical forces during a fault are required, and the d.c. component also is taken into account. The transient reactance is often used during the period while the protective system is operating, and a circuit breaker is opening. It is often unnecessary to consider any variation in negative- and zero-sequence reactances of a synchronous machine during the sub-transient, transient and steady-state regions, so that one value for each is sufficient.

The process of calculating the unbalanced fault currents and voltages in any part of a power system may now be summarised as follows:

(*a*) determine the three sequence networks from generation to point of fault by inspection of the system diagram;

(*b*) insert the sequence impedances in each of the networks;

(*c*) reduce each network to a single impedance $\mathbf{Z}_1$, $\mathbf{Z}_2$ and $\mathbf{Z}_0$;

(*d*) connect these sequence networks together with any fault impedance in the way shown in section 8.4 to apply to that particular unbalanced fault;

(*e*) calculate the sequence components of current in phase *a* at the point of fault;

(*f*) calculate the sequence components of current in phase *a* in any branch of the sequence networks, by working back from the point of fault;

(*g*) calculate the actual current in any phase in any branch of the circuit using [*8.9*];

(*h*) calculate the sequence components of p.d. between phase *a* and earth at any point in the system from [*8.12*], [*8.13*] and [*8.14*];

(*i*) calculate the actual p.d. between any phase and earth at any point in the system from [*8.15*], [*8.16*] and [*8.17*].

8.3.5 PHASE-SHIFT THROUGH A DELTA/STAR TRANSFORMER

It was shown in section 6.3 and Fig. 6.4 that for a Dy1 transformer, the equivalent primary line-to-neutral voltage (i.e. the phase voltage of the star-connected alternator supplying the delta primary) was 30° ahead of the corresponding secondary line-to-neutral voltage.

It will now be shown that positive-sequence currents in the primary *lines* are also 30° ahead of the corresponding secondary positive-sequence *line* currents. The student should satisfy himself that if k equals the phase-to-phase turns ratio (N_s/N_p), the corresponding line voltage ratio is $\sqrt{3} \,.\, k$.

The positive-sequence currents in the secondary are $\mathbf{I}_{a1}$, $\mathbf{I}_{b1} = \mathrm{h}^2\mathbf{I}_{a1}$ and $\mathbf{I}_{c1} = \mathrm{h}\mathbf{I}_{a1}$. Multiplying these by k will give the corresponding primary phase currents and subtracting in pairs will give the line currents. Thus

$$\mathbf{I}_{A1} = \mathrm{k}\mathbf{I}_{a1} - k\mathrm{h}^2\mathbf{I}_{a1} = k\mathbf{I}_{a1}(1-\mathrm{h}^2) = \sqrt{3} \,.\, kI_{a1}\underline{/+30^\circ}$$

This shift of positive-sequence voltage and current by the same amount in the same sense is to be expected, since positive-sequence current represents load current, and the load power factor is not changed by an ideal transformer.

The student should now show that for negative-sequence currents

$$\mathbf{I}_{A2} = \sqrt{3} \,.\, kI_{a2}\underline{/-30^\circ}$$

Thus positive-, and negative-sequence currents are shifted 30° in opposite senses. It should be noted that the numerical transformation factor is the line-voltage ratio: this is to be expected since it is a secondary line to primary line current transformation. (The student should now show that for a Dy11 transformer (Fig. 6.5), the same results hold except that the phase shift is now -30° and $+30^\circ$ respectively.) The above results can be used when it is required to find the current distribution in a complex network on the primary side of a delta/star transformer.

Another illustration of application is to show that, if $\mathbf{I}_a$ is an earth-fault current on the secondary side, the primary fault current is $k\mathbf{I}_a$ (a result which is obvious from the definition of k). For an earth-fault it will be shown in section 8.4.1 that $\mathbf{I}_{a1} = \mathbf{I}_{a2} = \mathbf{I}_{a0}$ and $\mathbf{I}_a = \mathbf{I}_{a1} + \mathbf{I}_{a2} + \mathbf{I}_{a0}$ (equation [*8.8*]). Since there are no zero-sequence currents in the primary lines

$$\begin{aligned}
\mathbf{I}_A &= \mathbf{I}_{A1} + \mathbf{I}_{A2} \\
&= \sqrt{3}\, k\mathbf{I}_{a1}\underline{/+30^\circ} + \sqrt{3}\, k\mathbf{I}_{a2}\underline{/-30^\circ} \\
&= k\mathbf{I}_{a1}(1-\mathrm{h}^2) + k\mathbf{I}_{a2}(1-\mathrm{h}) = 3k\mathbf{I}_{a1} \\
&= k\mathbf{I}_a \\
\mathbf{I}_B &= \mathbf{I}_{B1} + \mathbf{I}_{B2} = \mathrm{h}^2\mathbf{I}_{A1} + \mathrm{h}\mathbf{I}_{A2} = 0 \\
\mathbf{I}_C &= -\mathbf{I}_A.
\end{aligned}$$

8.4 Asymmetrical Faults

An asymmetrical fault is any fault other than a 3-phase symmetrical fault: e.g. an earth-fault on one phase; a line-to-line fault (or phase-to-phase fault sometimes contracted to phase-fault); and a double-earth-fault (or line-to-line-to-earth fault). One notation is *a–e* for an earth-fault on phase *a* and *b–c* for a phase-fault across *b* and *c*. Problems involving asymmetrical faults are solved by resolving the asymmetrical fault currents into their symmetrical components. The power system is then analysed three times, once for each of the three sequence components, and the results added to give the actual fault current (or voltage).

During asymmetrical fault calculations it is usual (unless the contrary is explicitly stated) to make the same simplifying assumptions that were made during symmetrical 3-phase faults: namely, that all source e.m.f.s are equal (in per-unit) and in phase so that they can all be replaced by a single source e.m.f. If all the resistances can be neglected, the system reduces to a single reactance in the positive-sequence network with a single e.m.f. behind it.

In the analyses that follow, it should be noted that the reference phase denoted by *a* is always the symmetrical phase with respect to the fault: i.e. an earth-fault is on phase *a*, while phases *b* and *c* are healthy; and a phase-fault is across phases *b* and *c* while phase *a* is healthy. This is done to give the simplest results in mathematical terms. In applying the analysis in terms of phases *abc* to a practical problem in terms of phases *RYB*, the correct phase sequence must be maintained: i.e. if phase *a* is *B*, then *b* is *R* and *c* is *Y*.

8.4.1 EARTH FAULT ON PHASE *a*

Fig. 8.9 shows a simplified unloaded power system with an earth-fault on phase *a* at the point *f*. The source star-point is earthed via an impedance $\mathbf{Z}_n$ (usually a resistance) and the fault path is shown as having an impedance $\mathbf{Z}_f$ (usually a low resistance). The e.m.f.s and impedances of the power system are always assumed to be balanced up to but excluding the fault, so that there is no mutual coupling between the sequence networks as shown in sections 8.3.2 and 8.3.3. The fault current is $\mathbf{I}_{ae}$. Also shown in all three phases are the sequence components of current.

The procedure in this and in the following analyses will be to write down by inspection of the circuit diagram, the three equations for phases *a*, *b* and *c* which define the fault condition in terms of the actual phase currents and voltages. These currents and voltages will then be transformed into their symmetrical component form.

$$\mathbf{V}_{ae} = \mathbf{Z}_f\mathbf{I}_{ae} \qquad [8.18]$$

$$\mathbf{I}_{be} = 0 \qquad [8.19]$$

$$\mathbf{I}_{ce} = 0. \qquad [8.20]$$

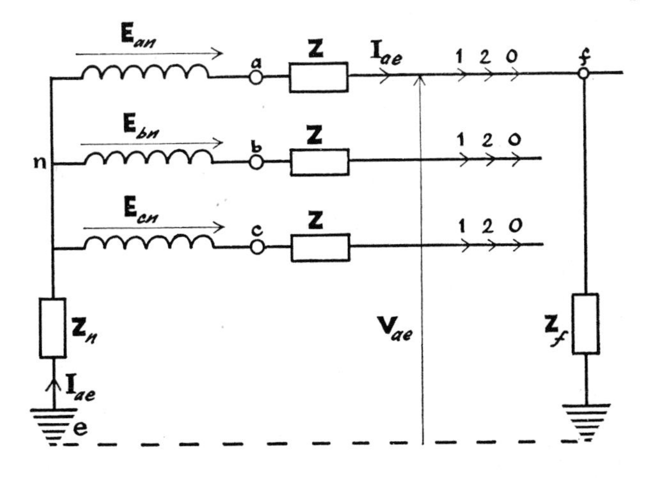

FIG. 8.9. *Earth-fault on phase* a.

Using [*8.9*], these voltages and currents may be replaced by the sum of their symmetrical components and [*8.18*], [*8.19*] and [*8.20*] become:

$$\mathbf{V}_{ae1}+\mathbf{V}_{ae2}+\mathbf{V}_{ae0} = \mathbf{Z}_f(\mathbf{I}_{ae1}+\mathbf{I}_{ae2}+\mathbf{I}_{ae0}) \qquad [8.21]$$

where $\mathbf{V}_{ae1}$, $\mathbf{V}_{ae2}$ and $\mathbf{V}_{ae0}$ are the symmetrical components of the p.d. between the reference phase and earth at the point of fault *f*.

$$\mathrm{h}^2\mathbf{I}_{ae1}+\mathrm{h}\mathbf{I}_{ae2}+\mathbf{I}_{ae0} = 0 \qquad [8.22]$$

$$\mathrm{h}\mathbf{I}_{ae1}+\mathrm{h}^2\mathbf{I}_{ae2}+\mathbf{I}_{ae0} = 0 \qquad [8.23]$$

Subtracting [*8.23*] from [*8.22*]

$$(\mathrm{h}^2-\mathrm{h})\mathbf{I}_{ae1}+(\mathrm{h}-\mathrm{h}^2)\mathbf{I}_{ae2} = 0$$

$$\mathbf{I}_{ae1} = \mathbf{I}_{ae2}.$$

Substituting this result in [*8.23*]

$$(\mathrm{h}+\mathrm{h}^2)\mathbf{I}_{ae1}+\mathbf{I}_{ae0} = 0$$

and since $h+h^2 = -1$

$$\mathbf{I}_{ae1} = \mathbf{I}_{ae0}$$

so that

$$\mathbf{I}_{ae1} = \mathbf{I}_{ae2} = \mathbf{I}_{ae0} = \mathbf{I}_{ae}/3. \qquad [8.24]$$

Equation [*8.24*] shows that for this particular fault the three sequence currents in phase *a* are all exactly equal in magnitude and in phase.

Re-arranging terms in [*8.21*]

$$(\mathbf{V}_{ae1}-\mathbf{Z}_f\mathbf{I}_{ae1})+(\mathbf{V}_{ae2}-\mathbf{Z}_f\mathbf{I}_{ae2})+(\mathbf{V}_{ae0}-\mathbf{Z}_f\mathbf{I}_{ae0}) = 0. \qquad [8.25]$$

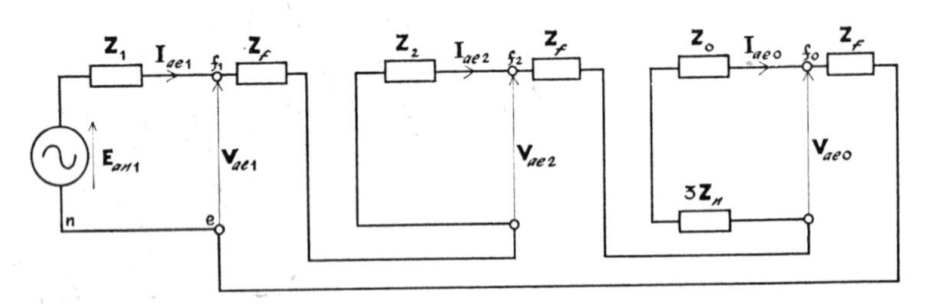

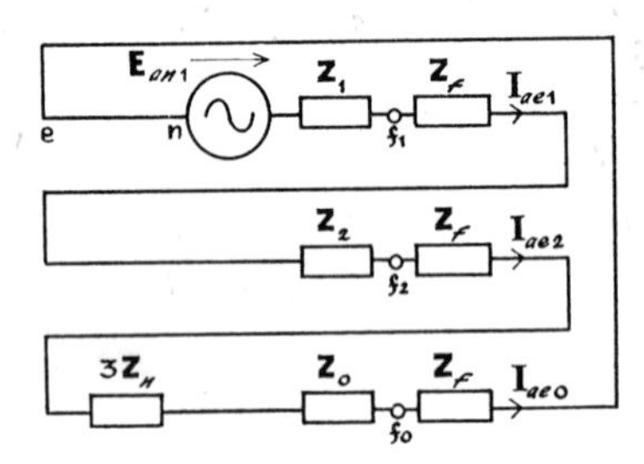

FIG. 8.10. *Interconnection of the sequence networks for an earth fault on the reference phase* a.

The interconnection of the sequence networks needed to satisfy [*8.24*] and [*8.25*] is shown in Fig. 8.10 where

$$\mathbf{I}_{ae1} = \mathbf{I}_{ae2} = \mathbf{I}_{ae0} = \frac{\mathbf{E}_{an}}{\mathbf{Z}_1+\mathbf{Z}_2+\mathbf{Z}_0+3\mathbf{Z}_n+3\mathbf{Z}_f} \qquad [8.26]$$

and the fault current

$$\mathbf{I}_{ae} = \frac{3\mathbf{E}_{an}}{\mathbf{Z}_1+\mathbf{Z}_2+\mathbf{Z}_0+3\mathbf{Z}_n+3\mathbf{Z}_f} \qquad [8.27]$$

If the system star points are solidly earthed then $\mathbf{Z}_n = 0$, and if the fault is a short-circuit (i.e. with negligible impedance) then $\mathbf{Z}_f = 0$, and Fig. 8.10 reduces to a simple series connection of the three sequence networks with a fault current given by

$$\mathbf{I}_{ae} = \frac{3\mathbf{E}_{an}}{\mathbf{Z}_1 + \mathbf{Z}_2 + \mathbf{Z}_0} \qquad [8.28]$$

The voltages to earth of the un-faulted phases b and c at the point of fault can be found from [*8.16*] and [*8.17*].

Worked example 8.4

A 3-phase, 33-kV, 37·5-MVA alternator is connected to a 33-kV overhead line which develops an earth fault on one conductor at the remote end. The positive-, negative-, and zero-sequence reactances of the alternator are 18, 12 and 10% (on rating), while those for the line are 6·3, 6·3 and 12·6 Ω/conductor. Calculate the fault current, and the phase voltages at the alternator terminals. Assume the alternator star point is solidly earthed.

SOLUTION

Base 37·5 MVA (3-phase), 33-kV (line).

$$\text{Base voltage} = 1 \text{ p.u.} = 33/\sqrt{3} = 19{\cdot}1 \text{ kV/phase}$$
$$\text{base current} = 1 \text{ p.u.} = 37{\cdot}5/(\sqrt{3} \times 33) = 0{\cdot}655 \text{ kA}$$
$$\text{base reactance} = 1 \text{ p.u.} = 19{\cdot}1/0{\cdot}655 = 29{\cdot}1\ \Omega$$

so $6{\cdot}3 \text{ ohms} = 6{\cdot}3/29{\cdot}1 = 0{\cdot}2165 \text{ p.u.}$

The total reactance of the equivalent network of the three sequence networks in series (Fig. 8.10) is

$$X_t = \text{j}\,[(0{\cdot}18+0{\cdot}2165)+(0{\cdot}12+0{\cdot}2165)+(0{\cdot}1+0{\cdot}433)]$$
$$= \text{j}\,1{\cdot}266 \text{ p.u.}$$
$$\mathbf{I}_{ae1} = 1\underline{/0^\circ}/\text{j}\,1{\cdot}266 = -\text{j}\,0{\cdot}791 \text{ p.u.}$$

and fault current $\mathbf{I}_{ae} = -\text{j}\,0{\cdot}791 \times 3 \times 0{\cdot}655 = -\text{j}\,1{\cdot}555 \text{ kA}$.

The voltages with respect to earth at the alternator terminals are calculated using [*8.12*], [*8.13*] and [*8.14*] where the reactances are those from the source to the alternator terminals.

$$\mathbf{V}_{ae1} = 1\underline{/0^\circ} - (-\text{j}\,0{\cdot}791)(+\text{j}\,0{\cdot}18) = 0{\cdot}858 + \text{j}\,0 \text{ p.u.}$$
$$\mathbf{V}_{ae2} = -(-\text{j}\,0{\cdot}791)(+\text{j}\,0{\cdot}12) = -0{\cdot}0948 + \text{j}\,0 \text{ p.u.}$$
$$\mathbf{V}_{ae0} = -(-\text{j}\,0{\cdot}791)(+\text{j}\,0{\cdot}1) = -0{\cdot}0791 + \text{j}\,0 \text{ p.u.}$$
$$\mathbf{V}_{ae} = \mathbf{V}_{ae1} + \mathbf{V}_{ae2} + \mathbf{V}_{ae0} = 0{\cdot}6841\underline{/0^\circ} \text{ p.u.}$$
$$= 0{\cdot}6841 \times 19{\cdot}1\underline{/0^\circ} = 13\underline{/0^\circ} \text{ kV}$$
$$\mathbf{V}_{be} = \text{h}^2\mathbf{V}_{ae1} + \text{h}\mathbf{V}_{ae2} + \mathbf{V}_{ae0} = 0{\cdot}945\underline{/-119{\cdot}16^\circ} \text{ p.u.}$$
$$= 18\underline{/-119{\cdot}16^\circ} \text{ kV}$$
$$\mathbf{V}_{ce} = 18\underline{/+119{\cdot}16^\circ} \text{ kV.}$$

The fault MVA $= 1{\cdot}555 \times 19{\cdot}1 = 29{\cdot}6$ MVA.

The student should check that this last result can be obtained from [*8.6*], i.e. $(\mathrm{VA})_f = 1/X$ p.u., providing that X is interpreted as X_t the total reactance of the fault equivalent network.

In practice alternators are usually earthed via a resistor of value about 100% on alternator rating. The corresponding value to be added to X_t is 300% = 3 p.u. Thus the total impedance of the fault equivalent network becomes $3+\mathrm{j}\,1{\cdot}266$ p.u.

8.4.2 LINE-TO-LINE (PHASE-FAULT) ACROSS LINES b AND c

Fig. 8.11 shows the system, and it should be noted that the conventional positive direction of current flow is taken as being away from

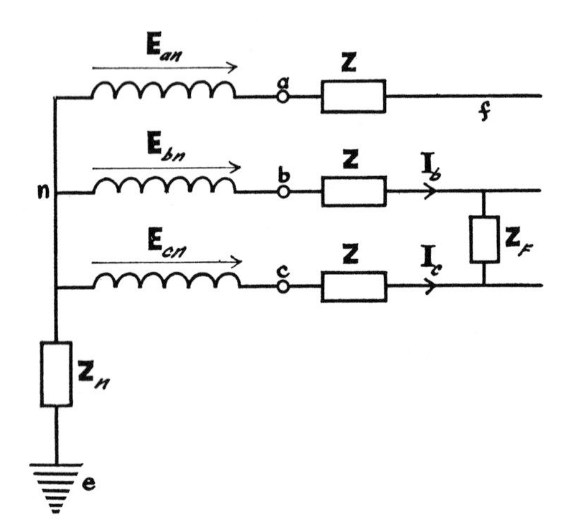

FIG. 8.11. *Line-to-line (phase) fault across* b *and* c.

the source. There is no voltage across $\mathbf{Z}_n$ and $\mathbf{I}_{a0}$ is zero since there is no path in which current can flow to the source star-point via earth.

By inspection of Fig. 8.11 we may write

$$\mathbf{I}_a = 0, \mathbf{I}_b = -\mathbf{I}_c \quad \text{and} \quad \mathbf{V}_{be} - \mathbf{Z}_F\mathbf{I}_b = \mathbf{V}_{ce}.$$

The student should show that by transforming the currents and voltages in these three equations into their symmetrical components, they may be solved to yield the results

$$\mathbf{I}_{a1} = -\mathbf{I}_{a2} \qquad [8.29]$$

$$\mathbf{V}_{ae1} - \mathbf{V}_{ae2} = \mathbf{Z}_F\mathbf{I}_{a1}. \qquad [8.30]$$

Equations [*8.29*] and [*8.30*] are satisfied by the interconnection of the sequence networks shown in Fig. 8.12. From [*8.29*] and [*8.30*] or by inspection of Fig. 8.12 it follows that

$$\mathbf{I}_{a1} = \frac{\mathbf{E}_{an}}{\mathbf{Z}_1+\mathbf{Z}_2+\mathbf{Z}_F} \qquad [8.31]$$

From [*8.9*], it is readily shown (though the student may wish to confirm this) that the fault current is given by

$$\mathbf{I}_b = -\mathrm{j}\,\sqrt{3}\,\mathbf{I}_{a1} = -\mathbf{I}_c \qquad [8.32]$$

Worked example 8.5

Using the data of example 8.4 but with a short-circuit between lines b and c at point f, calculate the fault current and the voltage of the healthy line to earth at the point of fault.

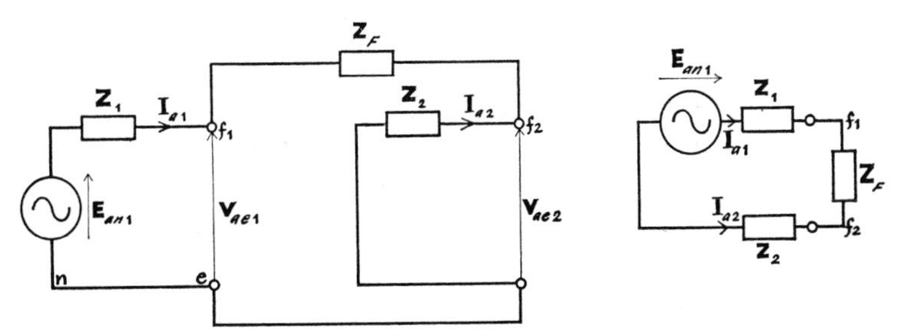

FIG. 8.12. *Interconnection of the sequence networks for a phase-fault.*

SOLUTION

The total reactance of the fault equivalent network, Fig. 8.12, is

$$\begin{aligned} X_t &= \mathrm{j}\,[(0{\cdot}18+0{\cdot}2165)+(0{\cdot}12+0{\cdot}2165)] \\ &= \mathrm{j}\,0{\cdot}733 \text{ p.u.} \\ \mathbf{I}_{a1} &= 1\underline{/0^\circ}/\mathrm{j}\,0{\cdot}733 = -\mathrm{j}\,1{\cdot}365 \text{ p.u.} \end{aligned}$$

The fault current is

$$\begin{aligned} \mathbf{I}_b &= -\mathrm{j}\,\sqrt{3}\times\mathbf{I}_{a1} = -\mathrm{j}\,\sqrt{3}(-\mathrm{j}\,1{\cdot}365) = -2{\cdot}36 \text{ p.u.} \\ &= 2{\cdot}36\times0{\cdot}655\underline{/180^\circ} = 1{\cdot}545\underline{/180^\circ} \text{ kA.} \end{aligned}$$

At the point of fault

$$\begin{aligned} \mathbf{V}_{ae1} &= (1+\mathrm{j}\,0)-(-\mathrm{j}\,1{\cdot}365)(+\mathrm{j}\,0{\cdot}3965) \\ &= 0{\cdot}459+\mathrm{j}\,0 \text{ p.u,} \end{aligned}$$

The student should check that

$$\mathbf{V}_{ae2} = 0-(+\text{j}\,1{\cdot}365)(+\text{j}\,0{\cdot}3365)$$
$$= \mathbf{V}_{ae1} \text{ (since } \mathbf{Z}_f = 0).$$

Thus

$$\mathbf{V}_{ae} = \mathbf{V}_{ae1}+\mathbf{V}_{ae2} = 0{\cdot}918\underline{/0^\circ} \text{ p.u.}$$
$$= 17{\cdot}5\underline{/0^\circ} \text{ kV.}$$

The fault MVA $= 1{\cdot}545\times 33 = 51$ MVA.

The student should check that this last result can be obtained from [*8.6*] providing that X is interpreted as the total reactance X_t of the fault equivalent circuit.

8.4.3 LINE-TO-LINE-TO-EARTH SHORT-CIRCUIT (DOUBLE-EARTH-FAULT) ON PHASES *b* AND *c*

Before proceeding to the general case of a double-earth-fault in section 8.4.4, the simpler case of a dead short-circuit to earth on two

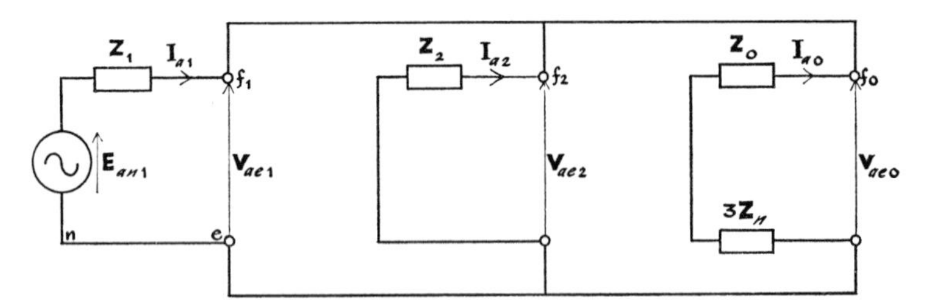

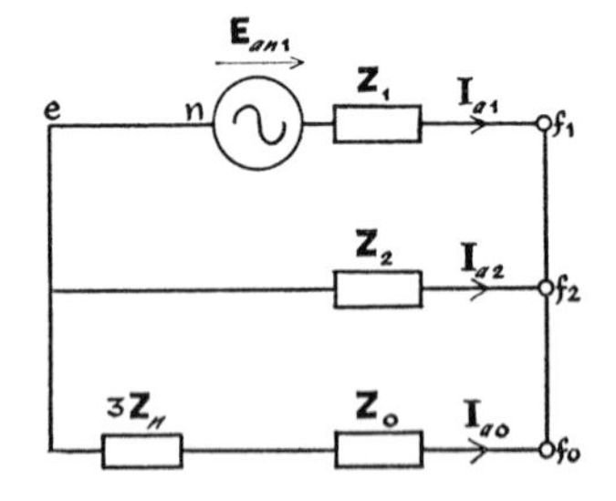

FIG. 8.13. *Interconnection of the sequence networks for a line-to-line-to-earth short-circuit on* b *and* c.

phases *b* and *c* is outlined in this section. For this fault, inspection of the circuit gives

$$\mathbf{I}_a = 0,\ \mathbf{V}_{be} = 0 \quad \text{and} \quad \mathbf{V}_{ce} = 0.$$

Transforming these three equations into their symmetrical components, gives (as students should check)

$$\mathbf{I}_{a1}+\mathbf{I}_{a2}+\mathbf{I}_{a0} = 0 \qquad [8.33]$$

and

$$\mathbf{V}_{ae1} = \mathbf{V}_{ae2} = \mathbf{V}_{ae0}. \qquad [8.34]$$

Equations [*8.33*] and [*8.34*] are satisfied by the circuit of Fig. 8.13 from which it can be seen that

$$\left.\begin{aligned} \mathbf{I}_{a1} &= \frac{\mathbf{E}_{an}}{\mathbf{Z}_1+\mathbf{Z}_2\mathbf{Z}_0'/(\mathbf{Z}_2+\mathbf{Z}_0')} \\ \mathbf{I}_{a2} &= -\mathbf{I}_{a1}\cdot\frac{\mathbf{Z}_0'}{\mathbf{Z}_2+\mathbf{Z}_0'} \\ \mathbf{I}_{a0} &= -\mathbf{I}_{a1}\cdot\frac{\mathbf{Z}_2}{\mathbf{Z}_2+\mathbf{Z}_0'} \end{aligned}\right\} \qquad [8.35]$$

It may be noted from [*8.14*] that if the system star points are earthed solidly (i.e. $\mathbf{Z}_n = 0$), then $\mathbf{Z}_0'$ becomes $\mathbf{Z}_0$ which is the single effective zero-sequence impedance corresponding to all the equipment connected between generation and fault. The currents in the two faulty lines may be obtained from [*8.35*] and [*8.9*], and the fault current in the earth path is $\mathbf{I}_b+\mathbf{I}_c = 3\mathbf{I}_{a0}$.

8.4.4 GENERAL CASE OF DOUBLE-EARTH-FAULT ON PHASES *b* AND *c*

The general case of a double-earth fault which is symmetrical with respect to the reference phase *a*, is a fault occurring between phases *b* and *c* with the mid-point of the fault impedance $2\mathbf{Z}_f$ between them connected to earth by an impedance $\mathbf{Z}_g$. It has already been noted that the method of symmetrical components is considerably simplified if it is applied to faults which are symmetrical with respect to the reference phase *a*. Again the system must be balanced from generation to the point of fault.

The method used here follows exactly the procedure used in section 8.4.1. The three equations which may be written for the actual voltages and currents at the point of fault are

$$\mathbf{I}_{ae} = 0 \qquad [8.36]$$

$$\mathbf{V}_{be} = (\mathbf{Z}_f+\mathbf{Z}_g)\mathbf{I}_{be}+\mathbf{Z}_g\mathbf{I}_{ce} \qquad [8.37]$$

$$\mathbf{V}_{ce} = \mathbf{Z}_g\mathbf{I}_{be}+(\mathbf{Z}_f+\mathbf{Z}_g)\mathbf{I}_{ce}. \qquad [8.38]$$

Writing these currents and the potential differences between the line conductors and earth at the point of fault, as the sum of their symmetrical components, gives

$$\mathbf{I}_{ae1}+\mathbf{I}_{ae2}+\mathbf{I}_{ae0} = 0 \qquad [8.39]$$

$$\mathrm{h}^2\mathbf{V}_{ae1}+\mathrm{h}\mathbf{V}_{ae2}+\mathbf{V}_{ae0} = (\mathbf{Z}_f+\mathbf{Z}_g)(\mathrm{h}^2\mathbf{I}_{ae1}+\mathrm{h}\mathbf{I}_{ae2}+\mathbf{I}_{ae0})+ \mathbf{Z}_g(\mathrm{h}\mathbf{I}_{ae1}+\mathrm{h}^2\mathbf{I}_{ae2}+\mathbf{I}_{ae0}) \qquad [8.40]$$

$$\mathrm{h}\mathbf{V}_{ae1}+\mathrm{h}^2\mathbf{V}_{ae2}+\mathbf{V}_{ae0} = \mathbf{Z}_g(\mathrm{h}^2\mathbf{I}_{ae1}+\mathrm{h}\mathbf{I}_{ae2}+\mathbf{I}_{ae0})+ (\mathbf{Z}_f+\mathbf{Z}_g)(\mathrm{h}\mathbf{I}_{ae1}+\mathrm{h}^2\mathbf{I}_{ae2}+\mathbf{I}_{ae0}). \qquad [8.41]$$

Subtracting [*8.41*] from [*8.40*] gives

$$(\mathrm{h}^2-\mathrm{h})\mathbf{V}_{ae1}+(\mathrm{h}-\mathrm{h}^2)\mathbf{V}_{ae2} = \mathbf{Z}_f(\mathrm{h}^2\mathbf{I}_{ae1}+\mathrm{h}\mathbf{I}_{ae2}+\mathbf{I}_{ae0})- \mathbf{Z}_f(\mathrm{h}\mathbf{I}_{ae1}+\mathrm{h}^2\mathbf{I}_{ae2}+\mathbf{I}_{ae0}).$$

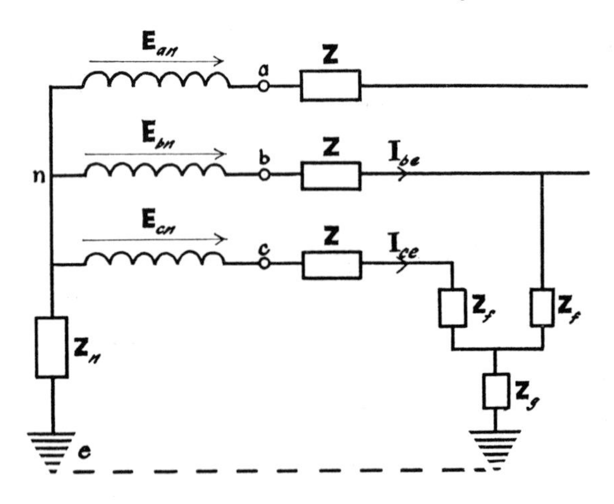

FIG. 8.14. *General double-earth fault on phase* b *and* c.

Collecting together positive-sequence terms on the left-hand side and negative-sequence terms on the right-hand side gives

$$(\mathrm{h}^2-\mathrm{h})(\mathbf{V}_{ae1}-\mathbf{Z}_f\mathbf{I}_{ae1}) = (\mathrm{h}^2-\mathrm{h})(\mathbf{V}_{ae2}-\mathbf{Z}_f\mathbf{I}_{ae2})$$

$$\mathbf{V}_{ae1}-\mathbf{Z}_f\mathbf{I}_{ae1} = \mathbf{V}_{ae2}-\mathbf{Z}_f\mathbf{I}_{ae2}$$

and substituting this result in [*8.41*] gives

$$\mathrm{h}(\mathbf{V}_{ae1}-\mathbf{Z}_f\mathbf{I}_{ae1})+\mathrm{h}^2(\mathbf{V}_{ae1}-\mathbf{Z}_f\mathbf{I}_{ae1})+(\mathbf{V}_{ae0}-\mathbf{Z}_f\mathbf{I}_{ae0}) = \mathbf{Z}_g[(\mathrm{h}^2+\mathrm{h})\mathbf{I}_{ae1}+(\mathrm{h}^2+\mathrm{h})\mathbf{I}_{ae2}+2\mathbf{I}_{ae0}]$$

and the last equation simplifies to

$$\mathbf{V}_{ae1}-\mathbf{Z}_f\mathbf{I}_{ae1} = \mathbf{V}_{ae0}-\mathbf{Z}_f\mathbf{I}_{ae0}-\mathbf{Z}_g(-\mathbf{I}_{ae1}-\mathbf{I}_{ae2}+2\mathbf{I}_{ae0}).$$

From [*8.39*], $(-\mathbf{I}_{ae1}-\mathbf{I}_{ae2}) = \mathbf{I}_{ae0}$
therefore

$$\mathbf{V}_{ae1}-\mathbf{Z}_f\mathbf{I}_{ae1} = \mathbf{V}_{ae2}-\mathbf{Z}_f\mathbf{I}_{ae2} = \mathbf{V}_{ae0}-(\mathbf{Z}_f+3\mathbf{Z}_g)\mathbf{I}_{ae0}. \qquad [8.42]$$

Equations [8.39] and [8.42] are satisfied by the interconnection of the sequence networks shown in Fig. 8.15.

From this general shunt fault, the results for a number of simpler faults which are more likely to occur in practice may be obtained,

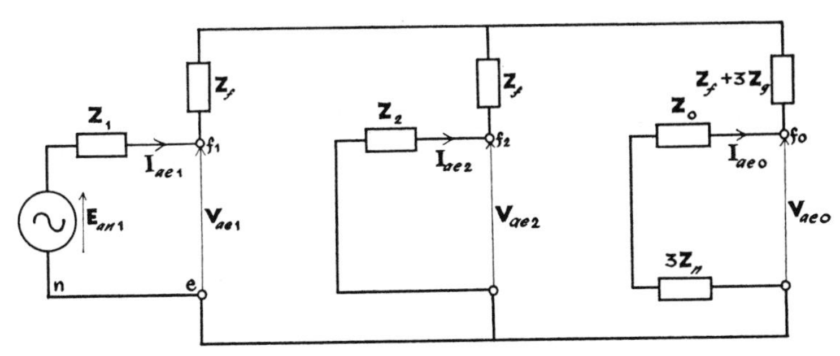

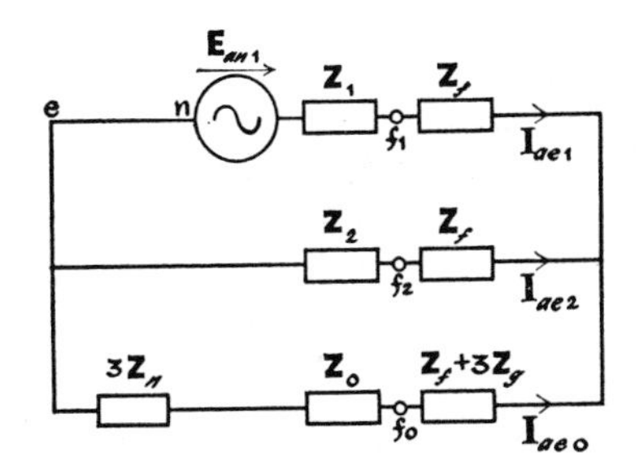

FIG. 8.15. *Interconnection of the sequence networks for a general double-earth fault on* b *and* c.

including those outlined in section 8.4.2 ($\mathbf{Z}_F = 2\mathbf{Z}_f$ and $\mathbf{Z}_g = \infty$), and in section 8.4.3 ($\mathbf{Z}_f = 0$ and $\mathbf{Z}_g = 0$).

8.4.5 GENERAL UNBALANCED THREE-PHASE-EARTH FAULT

Fig. 8.16 shows an earth fault on all three phases which is again symmetrical with respect to the reference phase *a*. The equations which may be written by inspection are:

$$\mathbf{V}_{ae} = \mathbf{Z}_f\mathbf{I}_{ae}+\mathbf{Z}_g(\mathbf{I}_{ae}+\mathbf{I}_{be}+\mathbf{I}_{ce})$$
$$\mathbf{V}_{be} = \mathbf{Z}_p\mathbf{I}_{be}+\mathbf{Z}_g(\mathbf{I}_{ae}+\mathbf{I}_{be}+\mathbf{I}_{ce})$$
$$\mathbf{V}_{ce} = \mathbf{Z}_p\mathbf{I}_{ce}+\mathbf{Z}_g(\mathbf{I}_{ae}+\mathbf{I}_{be}+\mathbf{I}_{ce}).$$

By transforming these equations into symmetrical components the student can deduce as an exercise that

$$\mathbf{V}_{ae1}-\mathbf{Z}_p\mathbf{I}_{ae1} = \mathbf{V}_{ae2}-\mathbf{Z}_p\mathbf{I}_{ae2} = \mathbf{V}_{ae0}-(\mathbf{Z}_p+3\mathbf{Z}_g)\mathbf{I}_{ae0} \qquad [8.43]$$

and

$$\mathbf{V}_{ae1}-\mathbf{Z}_p\mathbf{I}_{ae1} = (1/3)(\mathbf{Z}_f-\mathbf{Z}_p)(\mathbf{I}_{ae1}+\mathbf{I}_{ae2}+\mathbf{I}_{ae0}). \qquad [8.44]$$

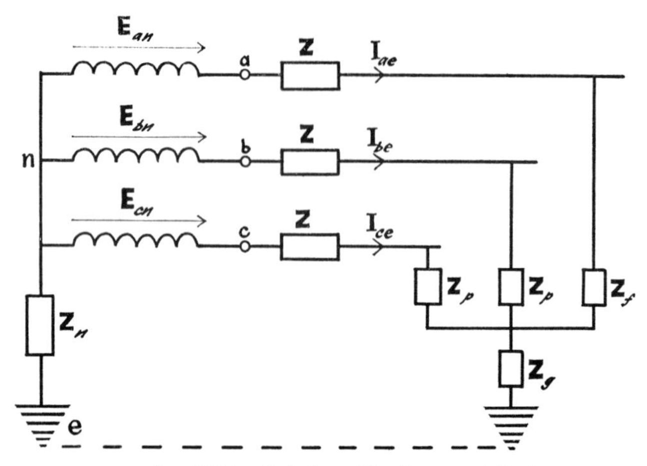

FIG. 8.16. *Unbalanced 3-phase earth fault.*

Equations [*8.43*] and [*8.44*] are satisfied by the interconnection of sequence networks in Fig. 8.17, from which the conditions for sections 8.4.2, 8.4.3 and 8.4.4, but not 8.4.1, may be deduced.

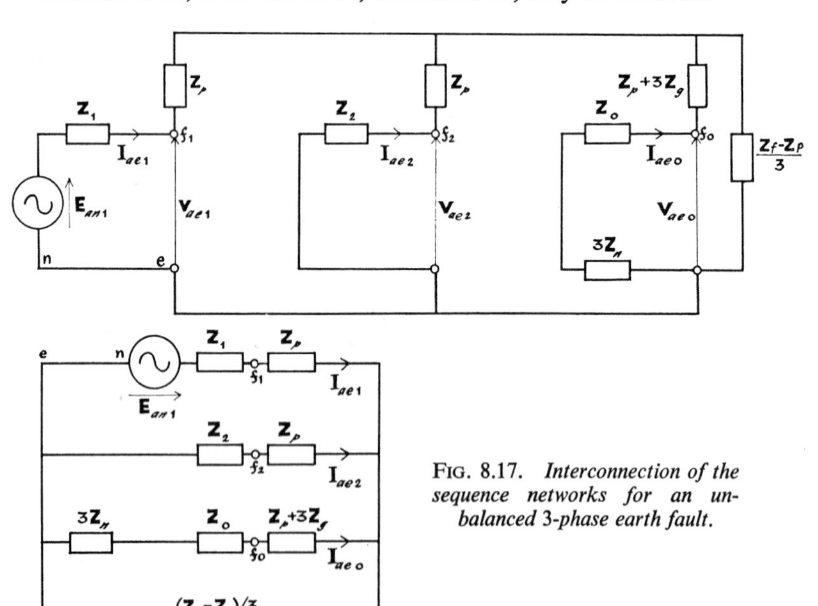

FIG. 8.17. *Interconnection of the sequence networks for an unbalanced 3-phase earth fault.*

8.4.6 SERIES FAULT ON PHASE *a*

If in one of the phase conductors a break occurs with an arc across the break, then this is equivalent to an impedance $\mathbf{Z}_r$ appearing in series with that phase with no corresponding impedance in the other

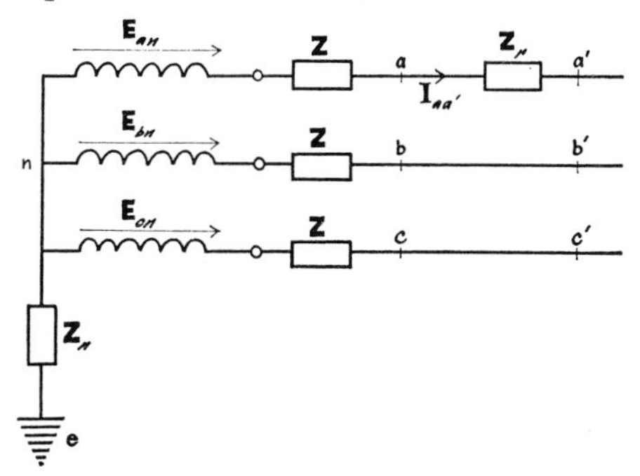

Fig. 8.18. *Series fault in phase* a.

two phases, as shown in Fig. 8.18. For this condition, we can write by inspection

$$\mathbf{V}_{aa'} = \mathbf{Z}_r\mathbf{I}_{aa'} \qquad [8.45]$$

$$\mathbf{V}_{bb'} = 0 \qquad [8.46]$$

$$\mathbf{V}_{cc'} = 0. \qquad [8.47]$$

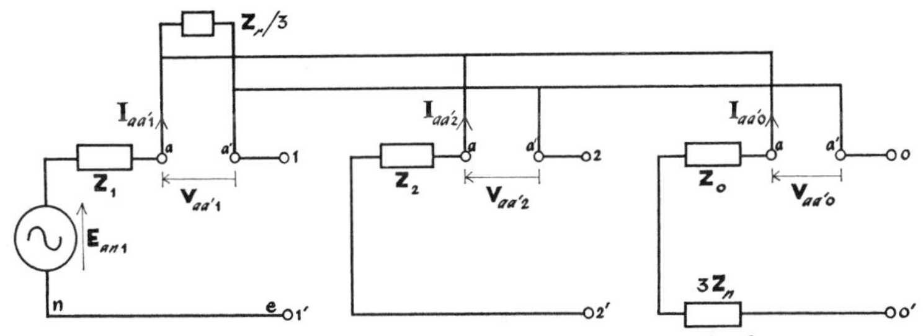

FIG. 8.19. *Interconnection of the sequence networks for a series fault in phase* a.

Transforming these equations into symmetrical component form gives

$$\mathbf{V}_{aa'1}+\mathbf{V}_{aa'2}+\mathbf{V}_{aa'0} = \mathbf{Z}_r(\mathbf{I}_{aa'1}+\mathbf{I}_{aa'2}+\mathbf{I}_{aa'0}) \qquad [8.48]$$

$$\mathrm{h}^2\mathbf{V}_{aa'1}+\mathrm{h}\mathbf{V}_{aa'2}+\mathbf{V}_{aa'0} = 0 \qquad [8.49]$$

$$\mathrm{h}\mathbf{V}_{aa'1}+\mathrm{h}^2\mathbf{V}_{aa'2}+\mathbf{V}_{aa'0} = 0. \qquad [8.50]$$

Equations [*8.49*] and [*8.50*] give $\mathbf{V}_{aa'1} = \mathbf{V}_{aa'2}$ and substitution of this result back in either equation gives

$$\mathbf{V}_{aa'1} = \mathbf{V}_{aa'2} = \mathbf{V}_{aa'0}. \qquad [8.51]$$

Substitution of [*8.51*] in [*8.48*] gives

$$\mathbf{V}_{aa'1} = (\mathbf{Z}_r/3)(\mathbf{I}_{aa'1}+\mathbf{I}_{aa'2}+\mathbf{I}_{aa'0}). \qquad [8.52]$$

Equations [*8.51*] and [*8.52*] are satisfied by the interconnection of the sequence networks as shown in Fig. 8.19. It should be noted that between terminals 1 and 1′, 2 and 2′, 0 and 0′, are connected the phase (line-to-neutral) positive-, and negative-, and zero-sequence impedances respectively of the circuit to the right of terminals *a*′, *b*′ and *c*′ in Fig. 8.18 (see Volume 2). If the arc at the broken conductor ends is extinguished then $\mathbf{Z}_r$ in Figs. 8.18 and 8.19 = ∞.

8.4.7 SIMULTANEOUS SHUNT AND SERIES FAULTS ON PHASE *a*

If the *a* phase conductor breaks it may touch earth, on the source side, with an impedance $\mathbf{Z}_f$ to earth, as shown in Fig. 8.20. For this

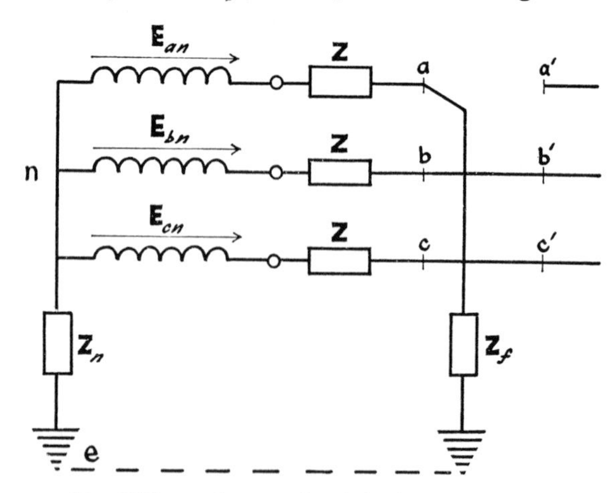

FIG. 8.20. a *phase conductor broken and earthed.*

condition [*8.24*], [*8.25*], [*8.51*] and [*8.52*] apply. In order to connect an equivalent circuit for the three sequence networks which satisfies all four equations, it is necessary to include ideal 1 : 1 isolating transformers as shown in Fig. 8.21.

Again, any power system connected to the right of *a′b′c′* is connected to the right of 11′, 22′, 00′. For example, if the right-hand system consists of live generation, its positive-, negative-, and zero-sequence networks should be connected across 11′, 22′ and 00′ respectively.

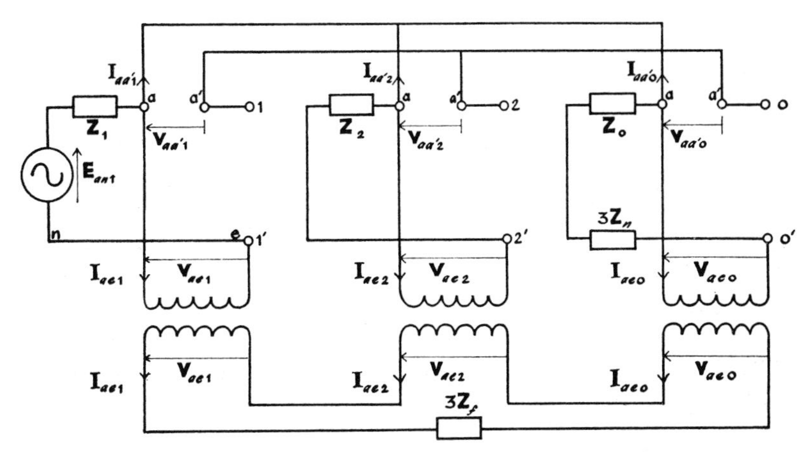

FIG. 8.21. *Interconnection of the sequence networks for a phase conductor broken and earthed.*

8.4.8 FAULT VOLT-AMPERES

The fault volt-amperes, usually termed the fault level and quoted in MVA, is the product of the fault current and the rated voltage of the system (not the voltage at the fault). In the case of an earth-fault, the voltage used is the line-to-neutral (or phase) voltage, while for a phase-fault the voltage used is the line-to-line voltage. In this context phase angles are ignored and only scalar values are considered.

It will now be shown that these volt-ampere fault-levels can be obtained directly from the fault equivalent networks, if the per-unit system is used. Let $\mathbf{Z}_t$ be the total impedance of the fault equivalent network so that for an earth-fault (Fig. 8.10)

$$\mathbf{Z}_t = \mathbf{Z}_1+\mathbf{Z}_2+\mathbf{Z}_0+3\mathbf{Z}_n+3\mathbf{Z}_f$$

and for a phase-fault (Fig. 8.12)

$$\mathbf{Z}_t = \mathbf{Z}_1+\mathbf{Z}_2+\mathbf{Z}_F.$$

Let V_r be the rated phase voltage of that part of the system on which the fault occurs. Then taking rated values as base values, the 3-phase base volt-amperes are

$$3V_rI_r = 3V_r^2/Z_r.$$

For an earth fault, [*8.28*] shows that

$$(\text{fault VA}) = 3V_r \,.\, E_{an}/Z_t$$

and in per-unit

$$(\text{fault VA})_{pu} = \frac{3V_r E_{an}}{(3V_r^2/Z_r)Z_t} = \frac{E_{an}/V_r}{Z_t/Z_r} = \frac{(E_{an})_{pu}}{(Z_t)_{pu}}. \qquad [8.53]$$

Similarly for a phase-fault, [*8.31*] and [*8.32*] show that

$$(\text{fault VA}) = \sqrt{3}\, V_r \,.\, \sqrt{3}\, E_{an}/Z_t$$

$$(\text{fault VA})_{pu} = \frac{3V_r E_{an}}{(3V_r^2/Z_r)Z_t} = \frac{E_{an}/V_r}{Z_t/Z_r} = \frac{(E_{an})_{pu}}{(Z_t)_{pu}}. \qquad [8.54]$$

If the e.m.f. behind the fault is taken to be equal to rated voltage (which is not far from the truth for the first few cycles after a fault), then as for a symmetrical 3-phase fault (section 8.2), [*8.53*] and [*8.54*] reduce to $1/(Z_t)_{pu}$. These fault volt-amperes are expressed relative to a 3-phase base volt-ampere value: this is logical since it is usual to quote base volt-amperes as a 3-phase value.

8.4.9 COMPARISON OF FAULT LEVELS

The relative magnitudes of fault currents and fault volt-amperes will now be compared, taking the 3-phase fault as the standard for comparison (1·0 p.u.). For simplicity it will be assumed that the system resistance is negligible and that $X_1 = X_2$: i.e. for alternators, the subtransient reactance is taken for X_1 (see section 7.5). Also, it will be assumed that the fault impedance is negligible. Since the main point of the exercise is to show that an earth-fault near a solidly-earthed source neutral-point can exceed that of a 3-phase fault, it will be assumed that the system is solidly earthed.

(*a*) 3-phase fault

$$\text{Fault current} = E_{an}/X_1$$
$$\text{Fault VA} = 3V_r E_{an}/X_1$$

(*b*) Earth-fault

$$\text{Fault current} = 3E_{an}/(X_1 + X_2 + X_0)$$

and expressed in per-unit of the 3-phase fault value

$$\text{fault current} = 3X_1/(2X_1 + X_0) \text{ p.u.}$$

If $X_0 = X_1$, then earth fault current $= 1{\cdot}0$ p.u.

If $X_0 < X_1$, then earth fault current $> 1{\cdot}0$ p.u.

$$\text{Fault VA} = 3V_r E_{an}/(X_1+X_2+X_0),$$

and again in per-unit of the 3-phase fault value we have

$$\text{fault VA} = X_1/(2X_1+X_0) \text{ p.u.}$$

This fault will be cleared on one pole of a 3-phase circuit breaker having three identical poles. Thus the 3-phase breaking VA rating of the circuit breaker will be

$$3X_1/(2X_1+X_0) \text{ p.u.}$$

Thus the conclusions are the same as for earth-fault current, namely, that if $X_0 = X_1$ the same circuit breaker will clear a 3-phase fault or an earth-fault, but if $X_0 < X_1$ an earth-fault requires a circuit breaker of larger rating. Since the zero-sequence network only extends from the point of fault back to the nearest transformer the primary of which is unearthed star-, or delta-connected, while the positive-sequence network extends right back to the source alternators, it can often happen that $X_0 < X_1$.

(*c*) Line-to-line short-circuit (phase-fault)

$$\begin{aligned} \text{Fault current} &= \sqrt{3}\, E_{an}/2X_1 \\ &= \sqrt{3}/2 = 0{\cdot}867 \text{ p.u.} \\ \text{Fault VA} &= \sqrt{3}\,.\,\sqrt{3}\, V_r E_{an}/2X_1 \\ &= 0{\cdot}5 \text{ p.u.} \end{aligned}$$

and since this fault is cleared on two poles of the circuit breaker, the corresponding 3-phase rating of the circuit breaker is

$$(3/2)0{\cdot}5 = 0{\cdot}75 \text{ p.u.}$$

This simple exercise illustrates why a phase-fault is relatively unimportant (in this context).

8.4.10 ASYMMETRICAL FAULT WHILST SYSTEM IS ON LOAD

Most elementary fault calculations assume that the fault occurs whilst the system is on no load with the alternators excited to the rated voltage of the system. If however the system is on load when the fault occurs at point f in Fig. 8.22, the fault current can be found by using the Thevenin (Helmholtz) theorem. The current in the fault path is found by open-circuiting the fault path and dividing the line-to-neutral voltage across the open-circuit (this will be $\mathbf{V}_f$ the (r.m.s.) voltage at the point of fault immediately prior to the fault) by the impedance of the passive network obtained by looking into the open-circuit (this will include the impedance of the fault path itself). The appropriate alternator reactance will be used—subtransient or transient. A symmetrical 3-phase fault on a loaded system has been dealt with in Section 7.5, and example 7.3 is an example of such a calculation. The polarity of $\mathbf{V}_i$ which creates the fault current should be carefully noted. $\mathbf{V}_i$ is superimposed upon the pre-fault $\mathbf{V}_f$ in order to reduce the total voltage at the fault to zero. Thus the conventional direction of the fault current is through the fault path from f to e. Assuming that the system was balanced prior to the fault, $\mathbf{V}_f$ will be a positive-sequence voltage only.

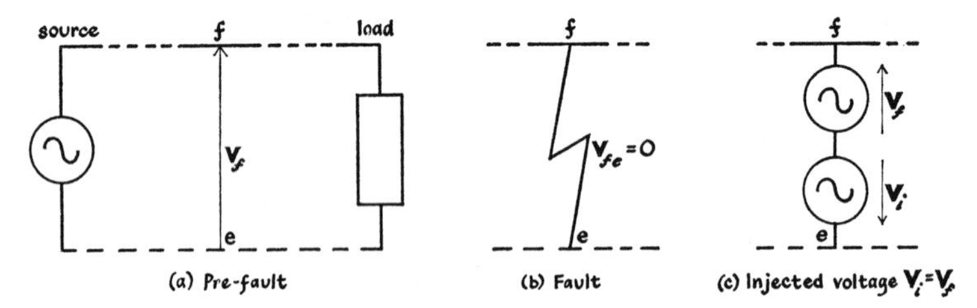

FIG. 8.22. *Application of Thevenin theorem at point of fault.*

The fault currents can be calculated by using the equations appropriate to the fault, e.g. [*8.26*] and [*8.27*] for an earth fault on phase a, [*8.31*] and [*8.32*] for a phase-fault, etc, with $\mathbf{V}_f$ in place of $\mathbf{E}_{an}$. This follows since Thevenin's theorem shows that for a fault on a linear system with balanced load, all the interconnections of the sequence networks (Figs. 8.10, 8.15, etc.) apply to their particular fault when $\mathbf{V}_f$ replaces $\mathbf{E}_{an}$ as the positive-sequence driving voltage. The resulting currents into the fault itself and through the system supplying it, may then be added by superposition to the original load currents in order to give the actual system currents during the fault period (subtransient or transient) under consideration.

Worked example 8.6

In Fig. 8.23a G is an exporting grid area (generator) having a 3-phase fault level of 20 000 MVA at the 400-kV busbars S. The corresponding data for the grid importing area M (motor) is 10 000 MVA at R. The short 400-kV line (L) has a series reactance of 40 ohms/phase. The load transfer is 150 MW, 0·8 power factor lagging, 360 kV at R. If a short-circuit to earth occurs on one phase at S, calculate the fault currents. For both areas during the subtransient period the negative-, and zero-sequence reactances may be assumed to be 100% and 50% respectively of the positive-sequence values

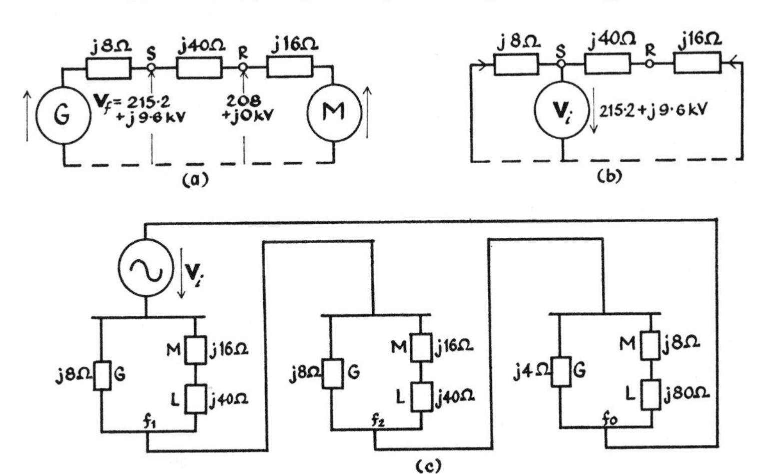

FIG. 8.23. *(a) Positive-sequence network before the fault. (b) Thevenin equivalent for positive-sequence. (c) Sequence interconnection.*

for the grid areas, and 100% and 200% for the line. Resistances may be neglected. Both G and M are solidly earthed.

SOLUTION

$$\text{Generator fault current} = 20\,000/(\sqrt{3}\times 400) = 28{\cdot}9 \text{ kA}$$

$$\text{Source reactance} = (400/\sqrt{3})/28{\cdot}9 = 231/28{\cdot}9 = 8\ \Omega/\text{phase}$$

$$\text{Internal reactance of motor} = 16\ \Omega/\text{phase}$$

Taking $\mathbf{V}_R$ as the reference phasor, $\mathbf{V}_R = 208\underline{/0^\circ}$ kV/phase

$$\text{Load current} = [150/(\sqrt{3}\times 360\times 0{\cdot}8)](0{\cdot}8-\text{j}\,0{\cdot}6)$$

$$= 0{\cdot}24-\text{j}\,0{\cdot}18 \text{ kA/phase}$$

$$\mathbf{V}_S = (208+\text{j}\,0)+(0{\cdot}24-\text{j}\,0{\cdot}18)(\text{j}\,40)$$

$$= 215{\cdot}2+\text{j}\,9{\cdot}6 \text{ kV/phase.}$$

This is the (balanced) voltage at the fault prior to the fault and is the value to be used for $\mathbf{V}_f$ in section 8.4.10.

Fig. 8.23(b) shows the Thevenin equivalent at the point of fault for the positive-sequence network of Fig. 8.23(a). The reactances of the positive-, negative-, and zero-sequence networks are j 7, j 7 and j 3·83Ω respectively, and their interconnection is shown in Fig. 8.23(c). Thus the fault current flowing from the *a* phase conductors into the fault at *S* is given by [*8.28*], with $\mathbf{E}_{an}$ replaced by $\mathbf{V}_f$ as discussed in section 8.4.10, as

$$\mathbf{I}_{ae} = \frac{3(215{\cdot}2+\mathrm{j}\,9{\cdot}6)}{\mathrm{j}\,7+\mathrm{j}\,7+\mathrm{j}\,3{\cdot}83} = 1{\cdot}62-\mathrm{j}\,36{\cdot}3 \text{ kA}.$$

The positive-sequence current in the *a* phase fed into the fault from the generator is

$$\frac{1{\cdot}62-\mathrm{j}\,36{\cdot}3}{3}\times\frac{56}{64} = 0{\cdot}472-\mathrm{j}\,10{\cdot}56 \text{ kA}$$

whilst that from the motor is

$$\frac{1{\cdot}62-\mathrm{j}\,36{\cdot}3}{3}\times\frac{8}{64} = 0{\cdot}067-\mathrm{j}\,1{\cdot}51 \text{ kA}.$$

The negative-sequence currents are the same, while the zero-sequence current in the *a* phase fed into the fault from *G* is

$$\frac{1{\cdot}62-\mathrm{j}\,36{\cdot}3}{3}\times\frac{88}{92} = 0{\cdot}517-\mathrm{j}\,11{\cdot}57 \text{ kA}$$

and that from *M* is

$$\frac{1{\cdot}62-\mathrm{j}\,36{\cdot}3}{3}\times\frac{4}{92} = 0{\cdot}023-\mathrm{j}\,0{\cdot}527 \text{ kA}.$$

To these fault currents must now be added the load current, which is wholly positive-sequence, in order to give the actual current in any part of the system. The positive-sequence current in phase *a* flowing from the generator, is

$$(0{\cdot}472-\mathrm{j}\,10{\cdot}56)+(0{\cdot}24-\mathrm{j}\,0{\cdot}18) = 0{\cdot}712-\mathrm{j}\,10{\cdot}74 \text{ kA}.$$

The positive-sequence current in phase *a*, flowing from the motor, is

$$(0{\cdot}067-\mathrm{j}\,1{\cdot}51)-(0{\cdot}24-\mathrm{j}\,0{\cdot}18) = -0{\cdot}173-\mathrm{j}\,1{\cdot}33 \text{ kA}.$$

Thus the actual current in phase *a*, flowing from the generator is

$$(0{\cdot}712-\mathrm{j}\,10{\cdot}74)+(0{\cdot}472-\mathrm{j}\,10{\cdot}56)+(0{\cdot}517-\mathrm{j}\,11{\cdot}57) = 1{\cdot}701-\mathrm{j}\,32{\cdot}87 \text{ kA}.$$

The actual current in phase *a*, flowing from the motor is

$$(-0{\cdot}173-\mathrm{j}\,1{\cdot}33)+(0{\cdot}067-\mathrm{j}\,1{\cdot}51)+(0{\cdot}023-\mathrm{j}\,0{\cdot}527) = -0{\cdot}083-\mathrm{j}\,3{\cdot}367 \text{ kA}$$

(or 0·083+j 3·367 kA to the motor).

The sum of these actual currents in phase *a* is

$$(1{\cdot}701 - \mathrm{j}\,32{\cdot}87) + (-0{\cdot}083 - \mathrm{j}\,3{\cdot}37) = 1{\cdot}618 - \mathrm{j}\,36{\cdot}24 \text{ kA}$$

which is the actual fault current $\mathbf{I}_{ae}$ already found.

The currents in the *b* and *c* phases of the system can be calculated using [*8.9*].

An alternative solution, which gives the actual currents (load plus fault) directly, can be obtained by calculating the e.m.f.s of the generator and the motor, prior to the fault, and by assuming that these e.m.f.s remain unchanged during the fault. The e.m.f.s of the generator and motor are

$$\mathbf{E}_G = (208 + \mathrm{j}\,0) + (0{\cdot}24 - \mathrm{j}\,0{\cdot}18)(\mathrm{j}\,48) = 216{\cdot}6 + \mathrm{j}\,11{\cdot}52 \text{ kV/phase}$$

$$\mathbf{E}_M = 205{\cdot}1 - \mathrm{j}\,3{\cdot}84 \text{ kV/phase.}$$

The sequence-interconnection network for the earth fault is similar to that of Fig. 8.23(c) except that there are now two separate sources of positive-sequence e.m.f. $\mathbf{E}_G$ and $\mathbf{E}_M$, each feeding into their own section of the positive-sequence network. If the whole network is reduced to an equivalent delta network, the student can check that the reactance between the e.m.f. sources is j 105·1 Ω, the reactance across the generator e.m.f. is j 20·39 Ω and the reactance across the motor e.m.f. is j 142·8 Ω. The current from the generator in phase *a*, is

$$(216{\cdot}6 + \mathrm{j}\,11{\cdot}52)/\mathrm{j}\,20{\cdot}39 = 0{\cdot}566 - \mathrm{j}\,10{\cdot}6 \text{ kA}$$

and that from the motor is

$$(205{\cdot}1 - \mathrm{j}\,3{\cdot}84)/\mathrm{j}\,142{\cdot}8 = -0{\cdot}0268 - \mathrm{j}\,1{\cdot}44 \text{ kA.}$$

The sum of these two currents is clearly the zero-sequence curren flowing in Fig. 8.23(c), and three times this current equals the fault current $\mathbf{I}_{ae}$ already found.

8.5 Symmetrical Component Filter Networks

These networks are used in 3-phase power systems to measure symmetrical components of current or voltage. A zero-sequence current usually indicates an earth fault on the system but it could also be due to unbalanced leakage currents over the surface of dirty insulators or to unbalanced line-to-earth charging currents. Zero-sequence voltage is used to polarise directional earth-fault relays (see section 9.4.4). Negative-sequence currents are liable to cause excessive heating of alternator rotors (see section 7.6). Positive-sequence voltage is sometimes used to operate the automatic voltage regulator (AVR) controlling the excitation of an alternator (see section 7.9). Only a few of the many available circuits will be discussed.

8.5.1 SEQUENCE CURRENT

Zero-sequence current can be measured by connecting an ammeter (or relay) across the three identical line current transformers (C.T.s) connected in parallel (residually connected) (see equation [*8.10*]). If the primary circuit is a 3-core cable, only one C.T. is necessary and is fitted over the cable, but the cable sheath and armour must be cut and the ends bonded together outside the C.T. These residually connected C.T.s measure residual current which equals three times the zero-sequence current.

A circuit for measuring negative-sequence current is shown in Fig. 8.24. Thus

$$(\mathbf{I}_a-\mathbf{I})\mathbf{Z}_a+(\mathbf{I}_c-\mathbf{I})\mathbf{Z}_c-\mathbf{IZ}=0$$

$$\mathbf{I}=(\mathbf{I}_a\mathbf{Z}_a+\mathbf{I}_c\mathbf{Z}_c)/(\mathbf{Z}_a+\mathbf{Z}_c+\mathbf{Z})$$

$$=\frac{\mathbf{I}_{a1}(\mathbf{Z}_a+\mathrm{h}\mathbf{Z}_c)+\mathbf{I}_{a2}(\mathbf{Z}_a+\mathrm{h}^2\mathbf{Z}_c)+\mathbf{I}_{a0}(\mathbf{Z}_a+\mathbf{Z}_c)}{\mathbf{Z}_a+\mathbf{Z}_c+\mathbf{Z}}.$$

Since $(\mathbf{Z}_a+\mathbf{Z}_c)$ cannot be zero the zero-sequence current is suppressed by using auxiliary, or interposing, C.T.s. A simple method is to insert two C.T.s (all four C.T.s being identical) in the *b* line and connect each in parallel, but with reversed polarity, across one of the existing C.T.s. The student should now analyse this modified circuit and show that the ammeter current is independent of the positive-sequence current if $\mathbf{Z}_a$ and $\mathbf{Z}_c$ are numerically equal and their impedance angles differ by 120°.

This condition must be satisfied without involving a negative resistance. One possible solution is

$$\mathbf{Z}_a = Z'\underline{/-90^\circ} \quad \text{and } \mathbf{Z}_c = Z'\underline{/+30^\circ} \text{ ohms}$$

where Z' is any ohmic value less than the rated ohmic burden of the C.T.s. If the ammeter impedance Z is made very much less than Z', the student should show that the negative-sequence current is one-third of the ammeter reading: he should also consider how to modify this circuit so that the ammeter gives a reading proportional to the positive-sequence current, independent of the negative-, and zero-sequence currents. (In Fig. 8.24 the C.T.s were placed arbitrarily in the *a* and *c* lines with the polarities shown. Is there a better choice, especially one for which one of the two impedances is a pure resistance?). The action of the above circuit should be checked by assuming that there is an earth fault on line *a*.

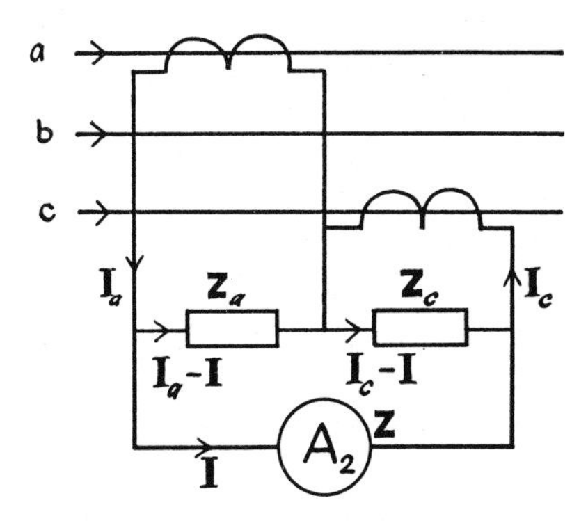

FIG. 8.24. *Circuit to measure negative-sequence current.*

8.5.2 SEQUENCE VOLTAGE

Zero-sequence voltage can be measured by using a 3-phase voltage transformer (V.T.) with a solidly-earthed, star-connected primary and an open (or broken) delta-connected secondary across which is connected a voltmeter. The voltmeter will measure the phasor sum of the three line-to-earth voltages (residual voltage) which equals three times the zero-sequence voltage. The V.T. must not be of the 3-limb, core-type construction (see section 6.10).

A circuit for measuring positive-sequence voltage is shown in Fig. 8.25. The two V.T.s are identical. Thus

$$\mathbf{V}_{ab} = \mathbf{I}_a\mathbf{Z}_a + (\mathbf{I}_a - \mathbf{I}_b)\mathbf{Z}$$
$$\mathbf{V}_{bc} = \mathbf{I}_b\mathbf{Z}_b - (\mathbf{I}_a - \mathbf{I}_b)\mathbf{Z}$$

Solving these equations, the current in the ammeter is given by

$$\mathbf{I}_a - \mathbf{I}_b = \frac{\mathbf{V}_{ab}\mathbf{Z}_b - \mathbf{V}_{bc}\mathbf{Z}_a}{\mathbf{Z}_a\mathbf{Z}_b + \mathbf{Z}(\mathbf{Z}_a + \mathbf{Z}_b)}$$

The three line voltages $\mathbf{V}_{ab}$, $\mathbf{V}_{bc}$, $\mathbf{V}_{ca}$ have no zero-sequence component. Taking $\mathbf{V}_{ab}$ as the reference phasor

$$\mathbf{I}_a - \mathbf{I}_b = \frac{\mathbf{V}_{ab1}(\mathbf{Z}_b - \mathrm{h}^2\mathbf{Z}_a) + \mathbf{V}_{ab2}(\mathbf{Z}_b - \mathrm{h}\mathbf{Z}_a)}{\mathbf{Z}_a\mathbf{Z}_b + \mathbf{Z}(\mathbf{Z}_a + \mathbf{Z}_b)}.$$

This equation is independent of the negative-sequence voltage when $\mathbf{Z}_b = \mathbf{Z}_a \underline{/120^\circ}$, i.e. the two impedances are numerically equal and their impedance angles differ by 120°. A possible solution is $\mathbf{Z}_a = Z' \underline{/-90^\circ}$ and $\mathbf{Z}_b = Z' \underline{/30^\circ}$. Assuming that the ammeter impedance can be neglected compared with Z' (which, being the burden on a V.T. will have a high ohmic value), the student should

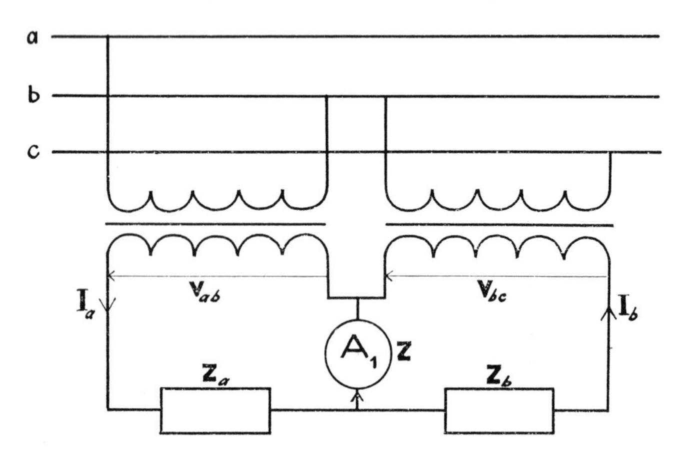

FIG. 8.25. *Circuit to measure positive-sequence voltage.*

show that the ammeter current is V_{ab1}/Z'. The action of the above circuit should be checked by assuming that the line voltages form a balanced 3-phase system, and that the ammeter impedance is zero.

REFERENCES

LACKEY, C. H. W. 1951. *Fault calculations.* Oliver and Boyd, Edinburgh.

MORTLOCK, J. R. & HUMPHREY DAVIES, M. W. 1952. *Power system analysis.* Chapman and Hall, London.

WAGNER, C. F. & EVANS, R. D. 1933. *Symmetrical components.* McGraw-Hill, New York.

WESTINGHOUSE (Ed.) 1964. *Electrical transmission and distribution reference book.* Westinghouse Electric Corporation, Pennsylvania, U.S.A.

Examples

In all the following examples, the student should assume, unless the contrary is stated, that all sources of e.m.f. are balanced, are excited to rated voltage, and are in phase, that resistance can be neglected, and that all per unit reactances for a machine are based on the rating of that machine and that the phase sequence is either *abc* or *RYB*.

1. Three 6·6-kV alternators of ratings 2000, 5000 and 8000 kVA and per-unit reactances 0·08, 0·12 and 0·16 respectively are connected to a common busbar. From the busbar, a feeder cable of reactance 0·125 Ω connects to a sub-station. Calculate the fault MVA, if a 3-phase symmetrical fault occurs at the sub-station.

(87·2 MVA)

2. A 3-phase transformer is connected star/star. It supplies a star load of 400+j 600 Ω/phase through a transmission line, each conductor of which has impedance 4+j 6 Ω. The secondary winding of the transformer has three times as many turns as the primary.

$$\text{Primary } R = 0{\cdot}5, \quad X = 2{\cdot}5\ \Omega/\text{phase}$$
$$\text{Secondary } R = 5{\cdot}0, \quad X = 25{\cdot}0\ \Omega/\text{phase.}$$

The transformer is fed from an alternator rated at 1500 kVA, 11 kV, 0·2 p.u. transient reactance. Find the transformer secondary terminal voltage when the primary terminal voltage is 11 kV.

If a dead short-circuit now occurs on all phases half way along the line find the alternator current. The alternator e.m.f. behind its transient reactance can be assumed constant.

(31 kV, 340 A)

3. Three star-connected 11-kV alternators are connected each in series with a similar current-limiting reactor to a common busbar. The alternators each have a rating of 10 MVA and sub-transient reactance/phase of 0·06 p.u. Two 11/33-kV transformers of 15 MVA rating, 0·03 p.u. reactance and 10 MVA 0·02 p.u. reactance respectively, connected in parallel to this busbar, supply a transmission line of impedance 0·2+j 0·7 Ω/km. At a sub-station 10 km from the generating station is a 25 MVA 33/11-kV transformer of 0·06 p.u. reactance. Calculate the reactances of the current-limiting reactors if each alternator is not to carry more than $2\frac{1}{3}$ times full-load current, when a symmetrical short circuit occurs on the 11 kV busbars in the sub-station.

(University of London) (0·837 Ω)

4. A power station contains three identical 3-phase 50-Hz alternators each of 200 MVA and 0·9 per unit synchronous reactance. These machines are running equally loaded and to each is connected an 11/132-kV transformer of 200 MVA and 0·09 per-unit reactance. The 132-kV busbar feeds a short 132-kV transmission line having a resistance of 0·5 Ω and an inductance of 8 mH for each phase conductor. The resistances of the machines and the transformer can be neglected.

A load of 400 MW at 0·8 p.f. lagging is supplied by the line. If one of the phase voltages at the load end of the line is taken as the reference of $1{\cdot}0\underline{/0^\circ}$ per unit, calculate the e.m.f. of the corresponding phase in any one of the alternators.

If a 3-phase short-circuit occurs at the load end of the transmission line, calculate the steady-state fault current at this point.

(University of Leeds) ($1{\cdot}703\underline{/24^\circ\ 20'}$, 4130 A)

5. Three 11-kV alternators rated at 10 000 kVA, with a resistance of 0·02 and subtransient reactance of 0·15 per unit, have their busbars connected by three 3-phase mesh-connected reactors, each having 0·015 per-unit resistance and 0·09 per-unit reactance/phase on a 10 000 kVA rating. A 20 000 kVA transformer of 0·01 per-unit resistance and 0·10 per-unit reactance is connected to the busbars of one machine and feeds at 132 kV, a transmission line which has 9 Ω resistance and 30 Ω reactance/phase. Find the currents which flow in each of the machines when all three phases are shorted together at the load end of the line.

If an oil circuit breaker to be installed at the point of the fault is to have its rating determined by this fault condition, what must be its minimum rating?

(University of London) (1725 A, 1075 A, 1075 A, 73·7 MVA)

6. A power station has three section busbars, with 11-kV generators, connected on the ring system as follows:

Section 1, two generators of 30 and 20 MVA with 0·15 and 0·12 per-unit reactance respectively;

Section 2, one generator of 60 MVA and 0·2 per-unit reactance;

Section 3, two generators of 30 and 60 MVA with 0·18 and 0·2 per-unit reactance respectively.

The sections 1 and 2, 2 and 3, and 3 and 1 are connected through reactors of ratings 60, 90 and 80 MVA with reactances of 0·2, 0·3 and 0·24 per unit respectively. Determine the fault current when a 3-phase symmetrical fault occurs at the far end of a feeder with a 0·05 per-unit reactance and 0·01 per-unit resistance and rating 30 MVA, connected to number three section busbar; and also the current supplied by the 60 MVA generator on section three busbar under this condition.

(University of London) (17 800 A, 6760 A)

7. The line currents in a 3-phase 3-wire system are $\mathbf{I}_a = 60+j\,40$, $\mathbf{I}_b = -80+j\,20$, $\mathbf{I}_c = 20-j\,60$ A. Calculate the symmetrical components of these currents for phase *a*.

$(\mathbf{I}_1 = 6{\cdot}9-j\,8{\cdot}87,\ \mathbf{I}_2 = 53+j\,48{\cdot}8,\ \mathbf{I}_0 = 0)$

8. A 3-phase system of voltages is given by $\mathbf{V}_a = 1000\underline{/35^\circ}$, $\mathbf{V}_b = 3000\underline{/100^\circ}$, $\mathbf{V}_c = 2000\underline{/270^\circ}$. Resolve these voltages into their symmetrical components for phase *a*.

$(\mathbf{V}_1 = 1077\underline{/186^\circ\,20'},\ \mathbf{V}_2 = 1800\underline{/5^\circ\,40'},\ \mathbf{V}_0 = 516\underline{/79^\circ})$

9. A 132/11-kV 5-limb transformer has an equivalent *T* circuit comprising a leakage reactance/phase on primary and secondary of 20 Ω and 0·2 Ω, and a magnetising reactance of 2160 Ω on the primary side. All resistances can be neglected. The supply has $X_1 = 15\ \Omega$, $X_2 = 10\ \Omega$, $X_0 = 7\ \Omega$/phase. Calculate the fault current when an earth fault occurs on one secondary terminal, for the following connections: (1) delta/star with earthed star-point, (2) star with earthed star-point/delta, (3) star with earthed star-point/star with earthed star-point, (4) star/star, and (5) star/star with earthed star-point.

(20 900 A, 0, 15 400 A, 0, 1185 A)

10. Two 3-phase alternators generating e.m.f.s unequal in magnitude but of equal phase difference are connected to the same busbars. Calculate the circulating current in the red phase and the voltage to earth at the star-point of alternator *B*. Alternator *A* has its star-point solidly earthed while that of *B* is earthed through a 1 Ω reactance.

Generated e.m.f./phase	*R*	*Y*	*B*
Alternator *A*	1000	900	850
Alternator *B*	900	1100	900

Sequence reactances Ω	X_1	X_2	X_0
Alternator *A*	4	2	1·0
Alternator *B*	6	5	1·5

(University of London) (19·35 A, 46·8 V)

11. A circuit breaker is to be installed to protect a 3-phase, star-connected, 11-kV alternator which has, at the instant of fault initiation, positive-, negative- and zero-sequence reactances of 10, 7 and 3 Ω. The alternator neutral is earthed through a 1 Ω reactance. If the breaker rating is to be determined by the greatest current that can flow under zero-impedance fault conditions, what will be this rating? What is the maximum voltage of the alternator neutral above earth?

(University of London) (15·8 MVA, 828 V)

12. If a line-to-line short-circuit occurs between phases *b* and *c* of a 132-kV system, calculate the fault current and the voltage to neutral of phase *a* at the point of fault. The positive-, negative- and zero-sequence reactances from generation to fault are respectively 12, 10 and 8 Ω. Resistance may be neglected.

(University of Leeds) (6000 A, 69·2 kV)

13. The positive-, negative-, and zero-sequence reactances/phase of a 3-phase synchronous generator are, respectively, 0·6 Ω, 0·5 Ω, and 0·2 Ω. The winding resistances are negligible. A short-circuit occurs between the *Y* and *B* lines and earth at the generator terminals. The sequence of the generator phase voltages is *RYB*, and the *R* phase e.m.f. is (6000+j 0) V.

Calculate from first principles the three sequence currents in the *R* phase, and the current in the earth return circuit, (*a*) if the generator neutral point is solidly earthed and (*b*) if the neutral point is isolated. Give in each case a phasor diagram to show the sequence currents in relation to the *R* phase e.m.f.

(University of London) (−j 8080, j 2310, j 5770, j 17 310 A: −j 5640, j 5460, 0, 0 A)

14. A single-line-to-earth fault of 0·05 Ω resistance occurs on a three-phase system supplied by a synchronous generator with a generated e.m.f. of 11 kV between lines. The positive-, negative-, and zero-sequence reactances of the system from the fault to the generator

are 0·5 Ω, 0·2 Ω and 0·1 Ω respectively; the resistances are negligible. The generator neutral is earthed through a 0·2 Ω resistor.

From first principles calculate the fault current.

(University of London) (17 400 A)

15. An unloaded 3-phase, star-connected synchronous generator has constant balanced e.m.f.s giving 11·0 kV between the pairs of line terminals *RYB*. The positive- and negative-sequence impedances, considered constant, are respectively (0+j 0·9) Ω and (0+j 0·4) Ω/phase. Determine from first principles the current that will flow in a short-circuit between the *Y* and *B* terminals, taking the *R* phase e.m.f. as the reference phasor. Calculate also the potential differences between the pairs of line terminals.

(University of London) (8450 A, 5865 V)

16. A fault between the yellow and blue phases occurs on the busbars of an alternator which has, at the instant of fault initiation, the following sequence impedances/phase:

$$\mathbf{Z}_1 = 0+\mathrm{j}\,1{\cdot}0\,\Omega\text{: } \mathbf{Z}_2 = 0{\cdot}1+\mathrm{j}\,0{\cdot}2\,\Omega\text{: } \mathbf{Z}_0 = 0+\mathrm{j}\,1{\cdot}0\,\Omega.$$

The generator phase e.m.f. is 1000 V.

Calculate from first principles, the line currents and the line-to-neutral voltage of the unfaulted phase.

(University of London) (1440 A, 373 V)

17. An existing generating station has a generator of 5000 kVA with 8% sub-transient reactance and a generator of 4000 kVA with 6% sub-transient reactance, connected in parallel to 11-kV, 50-Hz, 3-phase busbars. Existing switchgear in feeders leading off from the busbars is rated at 180 MVA.

It is desired to extend the station by connecting a 'grid' supply on to the station busbars through a 10 000 kVA transformer having 10% leakage reactance. A current-limiting reactor is to be installed at the point of connection of this supply to the station busbars to safeguard the existing switchgear. Calculate the value, in ohms, required for this reactor.

Under symmetrical, zero-impedance fault conditions at the busbars, calculate the voltage drop across the reactor at the instant of short-circuit.

(University of London) (1·185 Ω, 3·15 kV)

18. A 3-phase, 50-Hz, 3000-kVA generator, having its star-point resistance-earthed, is connected to 6·6-kV busbars. From the busbars a star/star connected 6·6/33-kV step-up transformer T_1, having only its star-point on the 33-kV side solidly earthed, feeds into a cable. At the far end of the cable a sub-station is fed through a delta/star connected 33/6·6-kV step-down transformer T_2 having its star-point solidly earthed.

Calculate the fault current developing at the sub-station l.v. bars for (*a*) a line-to-line fault of zero impedance, (*b*) a line-to-earth fault of resistance 1 Ω.

Values of the sequence impedances in ohms are given below:

	$\mathbf{Z}_1$	$\mathbf{Z}_2$	$\mathbf{Z}_0$	
Generator	j 2	j 1·5	j 0·5	
Transformer T_1	j 1·8	j 1·8	j 18·0	} referred to the
Transformer T_2	j 1·6	j 1·6	j 1·6	} 6·6 kV side.
Cable	j 0·3	j 0·3	j 0·3	

(University of London) (640 A, 934 A)

19. The 11-kV busbars of a power station are in 3 sections, *A*, *B*, and *C*, and these are interconnected by a ring of reactors as follows: between *A* and *B*, 10% on 20 MVA; between *B* and *C*, 8% on 20 MVA; and between *C* and *A*, 12% on 20 MVA. On each section there is one alternator rated at 30 MVA and having a reactance of 20%. On section *B* there is a transformer rated at 45 MVA, 11/132 kV, having a reactance of 12·5%. Calculate the 3-phase symmetrical fault level on the h.v. side of the transformer.

(L.C.T.) (172 MVA)

20. The busbars of a 33-kV power station are in two sections *A* and *B* separated by a reactor of reactance 1·5 Ω/phase. On section *A* there are two 20 MVA sets each having a (transient) reactance of 18% on 20 MVA. Section *B* is fed from the 132-kV grid via two identical 25-MVA transformers, in parallel, each having a leakage reactance of 12·5% on 25 MVA. The 3-phase fault level infeed at the grid is 2500 MVA. Calculate the 3-phase fault level MVA at *A* and at *B*. What reactance would each 25-MVA transformer require to have to reduce the fault level at *B* to 500 MVA?

(L.C.T.) (454, 514 MVA, 19·7%)

21. An 11-kV busbar is in four sections, 1, 2, 3, and 4. Generators 1

and 4, each rated at 50 MVA and having a reactance of 16% are connected one to sections 1 and 4: similarly generators 2 and 3 each rated at 25 MVA and having reactances of 14% are connected one each to sections 2 and 3. The four sections are each connected by a reactor rated at 30 MVA and of reactance 5% to a tie-bar. Sections 2 and 3 are each connected to the 132-kV grid busbar by a transformer rated at 45 MVA and having reactances of 12·5%. The 3-phase fault level infeed at the grid busbar is 2000 MVA. Calculate the 3-phase fault level MVA at section 1 and at section 2; also the MVA outfeed from the station to a fault on the grid busbars.

(L.C.T.) (639, 782, 346 MVA)

22. The two sections of an 11-kV busbar are connected by a reactor with a through-load rating of 25 MVA and a reactance/phase of 10% based on 25 MVA. Each section is connected, by a 20-MVA transformer having a reactance of 10% on 20 MVA, to a 33-kV busbar which is not sectionalised. Connected to this 33-kV busbar are two generator-transformer sets each rated at 37·5 MVA and each having a total reactance of 25% on 37·5 MVA. The 33-kV busbar is connected to the 132-kV grid busbar by two identical transformers in parallel each rated at 45 MVA and having a reactance of 12·5% on 45 MVA. The 3-phase fault level infeed from the grid is 2000 MVA. Calculate the 3-phase fault level MVA at the 11-kV busbar with and without the reactor (reactor short-circuited), and at 33-kV busbar. Calculate the 3-phase fault MVA outfeed from the station to a fault on the 132-kV grid busbar.

(L.C.T.) (500, 685, 1028, 297 MVA)

23. An exporting grid area has a 3-phase, transient fault level of 20 000 MVA (at 400 kV) at the sending-end of a 400-kV interconnector. The importing grid area at the receiving end has a fault level of 10 000 MVA (at 400 kV). The interconnector has a series reactance of 40 Ω/phase. The load transfer is 150 MW, power factor 0·8 lagging, 360 kV measured at receiving end. If a 3-phase fault occurs at the sending-end of the interconnector, calculate the current in the fault path, from the exporting area and to the importing area, taking the 360 kV as the reference phasor. Assume the e.m.f.s behind the transient reactances are constant.

(L.C.T.) (1·372−j 30·8, 1·44−j 27·13, 0·068+j 3·67 kA)

24. An 11-kV star-connected generator of phase sequence *abc* is connected to a 132/11-kV star/delta, *Yd*1 transformer. The load on the 132-kV side consists of a balanced load of 40 MW at 0·8 power factor lagging across *A*, *B*, *C*, and a single phase load of 10 MW at 0·9 power factor lagging across *B* and *C*. Neglecting all losses, calculate the positive-, and negative- sequence components of current, and the line current in the generator phase *a*.

(I.E.E.) (3,200; 583; 3,030 A)

25. In a 3-phase 4-wire supply of phase sequence *RYB*, the positive-, and negative-sequence components of current in the *R* line, flowing towards the load, are respectively $200\underline{/0^\circ}$ A and $100\underline{/60^\circ}$ A. The current in the neutral flowing towards the supply, is $300\underline{/300^\circ}$ A. Calculate the three line currents in magnitude and phase. If the supply e.m.f.s form a symmetrical system, and if the power factor of the *R* phase load is 0·867 leading, calculate the power factors of the other two line currents.

(I.E.E.) ($300\underline{/0^\circ}$, $300\underline{/-120^\circ}$, 0 A; 0·867 leading, —)

26. A 3-phase 132-kV system may be resolved into a solidly-earthed source feeding a 132/33-kV star/delta transformer whose star point is also solidly earthed. If an earth fault occurs on one of the 132-kV transformer terminals when the 33-kV side is not connected to a load, determine the fault current to earth and the current in the transformer delta winding. The sequence reactances based on 100 MVA are as follows:

	Source	Transformer
Positive:	20%	15%
Negative:	15%	
Zero:	10%	

(L.C.T.) (3200, 985 A)

27. A 3-phase, 150-MVA, 13-kV alternator has positive-, negative-, and zero-sequence reactances of 20, 15 and 5% respectively based on machine rating. If it is running on no-load at rated voltage, calculate the value in ohms of the earthing resistor to limit a terminal earth-fault to 75 A. The alternator is now excited to 1·1 p.u. voltage and connected to a 140-MVA, 12/275-kV delta/solidly-earthed star transformer having a leakage reactance of 12% on its rating. The

star-point is connected to an earth-bar by a copper strap of negligible impedance but the connection between the earth-bar and true earth will be deemed to have a resistance of one ohm. Calculate the voltage of the earth-bar to true earth during an earth-fault on a 275-kV terminal.

(L.C.T.) (1·41 kV)

28. A 33-kV busbar has a 3-phase fault infeed level of 1000 MVA. The negative-, and zero-sequence source reactances are 2/3 and 1/3 of the positive-sequence reactance: the zero-sequence source resistance is 60 Ω. A 30-MVA 132/33-kV solidly-earthed star/delta connected transformer having a reactance of 0·1 per unit is fed from the 33-kV busbars. Calculate the fault current in kA and fault MVA at the 132-kV terminals for the following types of fault: (i) 3-phase fault, (ii) earth-fault, (iii) phase-to-phase fault, and (iv) phase-to-phase to earth (double earth) fault. Calculate the MVA ratings of the circuit breakers needed to clear each fault.

(L.C.T.) (1·01, 231; 1·12, 85·8; 0·91, 120;
$I_b = I_c =$ 1·07, $I_e =$ 1·164, 163; 231, 257, 180, 246)

29. A 3-phase, star-connected alternator is earthed by a 100% resistor, has a rating of 37·5 MVA, 11 kV and positive-, negative-, and zero-sequence reactances of 20, 15 and 5% respectively. It is connected to a solidly-earthed star/delta, 66/11-kV Yd1 transformer rated at 25 MVA and having a leakage reactance of 10%. If an earth fault occurs on a 66-kV terminal, calculate the fault current and hence the fault current on the 11-kV side. Check the answer by transferring the sequence components of the (66 kV) fault current through the transformer.

(L.C.T.) (1230, 4260 A)

30. The 3-phase fault level at the 33-kV busbars of a power station is 1500 MVA and it may be assumed that $X_1 = X_2 = X_0$. Two identical solidly-earthed star/delta transformers rated at 45 MVA, 132/33 kV operating in parallel are now connected to the 33-kV busbars. Each transformer has a leakage reactance of $12\frac{1}{2}$% (on rating). (*a*) For a 3-phase fault on the 132-kV busbars, calculate the fault MVA and fault current. (*b*) For an earth-fault on the 132-kV busbars calculate the fault current. (*c*) If the two 132-kV star points are solidly connected to an earth bar which is then earthed via an

earth electrode system having a resistance to true earth of 10 ohms, recalculate the earth fault current assuming (i) both star-points earthed and (ii) only one star-point earthed (accidental open-circuit on the other).

(L.C.T.) (487 MVA; 2·13, 2·38, 2·27, 1·85 kA)

31. A solidly-earthed 33-kV source has a 3-phase fault level of 500 MVA and it may be assumed that it has equal reactances to all three sequences. It feeds a star/solidly-earthed star/tertiary delta, 33/6·6/1·5-kV, 5-MVA transformer having leakage reactances between pairs of windings as follows: primary-secondary 8%, primary-tertiary 13%, secondary-tertiary 4%, all based on 5 MVA. The primary is connected to the 33-kV source. If an earth fault occurs on the 6·6-kV busbars, calculate the earth fault current if the primary star point is (*a*) isolated, (*b*) solidly earthed.

(L.C.T.) (5·72, 6·4 kA)

32. Two identical star/star/tertiary delta transformers operating in parallel have phase turns ratios of 1 : 1 : 1. Their primaries are unearthed. Their leakage reactances per phase between pairs of windings are: primary-secondary 1/2 Ω, primary-tertiary 3/4 Ω, secondary-tertiary 1 Ω, all referred to the first-mentioned winding. For an earth fault on the *R* busbar on the secondary side, calculate the fault current distribution throughout the system for (*a*) only one secondary star point earthed, and (*b*) both secondary star points earthed but one primary circuit breaker open. The supply e.m.f. is 6 kV/phase and has zero internal impedance. The answers given are for (*a*) the current to earth and the current in the faulty secondary phase of the earthed transformer, and for (*b*) the current to earth and the current in the faulty secondary phase of the transformer whose primary circuit breaker is open.

(L.C.T.) (12, 8; 12, 2 kA)

33. A 3-phase, star-connected, solidly-earthed alternator is excited to 10 kV/phase and has positive-, negative-, and zero-sequence reactances of 10, 8 and 6 Ω/phase respectively. It is connected by a short cable of zero impedance to the primary of a star/star/tertiary delta transformer whose star points are both solidly earthed and having leakage reactances between pairs of windings of: primary-secondary 15, primary-tertiary 12, and secondary-tertiary 5 Ω/phase,

all on a 10 kV/phase voltage level. If an earth fault occurs on phase *a* of the cable, calculate (*a*) the fault current distribution throughout the 10 kV/phase system assuming no infeed from the secondary side of the transformer, and (*b*) the fault current to earth if the secondary is connected to a solidly-earthed infinite busbar excited to 10 kV/phase. (All the answers are reactive lagging currents in amperes flowing towards the fault, in the order *a*, *b*, *c*, and relative to $\mathbf{E}_{an}$.)

(L.C.T.) ((*a*) 1365 to fault, alternator 1213, −152, −152, 909 from earth transformer 152, 152, 152, 456 from earth; (*b*) 1971)

34. With the same data as in example 33, but with a double earth fault ($b-c-e$) on the cable, recalculate the current distribution assuming no infeed from the secondary side. The answer gives the total fault current flowing towards earth, relative to $\mathbf{E}_{an}$.

(L.C.T.) (+j 1581 A)

35. A star/star/tertiary delta, 275/132/33-kV transformer has a primary and secondary each rated at 180 MVA: the tertiary has a short-time fault rating but no load rating. Three-phase short-circuit tests carried out on pairs of windings, with the third winding on open-circuit, gave the following results, the input being into the first-mentioned winding:

primary-secondary: 22·08 kV, 300 A
primary-tertiary: 32·85 kV, 300 A
secondary-tertiary: 14·03 kV, 700 A.

Calculate the reactance of the primary, secondary and tertiary windings of the equivalent circuit, on a base of 100 MVA, and estimate the rating of the test plant needed to carry out the above tests. Neglect resistance.

(L.C.T.) (3·66, 1·96, 4·71 %; 20·2 MVA)

36. A 3-phase generator having a transient reactance of 0·2 p.u. feeds a motor having a transient reactance of 0·5 p.u., via a line having a reactance of 0·05 p.u. The load is $0{\cdot}95\underline{/0^\circ}$ p.u. voltage, 0·75 p.u. current at unity power factor, all measured at the terminals of the motor. If a 3-phase short-circuit occurs at the motor terminals, calculate the total current from the generator and to the motor. Assume the e.m.f.s behind the transient reactances are constants.

(L.C.T.) (0·75 − j 3·8, 0·75 + j 1·9 p.u.)

37. The following data refers to an alternator-transformer set: Alternator rated 120 MW, power factor 0·8 lagging, $X_d = 188\%$, $X_d' = 23\%$, $X_d'' = 16\%$, (all on rating). Transformer rated 135 MVA, $X = 15{\cdot}3\%$ (on rating). With the alternator on full-load current with a power factor of 0·8 lagging measured at the h.v. terminals of the transformer, and the voltage at the same point equal to $1{\cdot}06\underline{/0^\circ}$ p.u., calculate the e.m.f.s behind the synchronous, transient and sub-transient reactances; also the steady-state short-circuit current (3-phase short-circuit at h.v. terminals) and the corresponding e.m.f. behind the transient reactance. Given that the e.m.f. behind the transient reactance decays exponentially between the two values already calculated, with a time-constant of 1·75 seconds, calculate the e.m.f. behind the transient reactance 0·12 second after fault initiation with the machine on load and hence find the corresponding fault current (r.m.s. p.u.). Assume the e.m.f. behind the synchronous reactance is constant; neglect d.c. transient.

(L.C.T.) (2·29+j 1·63, 1·3+j 0·32, 1·26+j 0·26; 0·8−j 1·12, 0·44+j 0·31; 1·27, 3·2)

38. A 3-phase synchronous generator is connected via a 4-wire feeder to a 10-MW 11-kV balanced heating load at unity power factor. The generator and feeder together have impedances of (0+j 0·4), (0+j 0·2), (0+j 0·2) p.u. to positive-, negative-, and zero-sequence currents respectively on a base of 11 kV, 10 MVA. A break occurs in the *R* conductor at the load. Determine (*a*) the voltage across the break, and (*b*) the currents in the *Y*, *B*, and neutral conductors. Assume the generator's e.m.f.s remain constant.

(I.E.E.) (6·64 kV: 536, 536, 536 A)

Chapter 9

PROTECTION

9.1 Introduction

If a fault (or any other abnormal condition) arises in an electrical power system, that fault must be isolated as quickly as possible from all live supplies in order (*a*) to retain system stability (Chapter 7 and Volume 2) i.e. to keep all generators running in synchronism, and (*b*) to reduce the damage both at the point of fault, due to fire and explosion, and in the parts of the system carrying the fault current, due to overheating.

Protection is the art or science of detecting the presence of a fault and initiating the correct tripping of the circuit breakers (C.B.s) (see Volume 2). A protective system comprises protective current transformers (C.T.s), voltage transformers (V.T.s) and protective relays with their associated wiring (referred to as the a.c. part of the system), and the relay contacts which close a circuit from the station battery to the C.B. trip coil (referred to as the d.c. part of the system).

A complete power system is divided into zones—associated, for example, with an alternator, a transformer, a busbar section, or a feeder—and each zone has one or more co-ordinated protective systems connected to it (referred to altogether as a protective scheme). Protection is sometimes likened to an insurance premium, the capital and maintenance costs of the scheme being the premium, while the return is the possible prevention of loss of system stability and the minimisation of damage. For very expensive plant, this premium can easily be justified, but for remote lightly loaded zones where the cost of protection could represent several years' profit from the revenue, the reverse is the case. Nevertheless, consideration must be given to loss of consumer revenue and goodwill if he loses supply for long periods. A major factor is possible danger to human life.

The modern tendency is to make protection systems static, i.e. to remove all moving parts by using, for example, transductors and solid-state electronic devices. Some of the systems described in this chapter are semi-static in the sense that the only moving part is the final output relay (which trips the circuit breaker) which is usually an attracted-armature (Post Office type) relay.

In this chapter the treatment of protection is mainly descriptive and explains basic principles rather than the details of existing systems —for these latter the student is referred to manufacturers' pamphlets. Also, for the many practical details which cannot be dealt with in this text, the student is referred to the appropriate British Standard (B.S.) Specifications listed at the end of this chapter.

9.2 Current Transformers (C.T.s.)

An important component in all protective systems is a C.T., which reduces the large current in the primary circuit to a much smaller value in the secondary circuit. A C.T. also insulates the secondary

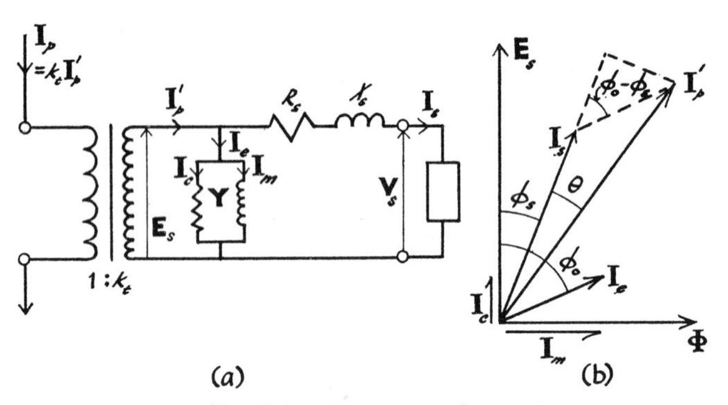

FIG. 9.1. *Current transformer.*

(*a*) *Equivalent circuit* (*b*) *Phasor diagram*

circuit, including the protective relays, from the high voltage of the primary circuit. The secondary circuit is earthed at one point. The rated primary and secondary currents of a C.T. are those values marked on the rating plate (this applies to any rated quantity). The performance of the C.T. is given at stated fractions or multiples of the rated current (e.g. at 20 times rated current). B.S.S. 3938 specifies rated primary currents up to 75 kA and rated secondary currents of

5 A or 1 A. The burden of a C.T. is the load (including leads) connected across its secondary terminals and is given in ohms, or in voltamperes at rated secondary current and usually at a power factor of 0·7 lagging. In this latter form, burden is equal to the output of the C.T. A large C.T. is rated at 15 VA, so the corresponding burdens are 15 ohms for a 1 A C.T. or 0·6 ohm for a 5 A C.T.

The construction and theory of a C.T. is similar to that of a power transformer, but an essential difference is that a power transformer is a constant-voltage, shunt-operated device while a C.T. is a constant-current, series-operated device (for any given operating conditions). The VA input to a C.T., as the burden varies between the extremes of open-circuit and short-circuit on the secondary side, is insignificant compared with the MVA behind the primary current. Thus the primary current is not altered by the insertion of the C.T. Consider now the equivalent circuit of a transformer. The primary resistance and leakage reactance can be ignored (in this context) since they only affect the voltage across the primary terminals. It is convenient to put the exciting shunt circuit on the secondary side and to refer all quantities to that side, so that $\mathbf{I}'_p$ denotes the primary current on the secondary side, as shown in Fig. 9.1(a). From $\mathbf{I}'_p$ is deducted $\mathbf{I}_e$ to excite the core and induce $\mathbf{E}_s$ which circulates $\mathbf{I}_s$.

The exciting current is the cause of the current (ratio) error and the phase error of a C.T. For very large primary currents, the C.T. tries to circulate the corresponding secondary current, and this requires a large secondary e.m.f., core flux density and exciting current. A stage could be reached when any further increase in primary current is almost wholly absorbed in an increased exciting current so that the secondary current hardly increases at all. Such a C.T. is said to be saturated.

The simple looking circuit in Fig. 9.1(a) is in practice difficult to analyse. The relation between E_s and I_e is non-linear (see Fig. 9.2). The subtraction of $\mathbf{I}_e$ from $\mathbf{I}'_p$ should be done vectorially but the necessary phase relations are seldom known. $\mathbf{E}_s$ is the product of $\mathbf{I}_s$ and the total impedance of the secondary circuit and, for a given primary current, $\mathbf{I}_s$ is the unknown whose value is required. $\mathbf{I}_e$ cannot be found until $\mathbf{E}_s$ is known. The impedance of the burden is often not a constant but a function of $\mathbf{I}_s$. Thus most of the quantities are inter-related and some of the relationships are non-linear.

The open-circuit magnetisation curve of a C.T. is shown in Fig. 9.2 for three typical core materials. Considering the curve for C.R.G. O.S.S. (which is used for protective C.T.s), it will be seen that the permeability has a low value at low flux densities, rises to a maximum at M, and then decreases in the saturation region. The non-linear exciting circuit shunt admittance $\mathbf{Y} = \mathbf{I}_e/\mathbf{E}_s$ varies inversely with the

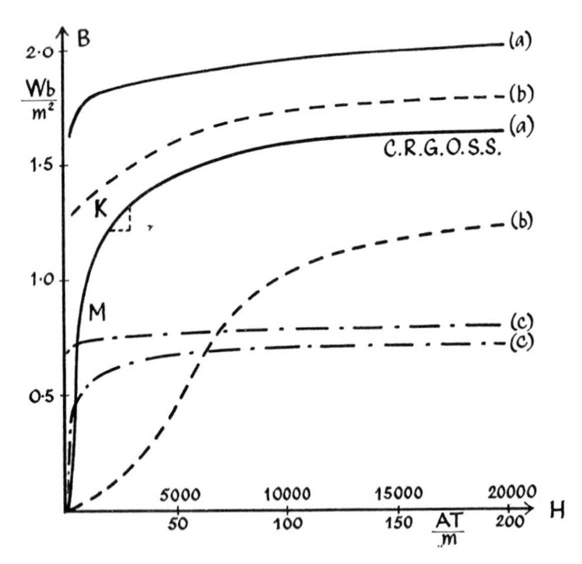

FIG. 9.2. *Magnetisation curves.*

(*a*) *Cold-rolled, grain-oriented, silicon steel* (3%)
(*b*) *Hot-rolled, silicon steel* (4%)
(*c*) *Nickel-iron* (77% *Ni*, 14% *Fe*)

permeability. The power factor of the exciting circuit, cos ϕ_o, varies from about 0·8 lagging at low excitation, to about 0·6 lagging at M and back to about 0·8 lagging at K: an average value could be taken as 0·7 lagging. For practical purposes the C.T. is deemed to be saturated at K, which is called the knee-point and defined as that point at which a 10% increase in applied voltage causes a 50% increase in exciting current. It is important to realise that in the saturation region, due to the non-linear shunt admittance, the exciting current contains harmonic components, especially third harmonics which give it its characteristic peaky waveform. Hence the secondary current is also non-sinusoidal. Under such conditions of saturation, the phasor diagram of Fig. 9.1(b) is not valid, since a phasor diagram can only represent a single frequency.

C.T.s can be classified into two types. Firstly, there are bar primary C.T.s for which the primary is a centrally situated straight conductor (extending some distance on both sides of the C.T.) which is a component part of the power system (e.g. a busbar) and is not strictly a part of the C.T. The secondary is wound uniformly over the whole periphery of the iron core, which is usually of the ring type (annular rings of steel). Thus the leakage flux of both the primary and secondary is negligible. Since there is only one primary turn, the rated primary current should be about 400 A or more, in order to provide enough exciting ampere-turns to give a reasonable output with reasonable percentage errors. A sub-class of the bar primary type is the bushing C.T. which fits over an insulator bushing enclosing a straight conductor. Due to the large diameter of the bushing, these C.T.s tend to have a relatively large magnetic path and exciting current. Secondly, there are wound-primary C.T.s the primary of which consists of one or more turns which, because of their small number, usually cannot be distributed uniformly over the iron core. Such C.T.s tend to have a high leakage flux, and usually have built-up cores (built up from E, I, T or L shaped steel stampings).

Measuring C.T.s are designed for use with instruments and meters and are required to give high accuracy for all load currents up to 125% of rated current. Nickel-iron cores are mainly used for measuring C.T.s, as the high permeability results in a low exciting current and low errors: also the almost absolute saturation at a relatively low flux density (and secondary voltage) prevents excessive currents being fed to the instruments during system faults.

Silicon steel, mostly of the C.R.G.O.S.S. type, is used for protective C.T.s which must not saturate even at high secondary currents, and yet must have reasonably small exciting currents and errors. The accuracy limit primary current is marked on the rating plate and is the maximum primary current at which the C.T. will maintain a specified accuracy when connected to rated burden. This current, when expressed as a multiple of rated current, is called the rated accuracy limit factor: the maximum value specified is 30. The accuracy limit current should at least equal the maximum fault current at the site of the C.T. Rated short-time current is the r.m.s. value of the a.c. component of primary current which the C.T. can carry, when connected to rated burden, without damage due to mechanical or thermal effects. The rated short-time current, when expressed as a multiple of rated current, is called the short-time factor. The mechanical forces are due to the peak current of the first major loop of current which is taken as 2·55 times the symmetrical r.m.s. value (see Fig. 7.9 and Volume 2). The thermal effect (I^2Rt) is due to rated short-time current being carried for rated time (largest specified value 3 seconds). Assuming that I^2t = constant which can be calculated, in kA^2-sec, for the short-time values of I and t, it is possible to estimate the time (>rated time) for which a fault current (<rated short-time current) can be kept on the power system without damage (other than at the point of fault). (This argument can be applied to most of the components of a power system during the short initial period for which the heat dissipation is negligible.)

The errors of a protective C.T. are specified at rated current and at the accuracy limit current.

$$\text{Current (ratio) error} = \frac{\text{nominal ratio} - \text{actual ratio}}{\text{actual ratio}} \times 100\%$$

$$= \frac{k_n - (k_t I'_p / I_s)}{k_t I'_p / I_s} \times 100\%$$

$$= \frac{k_n I_s - k_t I'_p}{k_t I'_p} \times 100\% \qquad [9.1]$$

where k_n is the nominal ratio (rated current transformation ratio) and k_t is the actual turns ratio. Assuming that $k_n = k_t$ then current (ratio) error

$$= \frac{I_s - I'_p}{I'_p} \times 100\% \qquad [9.2]$$

$$= \frac{I_e}{I'_p} \cos(\phi_o - \phi_s) \times 100\%. \qquad [9.3]$$

The current (ratio) error is typically negative since $I_s < I'_p$. If the secondary turns are reduced slightly (called turns compensation) then I_s is increased and the current error is reduced.

Phase error $= \theta \simeq \sin\theta$

$$\simeq \frac{I_e}{I'_p} \sin(\phi_o - \phi_s) \text{ radians} \qquad [9.4]$$

and is defined as positive if $\mathbf{I}_s$ (reversed) leads $\mathbf{I}'_p$.

It has already been mentioned that ϕ_o and ϕ_s are both typically about 45°. Thus the phase error tends to be small.

Equations [9.3] and [9.4] can be deduced from

$$\mathbf{I}_e = \mathbf{E}_s\mathbf{Y} = \mathbf{I}_s\mathbf{Y}\mathbf{Z}_s = \mathbf{I}_s YZ_s \underline{/\phi_s - \phi_o} \qquad [9.5]$$

where $\mathbf{Y}$ is the admittance of the shunt exciting circuit and $\mathbf{Z}_s$ is the total impedance of the secondary circuit including leads and burden. Thus the errors can be reduced by reducing $\mathbf{Z}_s$ and $\mathbf{Y}$. $\mathbf{Y}$ could be reduced by using a core of high permeability, large section and short magnetic length. In practice the need for a high saturation voltage (i.e. use C.R.G.O.S.S.) usually takes priority over the desirability of reducing the errors (i.e. use a nickel-iron steel); and the physical dimensions of the steel must be kept as small as possible since it must be fitted into a restricted space in the circuit breaker.

I_e is called the composite error of the C.T. when it is measured at, and expressed as a percentage of, the accuracy limit secondary current, (I_{ALS}). Since the current and phase errors are both due to components of $\mathbf{I}_e$, neither of these errors can exceed the composite error. Since the power factor of the exciting circuit (cos ϕ_o) is often not known, a pessimistic approximation to the errors can be obtained from

$$\text{current error} \leqslant (I_e/I_{ALS})\ 100\% \qquad [9.6]$$

$$\text{phase error} \leqslant (I_e/I_{ALS}) \text{ radians} \qquad [9.7]$$

Thus if the exciting current is 10% of I_{ALS} (measured for $I_s = I_{ALS}$) then the composite error is 10%, the current error cannot exceed 10% and the phase error cannot exceed 0·1 radian or 5·73° (and this last is very pessimistic). Equation [9.6] assumes $\mathbf{I}_e$ to be in phase with $\mathbf{I}'_p$: [9.7] assumes $\mathbf{I}_e$ to be in quadrature with $\mathbf{I}'_p$. It must be pointed out that the definition of composite error given above ignores the effect of harmonics due to saturation (see B.S. 3938).

The leakage fluxes of the primary and secondary windings, when added to the mutual flux linking both windings, increase the flux density in certain parts of the iron core, thus increasing the exciting current: in this context the primary leakage flux cannot be ignored. This increase in $\mathbf{I}_e$ when the C.T. is on load, and especially if it is nearly saturated, is such that the errors cannot be obtained accurately from a no-load excitation test. A C.T. having appreciable leakage flux is referred to as a high-reactance C.T. By contrast, a C.T. having negligible leakage flux (a bar primary, ring core type which is used in most protective systems) is referred to as a low-reactance C.T. and the errors of such a C.T. can be obtained accurately from a no-load excitation test (together with the secondary winding resistance and the turns ratio). If a C.T. is open-circuited, the whole of the primary current becomes an exciting current, the C.T. usually saturates and the secondary voltage becomes so high as to be a danger to life and liable to overstress the insulation of the C.T. C.T.s can be designed to have approximately linear characteristics by using iron cores with small air gaps. These are called distributed air-gap C.T.s (D.A.G.C.T.), but they are seldom used now. Fault current is liable to contain a d.c. component (see Fig. 7.9). The effect which this component has on the operation of a C.T. is very important when discussing high-speed, unit (or differential) protective systems.

9.3 Explanation of Protective Terms

The following explanations will be given, for simplicity, on the assumption that the relay is a current-operated device with normally-open contacts (i.e. the trip contacts are open when the relay is not energized).

Reliability. In the event of a fault in a zone, the protection of that zone should initiate the tripping of the necessary circuit breakers to isolate that zone, and only that zone, from all live supplies. If it does not do so, the protection is said to *maloperate*: and it should be noted that, for statistical purposes, the operation is either perfectly correct, or wrong. Reliability covers the correct design, installation and maintenance of all C.T.s, V.T.s (voltage transformers), relays, a.c. and d.c. wiring; also the accidental tripping of relays due to mistakes by personnel (this is a maloperation). For the whole of the electricity supply system in the U.K., protection is at least 95% reliable.

Discrimination. The protection of any zone is said to discriminate when it can distinguish between an internal fault in that zone and an external (through) fault in any other zone. The protection should trip for an internal fault but restrain from tripping for an external fault. The protection should not trip for any load current.

Unit protection is a system which, by virtue of its design, is inherently discriminative. Unit protective systems can have settings below the load current of the zone but non-unit phase-to-phase fault protective systems must have settings greater than the maximum load of the zone.

Main protection is the system which is normally expected to operate in the event of an internal fault.

Back-up protection is a second (often cheaper, slower) protective system which supplements the main protection should the latter fail to operate. The trip contacts of the relays are in parallel. The failure to clear the fault could be due to some component common to both systems (e.g. the circuit breaker), so most schemes provide overall back-up to clear the fault at another circuit breaker.

Check protection is a system whose trip contacts are in series with those of the main protection, thus guarding against the inadvertent operation of either system. Check operation is used where a mal-operation cannot be tolerated, e.g. the busbars of a major power station.

Sensitivity. A protective system is said to be sensitive when it will operate for very small internal fault currents. If an overhead conductor breaks and falls on dry ground or hedges, the fault current can be very small, and it is quite a problem to provide a protection sufficiently current-sensitive to detect this fault condition.

Stability. A protective system is said to be stable when it will restrain from tripping for a large external fault current. The system should be stable up to the maximum fault current liable to flow through the zone. A typical specification might require stability up to 20 times the rated primary current of the C.T.s. Rated stability limit is given in terms of the r.m.s. value of the symmetrical component of external fault current.

The terms stability and sensitivity are relative terms. A protective system can be very sensitive but not stable enough. Ideally it should be very sensitive and very stable (up to reasonable limits) but often the conditions for achieving these two ideal states are mutually incompatible.

Characteristic quantity is that which characterises the operation of the relay (e.g. current for an over-current relay, time for a time-lag relay).

Energising quantity is the current and/or voltage which, when applied to a relay, tends to make it operate.

Rated value is the value of the energising quantity, marked on the rating plate, on which the performance of the relay is based. In the case of a current-operated relay, its rated current will normally be the rated secondary current of the C.T. to be used with the relay (i.e. 1 A or 5 A).

Setting value is the nominal value(s) (usually as a percentage of rated value), marked on the setting plug (or dials) of the relay, at which the relay is designed to operate (e.g. 40% of 5 A). Since a protective relay and its C.T. cannot be considered separately, the setting of a protective system is often quoted as a percentage of the rated primary current of the C.T. The power system must be so designed and operated that the minimum fault current likely to occur in any zone is at least twice the setting of the protection of that zone. For some relays (those which operate for a fixed voltampere input at all current settings e.g. I.D.M.T.L. relays), minimum relay current setting corresponds to maximum relay impedance, which impedance is the burden on the C.T. secondary. It is thus possible for minimum relay current setting to cause the C.T. to saturate. Hence minimum relay-current setting does not necessarily result in minimum fault current setting in terms of C.T. primary current.

Pick-up (operating) value is the minimum value of the energising quantity which is just sufficient to make the relay pick up (close its contacts). For an ideal relay, the pick-up and setting values would be equal.

Operating time is the time between the application of a specified value of the characteristic quantity (e.g. 5 times setting) and the operation of the relay.

Reset (drop-off) value is the maximum value of the energising quantity which is just insufficient to hold the relay contacts closed after operation. A good relay has a drop-off/pick-up ratio only slightly less than unity (and a small resetting time when the energising quantity is removed).

9.4 Time-graded (Non-unit) Protective Systems

A time-graded (non-unit) protective system consists of several protective devices, each installed at one point, at intervals, along a (radial) power system. These devices are not inter-connected electrically. If a fault occurs on the power system, each device carries the same current. The problem is to obtain correct discrimination between the devices. The solution is to adjust their operating times (time-grading) so that the device nearest the fault, and on the source side of it, clears the fault before any of the other devices operate. Time-graded protective systems in the form of fuses and miniature circuit-breakers (M.C.B.) are used for 415/240-V distribution systems: fuses are used to a limited extent at 11 kV. Time-graded induction relays are used as main protection at 11 kV and as back-up protection at higher voltages.

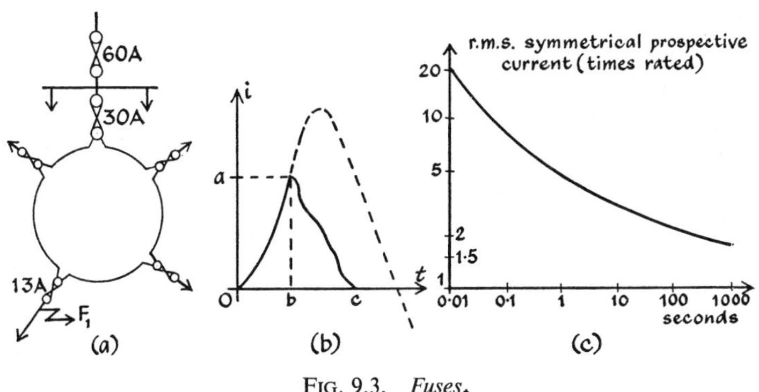

FIG. 9.3. *Fuses.*

9.4.1 FUSES

Fig. 9.3(a) shows a typical 240-V 13-A ring main circuit. If a fault occurs at F_1, correct discrimination can only be obtained if the 13-A fuse clears the fault before the 30-A fuse melts. Fig. 9.3(b) shows the first major loop of fault current (see Fig. 7.9). The prospective current (shown dotted) is the current which would have occurred had the fuse not been present in the circuit: it is measured in terms of the r.m.s. value of the a.c. (symmetrical) component. The current at which the fuse element melts is called the cut-off current (*oa*) and is measured as an instantaneous value. It is thus possible for a prospective current to be numerically less than the cut-off current. *ob* is called the pre-arcing time, *bc* the arcing time, and *oc* the total operating time.

The rated current of a fuse (e.g. 13 A) is the current it can carry indefinitely without fusing. The minimum fusing current is the smallest current, r.m.s. value, at which the fuse will melt. The fusing factor is the ratio of minimum fusing current to rated current. Fig. 9.3(c) shows a typical time-current characteristic for an 11-kV fuse, the current scale being in multiples of rated current (= 1). In practice this graph is usually given in terms of pre-arcing time and prospective, current, together with information relating arcing-time to current and cut-off current to prospective current. For small currents (overloads) arcing-time is very much less than pre-arcing time, but the converse is true for large currents (short circuits).

To obtain discrimination between two adjacent fuses carrying the same current, the pre-arcing time of the major fuse (nearer the source) must exceed the total operating time of the minor fuse. In the current range between rated current and minimum fusing current, a fuse gives no protection, but the excessive heating can cause deterioration of the fuse. The initial cost of a fuse is small but its installation, maintenance (against corrosion and deterioration) and replacement (periodically and after blowing) can be quite expensive (for example, on remote rural 11-kV networks).

A fuse is a combination of a protective device and a circuit breaker. The time/current characteristic of a fuse is approximately I^2t = constant for large currents, so that its operating time can be very small (fraction of a cycle). Thus the time grading of fuses with other protective devices is not a simple matter.

9.4.2 INVERSE DEFINITE MINIMUM TIME LAG (I.D.M.T.L.) INDUCTION RELAYS

This time-graded system is used to protect 11-kV distribution systems. It will be assumed that the reader has already studied the principle of the induction instrument (or relay). (Most textbooks on electrical instruments cover this subject: references Warrington and Shotter give detailed treatments.) For our present purpose the following summary will be sufficient. The construction of a wattmetric induction relay is shown in Fig. 9.4(a): it is similar to that of a kWh meter except that the lower magnet is energised, by mutual coupling, from the upper magnet. The shaded pole type is shown in Fig. 9.4(b). The trip contacts are shown schematically: in fact the moving contact is mounted on the spindle of the relay.

The basic principle of the induction relay is that its disc of copper or aluminium is cut by two alternating fluxes not in time phase, which induce eddy currents in the disc. Each flux interacts with the eddy currents of the other flux to produce torques which rotate the disc. The driving torque is given by (neglecting saturation of the relay electromagnets)

$$\text{torque} = KfI^2 \sin\theta \cos\lambda \qquad [9.8]$$

where I is the relay current, θ is the angle between the fluxes ($\mathbf{\Phi}_2$ leads $\mathbf{\Phi}_1$ by about 60°) and $\cos\lambda$ is the power factor of the eddy current circuit in the disc (nearly unity). The dependence of the torque on frequency f should be noted.

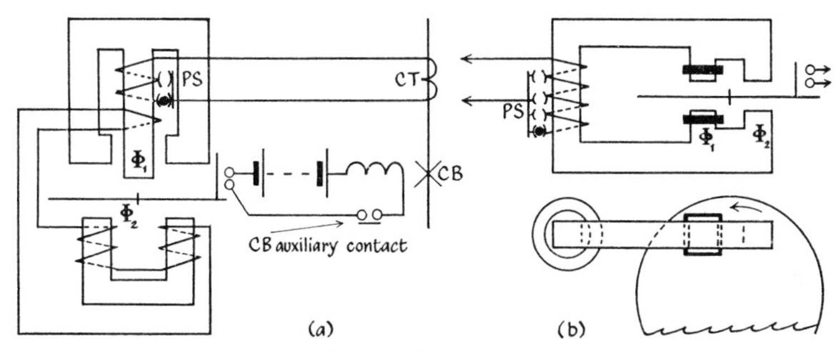

FIG. 9.4. *Induction relay (PS = plug setting).*
(a) Wattmetric (or double wound) type (b) Shaded pole type

The operation of the relay requires a certain flux and ampere turns. The current settings of the relay are chosen by altering (by means of a plug P.S. in Fig. 9.4) the number of turns of the exciting coil in use: e.g. for a 5 A (rated current) earth-fault relay, 1 A in the whole coil or 2 A in half of the coil or 4 A in a quarter of the coil will create (with good design) the same ampere turns, flux, torque and operating time. Thus only one time calibration curve is given and is applicable to all settings. The plug-setting (P.S.) can either be given directly in amperes or indirectly as percentages of rated current. Typical settings for an earth-fault relay are 20 to 80% in steps of 10%, and for an over-current relay (phase-to-phase fault relay) are 50 to 200% in steps of 25%.

The actual r.m.s. current in a relay, expressed as a multiple of the setting current is called the plug setting multiplier (P.S.M.): e.g. if a 5 A (rated current) over-current relay is set at 200% = 10 A and if the relay current is 150 A then the plug setting multiplier = 15, and if the C.T. is rated at 400/5 A, then the fault current is 12 kA.

A given relay current creates a driving torque which accelerates the disc to a speed such that the permanent magnet brake (not shown in Fig. 9.4) creates a braking torque slightly less than the driving torque, the difference being the torque to overcome the resetting spring and friction. The difference between the minimum current required to start the disc rotating, and the pick-up current should be small (5% of setting). A good relay has a large driving torque, a large braking torque, and low friction: also the disc is light so that the initial acceleration interval can be neglected and the disc rotates at constant speed (for a given current). Thus the time to operate the relay is proportional to the angle of travel. The large driving torque gives a good pressure at the trip contacts, when these close, thus minimising arcing. When the fault has been cleared by the circuit breaker, the resetting spring returns the disc to its initial position against the back-stop. The large braking torque gives a small over-shoot time (the time the disc continues to rotate after the removal of the driving torque) but gives a longer resetting time during which the timing function of the relay is incorrect should another fault occur.

The time setting of the relay marked 0 to 1 and called the time multiplier (T.M.), adjusts the position of the movable back-stop. With the time multiplier set at 1 the back-stop is as far back as it can go while with the time multiplier set at 0 the back-stop is so positioned that the relay contacts are almost closed. The time multiplier is very nearly proportional to the angle of travel of the relay contacts, but is calibrated to allow for the initial acceleration and the resetting spring. The minimum practical setting is about 0·1. With any lesser setting the relay contacts might close accidentally due to vibration etc.

B.S. 142 gives, for the standard I.D.M.T.L. relay, the relation between the plug setting multiplier and the relay operating time for a time multiplier of 1. The data is given in Fig. 9.5. The relation is an inverse one except that the time tends to become constant between P.S.M. 10 and 20 (the so-called definite minimum time). A P.S.M. = 1 means that the relay current equals its setting current. Clearly the relay must not operate for a current less than its setting. The manufacturer is allowed a tolerance of 30% on the pick-up current so that the relay must operate (pick-up) between P.S.M. = 1 and P.S.M. = 1·3, and take about 30 seconds to operate. To avoid working in this uncertain region, it is usual to arrange that, for minimum fault current the P.S.M.>2. In the case of over-current relays it is necessary to check that for maximum load currents the P.S.M.<1 by a reasonable margin of safety, since protection is not intended to deal with temporary overloading of the system.

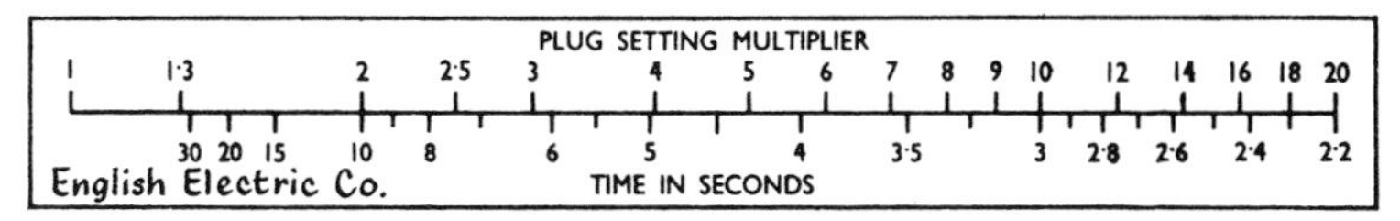

FIG. 9.5. *I.D.M.T.L. relay characteristic.*

The definite minimum time part of the curve can be obtained for a double-wound type of relay by connecting a saturable reactor (closed iron core) across the lower, U-magnet; and for a shaded-pole type of relay, by decreasing the iron section in steps from the poles to the operating coil so that the leakage flux of the coil is largest at the largest currents. The purpose of the definite-minimum-time part of the curve is discussed in section 9.4.3.

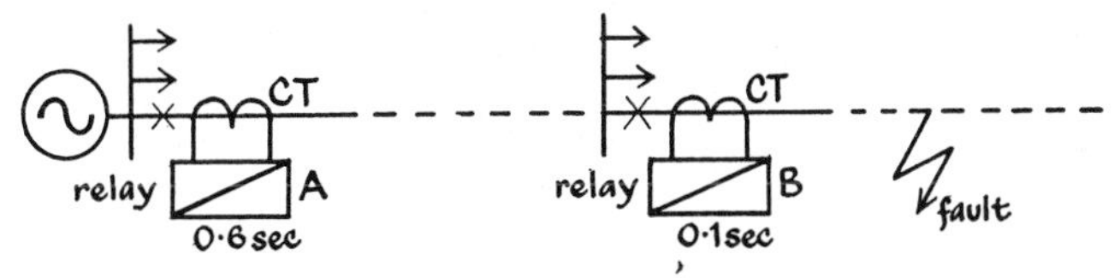

FIG. 9.6. *Time-graded protection of a radial feeder. The times shown are relay operating times (definite minimum for I.D.M.T.L. relays).*

The application of time-graded relays to the protection of a radial feeder is shown in Fig. 9.6. The time-grading margin between adjacent relays is shown as 0·5 second a typical value. 0·1 second after fault initiation, relay B closes its trip contacts. The 0·5 second margin allows time for the C.B. at B to clear the fault from the system before relay A closes its tripping contacts. It would be a maloperation of the protective system if the C.B. at A opened for a fault beyond the C.B. at B.

The 0·5 second margin is made up as follows:

Circuit breaker total time, 0·12 second.

Relay errors, 0·1 second each (A fast, B slow), 0·2 second.

Safety margin, 0·1 second.

Total 0·42 second.

With modern fast C.B.s and modern accurate relays it might be possible to reduce the total grading time to 0·4 second.

The 0·5 second margin can be obtained by a proper choice of:

1. The rated primary current of the C.T.
2. The plug setting. This affects the P.S.M. Check that
   (*a*) for maximum load the P.S.M. < 1 (overcurrent relays only);
   (*b*) for minimum fault current the P.S.M. > 2 (maximum source impedance and the fault at the remote end of the next section).
3. The time multiplier (T.M.) = the actual operating time/time from Fig. 9.5:
   (*a*) for remotest relay set the T.M. $\nless 0{\cdot}1$,
   (*b*) for maximum fault current to a fault just beyond the C.T. of the remotest relay set the T.M. of the next relay to give a time grading margin of 0·5 second.

This procedure is repeated for each pair of relays. For all fault currents less than the maximum, all the plug setting multipliers will be reduced and, because of the inverse shape of the time curve (Fig. 9.5) the time intervals between the relays will tend to increase. Since each piece of equipment in a power system has a short-time fault rating (I kA for t seconds), it is necessary to check, for the setting chosen above, that I^2t for the maximum fault current is less than I^2t corresponding to the short-time data.

The I.D.M.T.L. relay, whose time characteristic is shown in Fig. 9.5, is the one normally used to protect power systems. Its time characteristic cannot be expressed by a simple formula. Relays can be designed to have a time characteristic of the form It = constant, or I^2t = constant: the latter have an application where it is desired to protect against over-heating (rather than over-current). The general time characteristic could be written as I^nt = constant. Static (electronic) relays can be made having values of n up to about 8 (Warrington 1962). Such relays have specialised applications.

Three-phase power systems are protected by three C.T.s connected in star, and three I.D.M.T.L. relays with their trip circuits in parallel. They can be set as three over-current relays as in Fig. 9.7(a), or as two over-current and one earth-fault relay as in (*b*).

In (*a*), all three relays must be set above maximum load current and hence the minimum fault current must be greater than the maximum load current. For earth-faults and phase-to-phase faults the burden on each C.T. is that of its own relay and secondary wiring, and since the settings are high, the ohmic burdens will be low (I.D.M.T.L. relays operate for an approximately constant volt-ampere input at all settings).

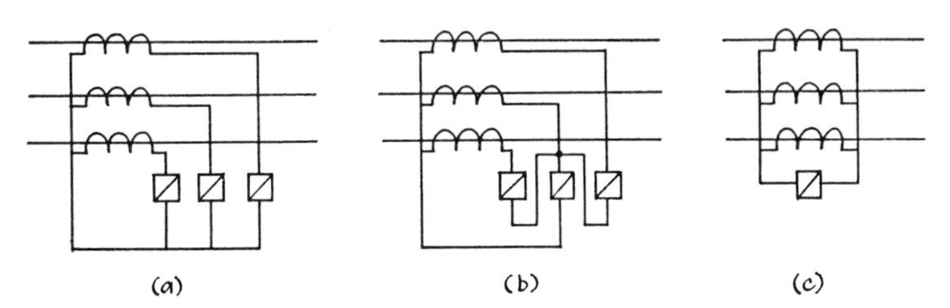

FIG. 9.7. (*a*) *Three over-current* (*b*) *Two over-current and one earth-fault* (*c*) *Residual* (*earth-fault only*)

In (*b*) the centre relay is the earth-fault relay since it can see only zero-sequence current (see Chapter 8). During normal, balanced load conditions this relay carries no current, so its earth-fault setting could in theory be any value greater than zero. In practice a minimum setting would be 10%. This circuit is used where the minimum earth-fault current is less than maximum load current. During an earth fault, the burden on the active C.T. is that of an over-current relay and the earth-fault relay in series. If the earth-fault relay setting is low, its ohmic impedance may be so high that the total burden on the C.T. may cause it to saturate. Thus if it is necessary to use the two over-current and one earth-fault connection, the C.T.s should have knee-point voltages sufficiently high to ensure operation of the relays under all fault conditions. If the three C.T.s are connected in parallel (referred to as residual connection) across a single relay (see Fig. 9.7(c)) this relay will see earth-fault currents only.

Worked example 9.1

A 20-MVA transformer, which may be called upon to operate at 30% overload, feeds 11-kV busbars through a circuit breaker; other circuit breakers supply outgoing feeders. The transformer circuit breaker is equipped with 1000/5 A current transformers and the feeder circuit breakers with 400/5 A current transformers and all sets of current transformers feed induction-type over-current relays. The relays on the feeder circuit breakers have a 125% plug setting and a 0·3 time setting. If a 3-phase fault current of 5000 A flows from the transformer to one of the feeders find the operating time of the feeder relay, the minimum plug setting of the transformer relay, and its time setting assuming a discriminative time margin of 0·5 second.

(I.E.E.)

SOLUTION

Feeder

Secondary current $= 5000(5/400) = 62{\cdot}5$ A
Plug setting multiplier $= 62{\cdot}5/(1{\cdot}25 \times 5) = 10$
Time from Fig. 9·5 $= 3$ sec
Operating time $= 0{\cdot}3 \times 3 = 0{\cdot}9$ sec.

Transformer

Overload current $= (1{\cdot}3 \times 20) \times 10^3/(\sqrt{3} \times 11) = 1360$ A
Secondary current $= 1360(5/1000) = 6{\cdot}81$ A
Plug setting multiplier $= 6{\cdot}81/(\text{PS} \times 5)$

where PS means plug setting. Since the transformer relay must not operate to overload current, its plug setting multiplier (PSM) must be less than 1. Thus plug setting (PS)$>6{\cdot}81/5>1{\cdot}36$ or 136%.

The plug settings are restricted to standard values (see text) in intervals of 25% so the nearest value is 150%.

Secondary fault current $= 5000(5/1000) = 25$ A
Plug setting multiplier $= 25/(1{\cdot}5 \times 5) = 3{\cdot}33$
Time from Fig. 9.5 $= 5{\cdot}7$ sec
Time setting $= (0{\cdot}9+0{\cdot}5)/5{\cdot}7 = 0{\cdot}246$.

9.4.3 DEFINITE-TIME RELAYS

A definite-time, time-graded protective system is one in which a relay, if it operates for a fault current, starts a timing relay which trips the circuit breaker after a pre-set time interval which is independent of the fault current. Discrimination is obtained by time-grading the relays. Thus a fault near the source end of the feeder is left on the system for the longest time, regardless of the magnitude of the fault current.

Referring to Fig. 9.6, if **E** is the source e.m.f. and $\mathbf{Z}_S$ its impedance up to the busbar A, and $\mathbf{Z}_F$ is the impedance of the feeder from A to the fault, then the fault current is $\mathbf{I} = \mathbf{E}/(\mathbf{Z}_S+\mathbf{Z}_F)$. If $\mathbf{Z}_S$ is small (high source fault level) then the fault current increases as the point of fault moves towards A. This argument is especially valid for an earth-fault fed from a transformer with a solidly-earthed secondary. Under such circumstances an inverse time relay is likely to clear a fault, especially one near the source end of each section, quicker than the corresponding definite-time relay.

If $\mathbf{Z}_S$ is large (or, in the case of earth-faults, if the source transformer is resistance earthed) the fault current tends to be a constant independent of the position of the fault. Thus all the relays in an inverse-time system would operate at fixed points on the time/plug-setting-multiplier curve for all positions of the fault: it would in effect be a definite-time system. Since definite-time relays are usually cheaper, there is an economic advantage in using them.

In a large, integrated power system (such as that of the U.K.) the source impedance is liable to vary over wide limits. Then the I.D.M.T.L. type of characteristic, which is inverse for plug-setting multipliers between 2 and about 10 and approximately definite-time for those above 10, has some advantages.

9.4.4 RING FEEDERS: DIRECTIONAL RELAYS

Compared with ring feeders, radial feeders and their protection are simple and cheap, but if a fault should occur in a radial feeder, all consumers beyond the fault would lose supply. A ring feeder is shown in Fig. 9.8(a). There are now two in-feeds to each substation and the design of the system is such that if any one section of the main cable is faulty, the loads can all be carried on the remaining cables (referred to as a firm supply). But the capital cost is high compared with that of two separate radial feeders, since the ring needs about twice as many circuit-breakers and more expensive protection. Thus ring systems tend to be used to supply large loads producing good revenue, continuous process load with penalty tariffs and where there is a danger to life as in mines and hospitals.

The relay times are set by considering each direction round the ring as a separate radial feeder. If all the relays were ordinary (non-directional) relays and if a fault were to occur between relays R_3 and r_2, then relays R_1 and r_1 would operate—a maloperation. R_1

and r_1 are therefore blocked off (prevented from operating) by directional control so that they operate only for fault current (strictly power) flowing in the direction of the arrow. By placing a fault in each section in turn, the following rule for directional control can be checked: directional control is required at each load substation on the relay with the lesser time setting (both if they are equal). Each directional relay should only operate when the fault current is flowing into its own section.

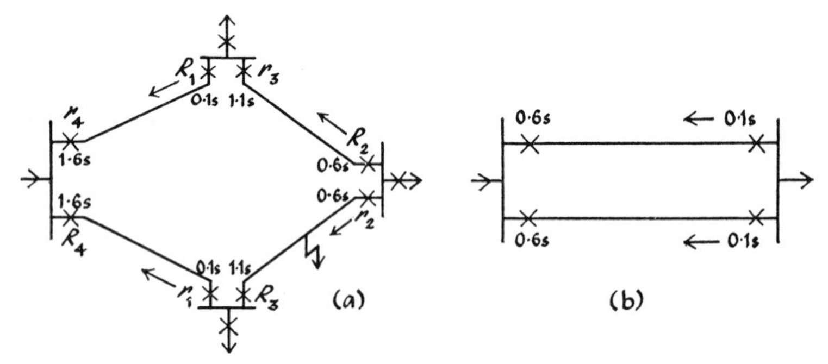

FIG. 9.8. (*a*) *Ring feeder* (*b*) *Duplicate feeder. The times shown are relay operating times*

The duplicate line system shown in Fig. 9.8(b) is a special case of a ring system. The same rule applies. If it is necessary to time-grade the duplicate feeders with other parts of the system, it may be necessary to modify the times shown in (b).

For an over-current relay the directional control can be obtained by using a wattmetrical relay, similar in construction and connections to a kWh energy meter. The disc will rotate in one direction when the angle between the voltage and current applied to the relay is less than 90°, and rotate in the reverse direction when the angle exceeds 90°. The voltage is referred to as the polarising quantity since the relay determines whether the current has a component in phase, or in anti-phase, with this voltage. The connections are so arranged that each directional relay closes its contacts (operates) when the direction of power flow is from the busbar into the section protected by that relay. The contacts of the directional relay are close together so that it operates in about 0·1 second. This time must be allowed for when calculating the time-grading, discriminative margins. The directional relay is specially designed (compensated) to ensure that it is sensitive to minimum fault current at a low power factor and low voltage (fault close to voltage transformer). This can take the form of a mutual coupling between the two electromagnets.

It is essential that the voltage and current at the directional relay should not be in quadrature or the directional control will fail. The characteristic angle of a relay is the angle between the voltage and current at the relay terminals at which the relay is most sensitive (creates maximum torque). For a normal, single-phase wattmetrical relay this angle is 0° (but other angles can be obtained). The relay connection angle is the angle between the voltage and current applied to the relay, assuming that the current is in phase with its own phase voltage on the primary sides of the V.T. (star/star) and C.T. (star). A characteristic angle of 0° and a connection angle of 30° is generally suitable for over-current protection of feeders. The directional relay is supplied with $\mathbf{V}_{RB}$ and $\mathbf{I}_R$ (the phase angle between these is 30° when $\mathbf{I}_R$ is at unity power factor with respect to $\mathbf{V}_{RN}$). Thus the relay is most sensitive when the power factor of the fault circuit is 0·866 lagging and fails when the fault power factor is 0·5 leading. When the fault power factor is zero lagging, the power factor at the relay is 0·5 lagging. To obtain a 90° connection angle the relay is supplied with $\mathbf{V}_{YB}$ and $\mathbf{I}_R$.

When a wattmetrical over-current relay is used with a wattmetrical directional relay, the contacts of the latter are connected in series with the circuit which couples the U-magnet to the E-magnet in the former (see Fig. 9.4a). To obtain directional control of a shaded-pole over-current relay (Fig. 9.4b) the contacts of the directional relay are connected in series with the shading coil (which is in this case a wound coil). Neither type of over-current relay is allowed to start timing until the directional relay operates. This is preferred to connecting the two sets of contacts in series in the tripping circuit of the circuit breaker.

For earth fault relays the current is obtained from residually-connected C.T.s (3 in parallel). The polarising quantity for the directional relay is usually the voltage output from a broken-delta tertiary winding on the V.T. which must have an earthed star primary and be of 5-limb construction (see sections 6.8 and 6.10). Because of the complications discussed above, 11-kV distribution systems are not often operated as ring systems. Sometimes the ring is installed but is operated with one circuit breaker open—the open-ring system. Closed ring systems usually operate at 33 kV and above, and unit protective systems are used (see section 9.5).

9.5 Unit Protective Systems

The non-unit systems so far discussed discriminate by virtue of time-grading. They are comparatively simple and cheap and require no pilot-wire circuits, but their fault clearance times are too slow for circuits operating at 33 kV and above, and for expensive plant such as large alternators and transformers. A unit protective system is one which comprises two sets of C.T.s, which may be some distance apart and which define the ends of the protected section (or zone). The C.T.s are inter-connected by a pilot-wire circuit incorporating protective relays, so that the protective system can inherently discriminate without the use of time-grading. Unit protection systems are instantaneous (0·1 second) and their time settings are independent of other protective systems.

The basic principle of unit protection (often referred to as Merz-Price protection) is that the currents into and out of a section are identical (neglecting complications to be discussed later) if the section is healthy. The relays are connected to measure the difference between these two currents (so that these systems are referred to as longitudinal differential, or balanced, protective systems). Since, for a healthy section, the relay current is ideally zero, unit protection is insensitive to load current and external (through) fault current. If an internal fault occurs in the section, the relay current is not zero and the sensitivity of the protective system must be such that the relay operates.

There are two sub-classes of differential protection:

(*a*) Circulating-current systems, used for short zones (alternators, transformers, busbars).

(*b*) Balanced-voltage systems, used for long zones (feeders, transmission lines), where the currents into and out of the section are converted to voltages before being compared (or balanced against each other).

9.5.1 CIRCULATING-CURRENT SYSTEMS

The basic circuitry is shown in Fig. 9.9 as a single-phase diagram. Assuming an ideal case, the currents into and out of the healthy zone are equal and as the C.T.s are nominally identical (matched) and the relay is at the electrical centre of the pilot-loop the currents in the two halves of the pilot loop are identical and the relay current is zero since there is no voltage across its terminals. This system is insensitive to load current and external (through) fault current. The relay is an attracted-armature type.

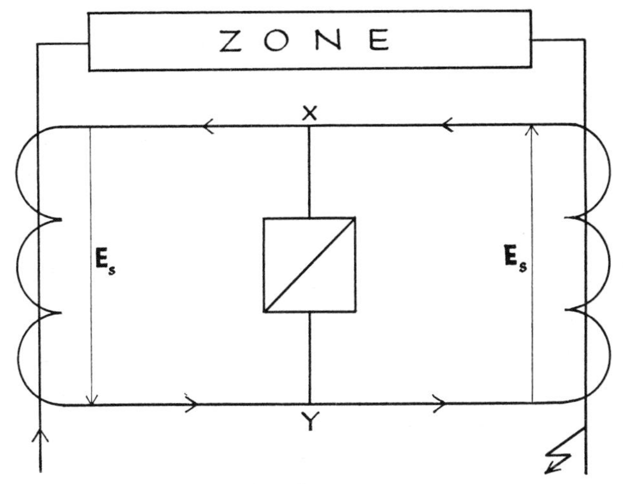

FIG. 9.9. *Circulating current system.*

Any departure from the ideal case will result in a voltage across, and hence a current in, the relay. The main cause of instability during an external fault is the asymmetry (mismatching) of the C.T.s while carrying large a.c. currents with d.c. components (see Fig. 7.9). Other causes of instability will be dealt with as they arise in this chapter.

Two methods are used to improve the through-fault stability of a protective system:

(*a*) Stabilising resistor.

(*b*) Biased relays.

If the zone has an internal fault fed from both ends then the right-hand currents in Fig. 9.9 are reversed, and both pilot currents (not necessarily identical) add and flow in the relay to cause tripping. If the zone is fed from one end only (conventionally from the left) then the right-hand currents are zero and only the left-hand current flows in the relay to cause tripping. For this reason, when considering the sensitivity of a protective system to an internal fault, a single-end infeed is usually assumed. The left-hand C.T. carrying the fault current is said to be active while the right-hand C.T. is said to be idling. Since, during an internal fault fed from one end only, a voltage exists across the relay, almost the same voltage exists across the idling C.T. which therefore takes a small exciting current.

Since an internal fault fed from one end only represents the worst case, the usual procedure for the study of a unit protective system is to assume a maximum plant external fault to check relay stability, then to delete the right-hand current to check tripping due to an internal fault fed from one end only.

Stabilising Resistor

This is a resistor (R) connected in series with the relay in Fig. 9.9. If, for the maximum external fault current, the largest (r.m.s.) voltage (due to C.T. asymmetry) across X and Y can be estimated as V, then R is chosen so that $V/(R+R_R)<$ setting current of the relay where R_R is the relay resistance (it can be treated as a resistance since its inductance is kept as low as possible, sometimes by using a tuned series capacitor).

The method usually adopted in practice to calculate R is to assume that one C.T. completely saturates while the other retains perfect current transformation. Since both C.T.s are nominally identical and carry the same primary current, this assumption may seem rather pessimistic. Referring to the equivalent circuit of a C.T., Fig. 9.1(a), the saturated C.T. is represented by a zero impedance shunt excitation circuit and a series impedance equal to the secondary resistance (leakage reactance can be ignored, especially for bar primary, ring-core C.T.s), while the other C.T. is represented by an open-circuited excitation circuit and a series secondary resistance. Thus if I_f is the maximum external fault current on the secondary side, via the perfect C.T., and r is the resistance of the half pilot loop and R_S is the C.T. secondary resistance, then

$$V = I_f(R_S+r)$$

Hence the setting current $\geqslant I_f(R_S+r)/(R+R_R)$.

This method of calculating the stabilising resistor can be justified on theoretical grounds. Modern instantaneous attracted armature relays, incorporating a stabilising resistor R, have their settings marked in volts (rather than in milliamperes). The stability limit current is the maximum (r.m.s. a.c. component) external fault current for which the protective system will remain stable. Having ensured that the relay will be stable to the maximum external fault current, it is now necessary to check that it will trip to the minimum internal fault current, assuming an infeed from one end only. If the voltage across the relay is its setting voltage, this is approximately the voltage across both C.T.s, and the corresponding exciting current(s) can be found from the magnetisation curve(s). By summating the exciting currents of all the C.T.s in the protective system (a bus zone system may have up to 30 C.T.s in parallel) and adding the relay setting current, the secondary current and hence the primary current of the active C.T. can be found. This primary current is called the primary fault setting (or primary operating current P.O.C.). It is usual to summate the currents arithmetically since they are approximately in phase. The calculated primary fault setting is then pessimistic and must be less than the available internal fault current. The stability ratio of a protective system is the ratio of the stability limit current to the primary fault setting.

Biased Relays

The attracted armature relay in Fig. 9.9 is now replaced by a biased relay, (Fig. 9.10). A biased relay consists of two shaded-pole relays

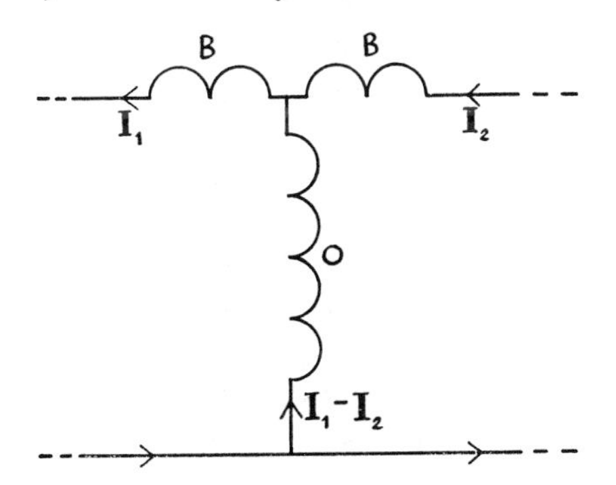

FIG. 9.10. *Biased relay (percentage differential).*

acting on the same disc but producing torques in opposite senses. The operating coil (O) which carries the out-of-balance (spill) current during an external fault, tends to trip the relay, while the bias coil

(B) which is centre-tapped, carries the secondary equivalent of the fault current and tends to restrain the relay.

During an external (through) fault, the two halves of the bias coil both carry large currents whose ampere-turns add, while the operating coil carries only a small out-of-balance current due to the imperfections of the protective system. During an internal fault fed from both ends, the right hand pilot current reverses so that the biasing action is weak and the operating action strong. During an internal fault fed from one end only, the right-hand pilot current is zero, so that the same current flows in the operating coil and in half the bias coil. The relay must be designed to trip under these conditions. For shaded-pole relays of similar design the operating coil must have more turns than half the bias coil.

The setting of the relay is the minimum current in the operating coil only (zero bias) which will operate the relay: it is expressed as a percentage of rated current. The setting current ($\mathbf{I}_0$) produces a driving torque to overcome the resetting spring, friction and inertia.

The bias (B) of a relay is the ratio of the number of turns in the bias coil to the number of turns in the operating coil, assuming the two shaded-pole magnetic circuits are identical. Typical values might be, for an alternator a setting of 10 to 20% and a bias of 10%, and for a transformer a setting of 20% and a bias of 20 to 40% (the higher bias values are used for tap-changing transformers).

If N is the number of turns in the operating coil

$$\text{then the operating ampere-turns} = (\mathbf{I}_1 - \mathbf{I}_2)N$$

$$\text{and the biasing ampere-turns} = \mathbf{I}_1(B/2)N + \mathbf{I}_2(B/2)N = \left(\frac{\mathbf{I}_1 + \mathbf{I}_2}{2}\right)BN$$

The relay will operate when (assuming linear relations)

$$|(\mathbf{I}_1 - \mathbf{I}_2)N| - |(\mathbf{I}_1 + \mathbf{I}_2)(BN/2)| > |\mathbf{I}_0 N|$$

The balance equation for operation is (Fig. 9.11)

$$|\mathbf{I}_1 - \mathbf{I}_2| = |\mathbf{I}_0| + |(\mathbf{I}_1 + \mathbf{I}_2)(B/2)| \qquad [9.9]$$

The relay will trip for operating currents greater than that given by the bias line. The out-of-balance (spill) current in the operating coil of the relay due to imperfections in the protective system is shown dotted and, for stability, must lie well below the bias line.

The scalar diagram of Fig. 9.11 is not very informative since the operating current $(\mathbf{I}_1 - \mathbf{I}_2)$ could be due to differences in magnitude and/or phase. The relay characteristic is better shown as the polar characteristic of $(\mathbf{I}_1/\mathbf{I}_2)$ in the complex plane.

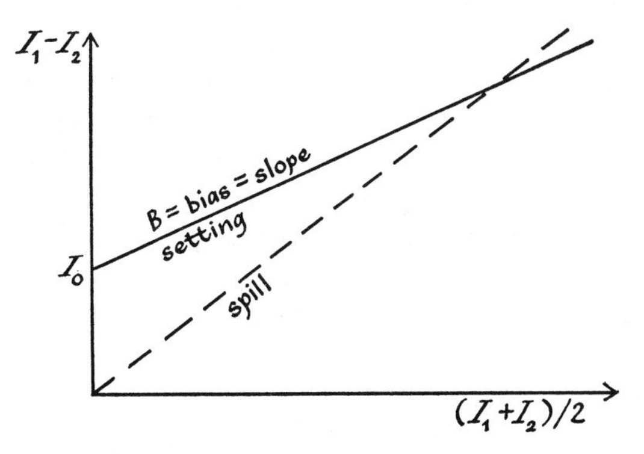

FIG. 9.11. *Biased relay characteristic.*

Generator Protective Schemes

The single-line diagram of Fig. 9.9 is now given in Fig. 9.12 as a 3-phase diagram for the protection of an alternator: for small alternators, the biased relays could be replaced by attracted-

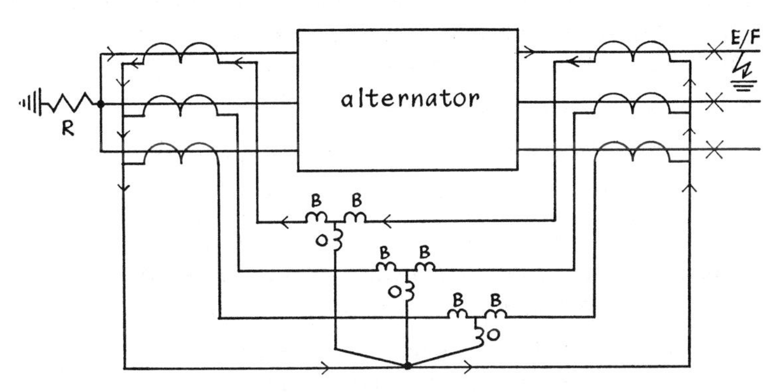

FIG. 9.12. *Biased differential protection of an alternator. All three relays are at the centre of the pilot loop.*

armature relays with stabilising resistors. The six C.T.s are identical (matched). The diagram shows that the system is stable for an external earth fault. The secondary fault current does not flow in the four idling C.T.s as their input impedances are very high (magnetising reactance). The student should check that the system is stable for an external phase-to-phase fault but should operate for internal earth faults and phase-to-phase faults on the stator winding. For internal faults it is usual to consider the worst case, i.e. no infeed from the busbars. The tripping contacts of the three relays are connected in parallel.

A typical alternator main-protection system might have a setting of 10%, a bias of 5%, an operating time of not more than 3 or 4 cycles at 5 times fault-setting and be stable up to at least 12·5 times rated current of the C.T.s.

If an internal fault occurs on the stator winding, it is not sufficient to open the main circuit breaker to clear the fault since the alternator itself can still supply power to the fault. For small alternators a 3-pole circuit breaker at the neutral end of the alternator would clear internal earth-, and phase-fault currents. Most alternators are now supplied with field suppression equipment, i.e. a make-before-break contactor which connects a field-suppression resistor across the rotor winding, to dissipate the electromagnetic energy, then opens the circuit to the excitation supply.

If the main protection operates, the main circuit breaker should be opened, the field-suppression equipment operated and the steam supply shut off, all as quickly as possible.

Large alternators are fitted with other protective devices, in addition to the main protection. Negative-sequence current relays (see section 8.5.1) are supplied by a set of C.T.s at the neutral end of the alternator. These negative-sequence currents, which are caused by unbalanced loading and by faults on the system, induce 100 Hz eddy currents in the rotor, and add to the heating of the rotor. Since rotor cooling is a serious problem, this additional heating cannot be tolerated for long. The relay has an I^2t = constant, type of characteristic. Alternators are also fitted with devices to indicate high temperature, under- or over-excitation, failure of cooling equipment, and loss of vacuum at the turbine.

If any non-urgent device operates, it first shuts down the steam supply and only opens the circuit breaker when the turbine has dissipated its trapped steam: this last condition is signalled when the alternator reverse-power relay operates.

A longitudinal, differential protective system will not detect a short-circuit between turns on one phase (except that such a fault will probably change to an internal earth fault).

If an earth fault occurs near the star-point of an alternator, the voltage behind the fault may be insufficient to create a relay operating current greater than the relay pick-up (setting) current.

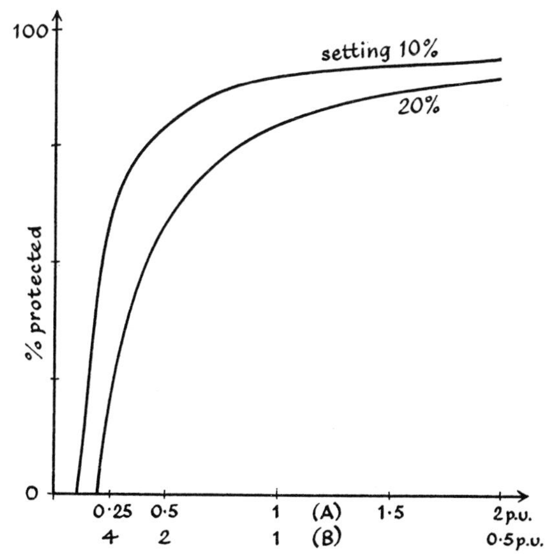

FIG. 9.13. *Percentage of alternator protected against earth-faults. Scale (A) earth-fault current. Scale (B) earthing resistance.*

For an alternator connected directly to a load (this excludes generator-transformer sets), the value of the earthing-resistor R in Fig. 9.12 is about 1 p.u. on alternator rating. Since the (transient) reactance of an alternator phase is about 0·2 p.u., the reactance of a small part of a phase near the star-point can be neglected compared with R: as usual, we consider the simple but pessimistic case of an alternator on no-load at rated voltage and speed.

Thus for $R = 0{\cdot}8$ p.u. (0·8 p.u. phase voltage across it when carrying rated current) and a differential protective system with an earth-fault setting of 20% of the rated current of the alternator, it follows that an earth-fault $0{\cdot}8 \times 20 = 16\%$ distant from the star-point would just be sufficient to trip the protection, so that only 84% of the phase is protected against internal earth-faults. In this way, Fig. 9.13 can be deduced. It is convenient to draw the abscissa to a linear scale (A) giving the current which rated phase voltage would create in the corresponding earthing resistor whose value is given in the non-linear scale (B). This current is referred to as the rated voltage, terminal earth-fault current: the alternator reactance is still neglected. Fig. 9.13 shows that for an earthing resistor of 1 p.u. and relay settings of 10 or 20% (of alternator rated current), only 80 or 90% of the phase is protected.

Since the insulation is uniform over the whole phase winding, and since the machine is totally enclosed, the probability of an earth fault near the star point is considered remote. Such a fault, if it does not operate the main protection, should operate the standby earth-fault relay. This is a sensitive non-unit relay, with a long inverse time characteristic, which must be time-graded with the I.D.M.T.L. relays on the distribution system.

The student should now analyse this problem and show that

$$x = kI_sR/V \qquad [9.10]$$

where x = the position of the fault from the star-point expressed as a fraction of the whole phase length (i.e. the fraction of the phase winding not protected),

I_s = relay setting current, amps,
k = C.T. current ratio,
V = phase voltage, volts,
R = earthing resistor value, ohms.

kI_s is the setting of the protective system measured on the primary side of the C.T. (primary fault setting). If all quantities are expressed as per-unit of the alternator rated current, then [*9.10*] becomes

$$x = I_{s\cdot pu}R_{pu}/V_{pu} \qquad [9.11]$$

Fig. 9.13 has been drawn for the case of $V_{pu} = 1$.

Transformer Protective Schemes

The circulating current (differential) protection system can be used to give overall protection for a 3-phase transformer (primary and secondary forming a single zone) but it must be realised that the comparison of the currents takes place at the relays, which are connected on the secondary side of the C.T.s.

Since the primary and secondary line currents are not equal when the transformer is on load, but are in the inverse ratio of the line voltage ratio, the rated primary currents of the two sets of C.T.s are also in the same inverse ratio.

When a delta/star transformer is on load, the primary and secondary line currents are 30° out of phase (see section 8.3.5). In order to bring the C.T. secondary currents into phase with each other, the C.T. secondaries must be connected in star on the delta side of the power transformer and in delta on the star side (this statement will be justified later). Since, for balanced 3-phase conditions, the line currents from a delta-connected set of C.T.s are $\sqrt{3}$ times the C.T.

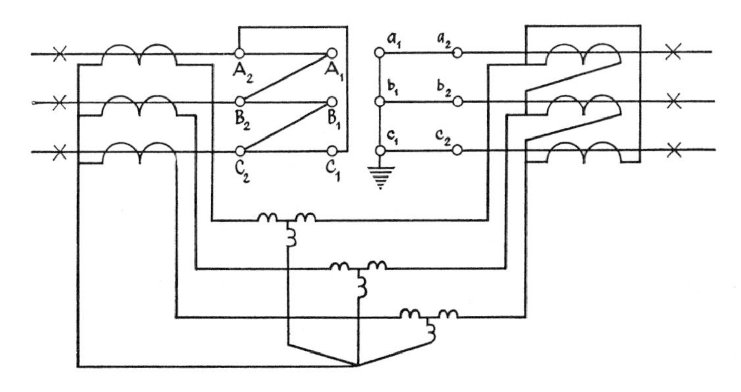

FIG. 9.14. *Biased, differential protection of a transformer.*

secondary currents, the rated secondary current of each delta-connected C.T. is $5/\sqrt{3} = 2{\cdot}89$ A for a relay rated at 5 A or $1/\sqrt{3} = 0{\cdot}577$ A for a relay rated at 1 A. Thus the C.T.s on the h.v. side of a 33/11-kV power transformer could be rated at 400/1 A and those on the l.v. side at 1200/0·577 A.

The student should check that an overall circulating-current protection scheme, using these C.T. ratios, will be stable when the transformer is on load with 1200 A on the secondary side (see Fig. 9.14): he should also check stability for an external earth fault and for an external phase-to-phase fault, and check that the protection should operate for an internal fault (other than an inter-turn fault on one winding). For an internal phase-to-phase fault on the secondary side, note that the current in one relay is twice that in the other two relays.

If an external earth-fault occurs on the star-connected secondary side of the power transformer, the secondary lines will carry zero-sequence currents (see Chapter 8) but no corresponding zero-sequence currents can flow in the primary lines. Thus, to obtain a current balance at the relay, the secondary zero-sequence currents must be prevented from reaching the pilot wires by connecting the secondary C.T.s in delta. Zero-sequence currents, in the form of single-phase currents, will flow round the secondaries of the delta-connected C.T.s, and round the delta-connected primary winding of the power transformer.

Similarly, if the power transformer is connected unearthed-star/earthed-star, the C.T.s must be connected delta/delta. The current distribution in such a transformer during an earth-fault is discussed in section 6.8. The student should now show that, if the fault is external to the (overall) protected zone, the protection is stable.

The relays used for overall transformer protective systems are of the biased type because:

(*a*) If overall protection is used (it is only used for large transformers) the transformer will probably be fitted with on-load tap changing gear. It is not practical to alter the C.T. ratio to match the varying ratio of the power transformer. The C.T. ratio is fixed and chosen to suit the nominal ratio of the power transformer. Thus for taps other than nominal, an out-of-balance current will flow in the relay during load and external fault conditions.

(*b*) Since the two sets of C.T.s are of different design (different current and voltage ratings), it is difficult to match them. Thus out-of-balance current is liable to flow in the relay during heavy external fault conditions.

When a transformer is on no-load, the no-load current is reproduced in the relay as if it were an internal fault condition. The relay setting current must be greater than the no-load current when both are expressed as percentages of rated primary current. For this reason, and reasons (*a*) and (*b*) above, transformer overall protective systems have higher settings than those for alternator protective systems, e.g. a setting of 40% and a bias of 20%.

When an unloaded transformer is switched on to a supply, the initial magnetising current can reach a value of several times rated current; and is called the magnetising surge or inrush current (see section 6.6 and Fig. 6.15. An overall protective system, using ordinary biased relays, would see this surge current as an internal fault. This problem used to be solved by time-delaying the protection until the magnetising current was less than the setting of the protection, a delay of about ½ second). But this delayed the clearance of a genuine internal fault and, as transformers became larger and more expensive, this delay could no longer be tolerated. The problem is solved by noting that the surge current has a large second-harmonic component (100 Hz) while a genuine fault has very little. A series-resonant circuit, tuned to 100 Hz, is mutually coupled into the operating coil circuit of Fig. 9.14, as shown in Fig. 9.15. The output

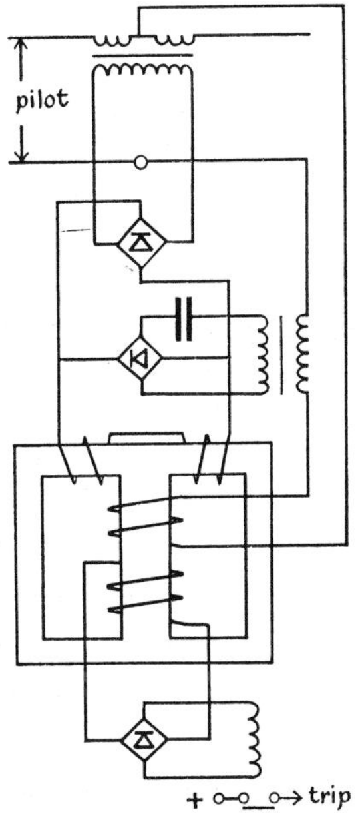

FIG. 9.15. *Harmonic bias.*

current from this coupling is rectified and fed to the bias or control winding of a transductor. The through-fault bias coil of Fig. 9.14 is replaced by a transformer, the output of which is also rectified and fed to the control winding. The input (primary) winding of the transductor takes the operating current of the protective system while the output (secondary) current is rectified and fed to an instantaneous attracted armature tripping relay. In the absence of d.c. bias the transductor behaves like a transformer and an internal fault should cause the tripping relay to operate. But under through fault, or magnetising surge conditions, the d.c. bias saturates the transductor, limits the pulsations of a.c. flux due to the transductor primary current, and so limits the secondary output that the tripping relay does not operate. This system is referred to as an harmonic-bias, overall, circulating-current protective system. It is used for large tap-changing transformers. Typical settings might be: setting 20%, bias 20 to 40%, maximum operating time 4 cycles at 5 times fault setting, and minimum stability limit 16 times rated current of the C.T.s.

Referring to Fig. 9.14 (but with resistance earthing replacing the solid earthing), if an earth fault occurs on the star-connected secondary winding, at a distance x from the star-point, expressed as a fraction of the whole phase length, the student should show that

$$\text{relay current} = \frac{x^2 V_s N_s}{R N_p k} \text{ A} \qquad [9.12]$$

where V_s = secondary voltage, volts/phase, N_s = secondary turns/phase, N_p = primary turns/phase, R = earthing resistor value, Ω, and k = ratio of C.T.s on primary side.

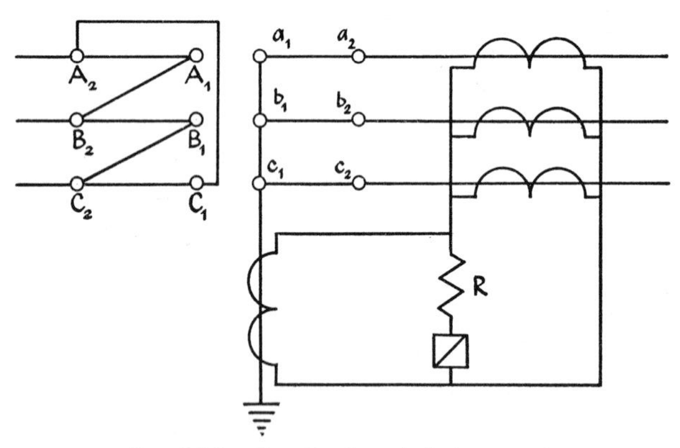

FIG. 9.16. *Restricted earth-fault protection.*

Equation [*9.12*] is derived on the assumption that the transformer is on no-load and that the fault is fed from the primary side only. Since $x<1$, it follows that for small values of x (the unprotected fraction of the winding), the relay current is very small, especially for a step-down transformer ($N_s<N_p$) with a high earthing resistor.

Thus the design of a satisfactory overall protective system for a transformer can be a difficult problem.

The student should convert [*9.12*] into a per-unit system, based on the rating of the transformer, and show that

$$x^2 = \sqrt{3}\,(PFS)_{pu}R_{pu}/V_{pu} \qquad [9.13]$$

where (PFS) = primary fault setting (primary current to operate the relay expressed as a fraction of the rated primary line current of the transformer),

R_{pu} = earthing resistor value, per-unit,

V_{pu} = transformer voltage, per unit.

Thus if $(PFS) = 0{\cdot}2$, $R_{pu} = 1$, $V_{pu} = 1$, then $x = 0{\cdot}588$ and $1-x = 0{\cdot}412$, i.e. only 41·2% of the winding is protected.

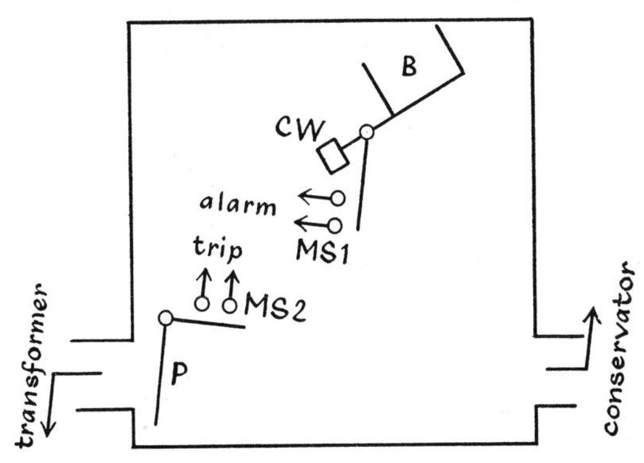

FIG. 9.17. *Buchholz (gas) relay.*

This problem of protecting an adequate fraction of the secondary winding is partially solved by adding a restricted earth-fault protective system, as shown in Fig. 9.16. The four C.T.s are identical (matched) and form a circulating-current protective system, which includes a stabilising resistor (R). During an internal earth-fault, fed only from the delta side, only the neutral C.T. is energised so the analysis is the same as that for alternator protection. The fault setting can be low (say 10%) since this system does not suffer from any of the disadvantages of an overall transformer protective system.

A protective device fitted to most transformers is the Buchholz or gas-actuated relay. This is fitted into the oil pipe between the transformer tank and the oil conservator. A sketch of its internal construction is shown in Fig. 9.17. Under healthy conditions the relay is full of oil, including the bucket B, but the counterweight CW is adjusted to hold the mercury switch MS1 open. If there is any partial failure of the insulation anywhere inside the transformer gas accumulates at the top of the relay. As the oil level falls, the bucket full of oil, being now heavier than the counterweight, tilts the mercury switch to the closed position, and operates the alarm. If there is a short-circuit inside the transformer, the explosion instantaneously forces the incompressible oil against the plate P and closes the mercury switch MS2 which trips the circuit breakers. An alternative design uses a ball-float instead of a bucket for the alarm element.

9.5.2 BALANCED VOLTAGE PROTECTIVE SYSTEMS

The circulating-current system of unit, differential protection is suitable for short zones such as alternators and transformers. If such a system were applied to feeders several km long, the C.T. secondary e.m.f.s would be required to circulate about 5 A or 1 A at full load, or several times the currents during external fault conditions, round a pilot loop of fairly high impedance. Such a burden is impracticable for any economic design of C.T.

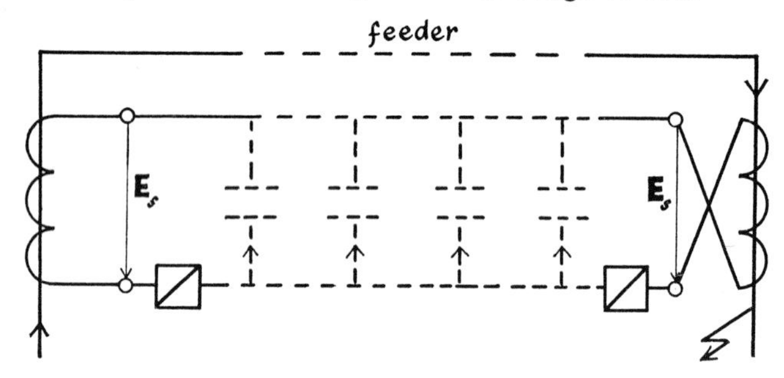

FIG. 9.18. *Balanced-voltage protective system.*

One solution of this problem was to reverse one of the C.T.s as shown in Fig. 9.18. During an external fault, the two secondary e.m.f.s opposed each other round the pilot loop, so that the relay operating currents were both zero. In practice the through-fault stability of this system was very poor for two reasons. Firstly, the C.T.s were operating under open-circuit conditions so that their secondary e.m.f.s were very high, were non-linear with respect to their primary currents, and were therefore difficult to match. In an attempt to solve this problem the C.T.s were designed with several small air-gaps (distributed air-gap C.T.s) but the slightest disturbance of the air-gaps mismatched them. Secondly, the pilot wires had a voltage across them along their whole length and the charging currents to the pilot-wire capacitance, flowing in the relays, tended to operate them.

Summation Transformers

Modern protective systems for feeders often use summation transformers (see Fig. 9.19), with the following advantages: (*a*) the line C.T.s are of conventional design operating into a low-burden circuit, (*b*) the output of the summation transformer is fed to a 2-wire pilot system and this output is available for any type of fault on any of the primary lines (with one exception which will be discussed later). As the two ends of a feeder protective system are identical, the pilot

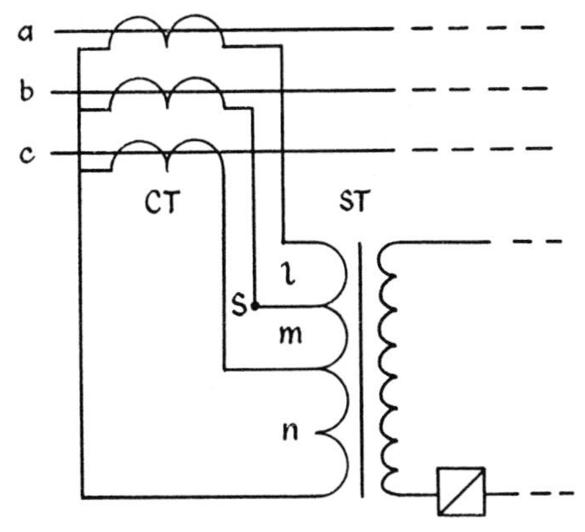

FIG. 9.19. *Summation transformer* (ST).

system can be operated as a balanced-voltage system. It must be clearly understood that the problem of the pilot charging currents tending to operate the relays has still to be discussed in terms of a practical solution.

The construction of a summation transformer is very similar to that of a C.T. except that the primary is tapped and it can operate with an open-circuited secondary.

The primary of a summation transformer consists of only a few turns of wire, and is so designed that, even with an open-circuited secondary, the burden which it imposes on the C.T.s is low enough for the C.T.s to be of conventional design.

Let l, m and n be the number of turns in the three sections of the primary. An earth-fault on phase a will create a secondary current in $(l+m+n)$ turns, while a phase-to-phase fault across phases b and c will circulate current in m turns. For an internal fault on a given system fed from one end only, and with the relay at that end on the point of operating, the output VA from, and hence the input ampere-turns to the summation transformer will be constants for all types of fault. (The corresponding magnetising, and iron-loss currents should be added to the operating current.) Thus the fault setting for a given type of fault will be inversely proportional to the number of turns (either actual or relative) of the summation transformer primary which are active for that fault. If one fault setting, and l, m and n are given, all other fault settings can be calculated. This inter-dependence of the fault settings is a disadvantage of the summation transformer, but not a serious one. The earth-fault settings can be reduced (made more sensitive) by increasing n without affecting the phase-to-phase fault settings which are independent of n. Since the protective system is a unit system it is not necessary to make the phase-to-phase fault settings greater than maximum load current.

In practice it is usual to use $l = m$, and the turns per section are defined in relative terms as $1:1:n$. The student should now show that the (relative) active number of turns corresponding to a symmetrical 3-phase fault is $\sqrt{3}$ (which is independent of n). (Hint: use superposition or take S in Fig. 9.19 as a star point). He should further show that if $n = 4$ (relative) and if the setting for an earth-fault on phase a is 40% (of the relay rated current) then the setting for an earth fault on phase c is 60%, for a phase-to-phase fault across a and c is 120% and for a 3-phase fault 139%.

The so-called blind spot of a summation transformer arises when the line currents are in the ratio of $1 : -2 : 1$ on phases a, b, c respectively. For this fault condition there is no output from the summation transformer. The student should check this statement and determine the circumstances under which this fault condition could arise. This problem might be solved by making $l \neq m$ but in practice this introduces other problems.

Pilots

To protect underground power cables, the protection pilot is usually laid at the same time as the power cables. Since there is little restriction on the size of wire used, and the route length is usually under 15 km, such pilots are referred to as low-impedance pilots.

Overhead power lines are a much more difficult problem. Suspended pilots on the same towers as the power line are usually regarded as a last resort. Buried pilots can be of two types—Post Office or private telephone.

Post Office telephone lines are designed for communication. If the telephone line is open-circuited, short-circuited or has its polarity reversed, it is inconvenient but as a protection pilot such happenings can be serious: the student should consider the effect of each on a balanced voltage system. The Post Office requires that there should be a 15 kV insulating barrier between the C.T.s and the telephone pilot, and that the pilot voltage be limited to 130 V (peak) and the current to 60 mA (r.m.s.). Post Office pilots are usually subjected to continual supervision by the protective system to detect (not prevent) interference with it.

With the development of supervisory systems for the remote control of a power system, using private multicore telephone cables, it may become economic to add protection pilot duties to the telephone system. Such private telephone systems are under the control of the electricity authority so that they are less liable to interference, are constant in their electrical characteristics and are usually homogeneous throughout their length.

Since telephone pilots consist of fine wire and can be up to 50 km route length, they are referred to as high-impedance pilots. The wires are lightly insulated and are very close together, so they have a large capacitance.

Feeder Protective Systems

The balanced-voltage protective system now to be described is normally used to protect ring feeders comprising underground cables rated at 33 kV or above and having route lengths of only a few km, so the pilots are of the low-impedance type. It can be used to protect overhead lines, using high-impedance, telephone type pilots, but with higher primary fault settings.

Both ends of the protective system comprise current-transformers and a summation transformer (as in Fig. 9.19), the latter giving an output voltage dependent on the fault ampere-turns in its primary. The relay comprises a transductor and an output tripping relay similar to that shown in the lower half of Fig. 9.15. The transductor primary is energised by the current circulating round the pilot loop due either to an internal fault on the feeder or to out-of-balance (spill) current during an external fault. To prevent tripping due to the latter, the d.c. for the transductor control winding is obtained from the pilot voltage via a high resistance and a full-wave rectifier, the relay design being such that the d.c. control current is large enough to restrain the relay from operating to the small pilot current. During an internal fault, the relatively large pilot current must be sufficient to operate the relay despite the restraint action of the pilot voltage. Thus the summation transformer should have a high voltage regulation, i.e. it is a high-impedance voltage source.

The summation transformer is designed to saturate, with a knee-point voltage of about 100 V, to limit the pilot charging current and so that the protective system can be used with telephone type pilots. This saturation results in harmonic currents flowing in the pilot and to prevent these from flowing in the transductor primary, the latter is shunted by a capacitor to form a parallel-resonant circuit tuned to 50 Hz.

The student should now draw the complete circuit diagram (both ends), with correct transformer polarities, such that the protective system restrains for an external fault and operates for an internal fault. It is conventional to indicate a balanced voltage protective system by showing the pilot wires crossing-over at the centre of the pilot loop: thus the two ends of the system are mirror images of each other. In the circuit used in practice, the voltage restraint is obtained from a tertiary winding on the summation transformer, and not directly from the pilot voltage. This protective system is a static system except for the output tripping relay.

Another balanced-voltage, ring-feeder protective system designed to protect overhead lines up to about 30 km route length, using high-impedance, telephone-type pilots also uses the circuit shown in Fig. 9.19 but the summation transformer is designed to operate like a current transformer by connecting a low-value resistor across its secondary terminals, the voltage across this resistor being the voltage applied via the relay to the pilots (see Fig. 9.20). Thus the summation transformer, when viewed from the pilot is a low-impedance source and when viewed from the current transformers is a low-impedance burden, even when the pilot loop is open-circuited. Across the above resistor is connected a non-linear resistor to limit the pilot voltage to about 100 V during maximum external fault (together with a filter circuit to prevent harmonics due to the non-linear resistor being passed on to the relay).

The pilot current is full-wave rectified and used to energise the relay operating coil (O). The relay is of the permanent-magnet moving-coil pattern similar to that of a loud-speaker.

During a through fault the charging current flowing to half the pilot capacitance (C), is supplied by the tuning choke (L) which together form a parallel resonant circuit. Thus the current in the operating coil can be reduced to a small value by tuning the choke to

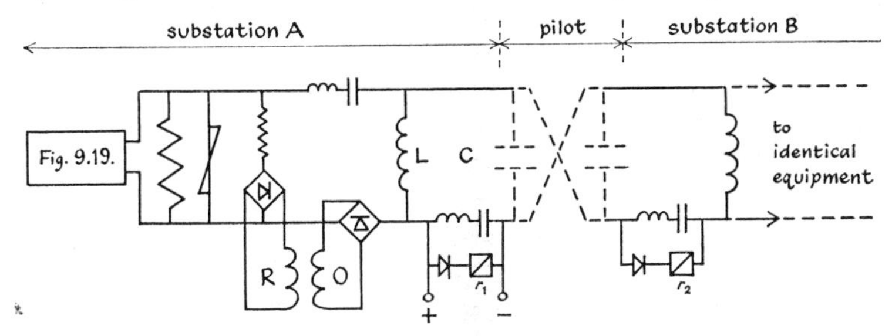

FIG. 9.20. *Feeder protection.*

match the pilot. The through fault stability is further improved by adding a voltage-operated restraint coil (R) which is designed to pull the relay against its back stop, against the action of the largest current likely to flow in the operating coil, during a through fault.

During an internal fault (fed from one end only) the current in the operating coil must be such as to trip the relay against the action of the restraint coil.

Fig. 9.20 also shows a typical form of pilot supervision. A low-voltage d.c. supply circulates about 6 mA round the pilot loop. The relay (r_1) detects failure of the supervising supply. The relay (r_2) detects failure of the pilot (open-circuit, short-circuit, reversal of connections). Both these relays operate the alarm via time-delay relays which prevent the alarm operating to transient disturbances in the pilot loop. The capacitors constrain the d.c. current to flow round the pilot loop and their a.c. impedances are cancelled by using series resonant circuits.

The above system is one of many pilot-wire feeder protective systems available. Volume 2 gives a general theory for such systems and discusses pilotless systems such as the impedance (distance) and carrier systems used for long overhead transmission lines.

REFERENCES

ATABEKOV, G. I. 1960. *The relay protection of high voltage networks.* Pergamon, London.

BUCKINGHAM, H. & PRICE, E. M. 1966. *Principles of electrical measurements.* English Universities Press, London.

ENGLISH ELECTRIC CO. 1966. *Applications guide to protection.* English Electric Co., Stafford, England.

FROST-SMITH, E. H. 1966. *The theory and design of magnetic amplifiers.* Chapman and Hall, London.

JENKINS, B. D. 1967. *Introduction to instrument transformers.* Newnes, London.

MASON, C. R. 1956. *The art and science of protective relaying.* Chapman and Hall (Wiley), London.

MATHEWS, P. 1955. *Protective current transformers and circuits.* Chapman and Hall, London.

SHOTTER, G. F. and TAGG, G. F. 1960. *Induction type integrating meters.* Pitman, London.

WARRINGTON, A. R. VAN C. 1962. *Protective relays.* Chapman and Hall, London.

BRITISH STANDARDS:

142. *Electrical Protective Relays.*

3938. *Current Transformers.*

3950. *Electrical Protective Systems for A.C. Plant.*

88. and 2692. *Fuses.*

I.E.E. Conference Publication No. 125, 1975, 'Developments in power-system protection'.

Examples

Note: The examples on time-graded protection using I.D.M.T.L. relays assume that the characteristic curve is given in Fig. 9.5 and that the plug settings available are those given in the text, section 9.4.2.

1. A 50-Hz current transformer has 20 primary turns and 60 secondary turns and the net iron cross-section of the core is 8·0 cm². The secondary current is 5·0 A. The impedance of the burden and the leakage impedance of the secondary winding are together equal to (0·7+j 0·4) Ω. The flux in the core is $\Phi\underline{/0^\circ}$. The magnetising current required in the primary winding to produce this flux if the secondary were open-circuited would be $0{\cdot}60\underline{/45^\circ}$A. Calculate the flux density in the core and estimate the phase-angle error and the ratio error of the transformer.

(I.E.E.) (0·377 pk tesla, 0·59°, −3·72%)

2. The secondary current in a 50-Hz current transformer is 3·0 A and the secondary-circuit total impedance is (0·60+j 0·45) Ω. The transformer has 4 primary and 40 secondary turns and the net cross-

section of the core material is 10·0 cm². Calculate the peak flux density in the core.

Under these conditions the primary current includes a core magnetising current consisting of a purely reactive component of 0·20 A and a core-loss component of 0·15 A. Determine the actual ratio and the phase-angle error.

(I.E.E.) (0·254 pk tesla, 10·08, 0·132°)

3. A current transformer with a bar-primary has 300 turns on its secondary winding. The resistance and reactance of the secondary circuit are 1·5 ohms and 1·0 ohm respectively (including the transformer resistance and leakage reactance). With 5 amperes flowing in the secondary circuit the magnetising magnetomotive force required is 100 ampere-turns and the iron loss is 1·2 watts. Determine the ratio and phase-angle errors.

(I.E.E.) (−5·92%, 2·3°)

4. An uncompensated bar-primary current transformer has a nominal ratio of 1000/5 A. When supplying 5 A to a relay, the relay burden is 1·4+j 2·45 Ω, the C.T. secondary impedance is 3·5+ j 0·35 Ω, the magnetising ampere turns are 95 and the core-loss ampere turns are 47·5. Estimate the current error, the phase error and the turns compensation to reduce the current error to a minimum (assume the data is unchanged).

(L.C.T.) (−8·43%, 3·1°, −17 turns)

5. A 3-phase, 11-kV cable *ABC*, fed at *A*, has 3-phase fault levels, during maximum load conditions, of 250, 225 and 200 MVA at *A*, *B* and *C* respectively (based on 11 kV). The 11-kV circuit breaker at *C*, controlling the 11 000/415-V, 1000-kVA transformer at *C*, is of the direct acting type with time-limit fuses which can be assumed to blow in 0·1 second for a 200 MVA fault on the h.v. side of the transformer. At *B* in the section *BC* and at *A* in the section *AB*, are fitted 400/5 A C.T.s, and I.D.M.T.L. relays. Assuming a discriminative time margin of 0·5 second and relay plug settings such that for a 3-phase fault immediately on the remote side of the relay, the relay plug setting multiplier has to be less than 20 by as small a margin as possible, calculate the plug settings and time setting multipliers of the relays at *B* and at *A*. If the maximum loads in sections *AB* and

BC are respectively 2 and 1 MVA, check that the protection will not operate during maximum load conditions.

(L.C.T.) (150%, 0·258; 175%, 0·452)

6. Assuming the circuit and the answers given in example 5 but with a minimum load 3-phase level in-feed at *A* of $\frac{3}{4}$ of its maximum load fault level, calculate the 3-phase fault levels at *B* and at *C* and the discriminative time margin between the relays at *A* and *B*, and comment on the last answer.

(L.C.T.) (173·2, 158 MVA; 0·564 s)

7. The circuit in example 6 is fed by a 33/11-kV, delta/resistance-earthed star-connected transformer such that the earth-fault level is 2000 A at all points along the cable *ABC*. Assuming that the I.D.M.T.L. relays are connected as three over-current relays with the settings given in the answers to example 5, and that the fuse still blows in 0·1 second, calculate the discriminative time margins between the protection at *B* and *C* and between *A* and *B*.

(L.C.T.) (1·37, 1·42 s)

8. Assuming the circuit and data in example 7 but with the I.D.M.T.L. relays now connected as two over-current and one earth-fault, the latter to have plug setting multipliers less than 20 by as small a margin as possible, calculate the plug settings and time settings of the earth fault relays at *B* and *A* to give a discriminative margin of 0·5 second between protective devices. Compare this 0·5 second margin with those for example 7.

(L.C.T.) (30%, 0·251; 30%, 0·461)

9. A transformer-feeder *AB* is fed at *A* at 33 kV where the 3-phase fault level is 1000 MVA (at 33 kV). The protection at the *A* end comprises 500/5 A C.T.s and overcurrent I.D.M.T.L. relays. The transformer at the *B* end is rated at 20 MVA, 33/11 kV, delta/resistance-earthed star-connected and is liable to be overloaded by 30%. The transformer has a leakage impedance of 10% based on its rating. The impedance of the 33 kV cable can be neglected. The protection on an outgoing 11-kV feeder comprises 400/5 A C.T.s and overcurrent I.D.M.T.L. relays whose plug settings are 150% and time settings 0·4. For 3-phase faults, calculate the fault level MVA at the 11-kV busbar and the operating time of the protection; also

the plug setting of the 33-kV protection which will keep the plug setting multiplier less than 20 by as small a margin as possible and its time setting multiplier to give a 0·5 second discriminative time with the 11-kV relays. Check that the 33-kV relays will not operate when the feeder is at maximum load.

(L.C.T.) (200 MVA, 0·932 s; 150%, 0·281)

10. An 11-kV busbar has duplicate incoming feeders each fitted with 1000/5 A I.D.M.T.L. relay having plug and time settings of 150% and 0·4. The corresponding data for an outgoing feeder is 400/5 A, 175%, 0·3. The substation is a firm supply for 25 MVA; explain the meaning of this statement. For a 3-phase 250-MVA fault, calculate the time margin between the relays for (*a*) two incoming feeders in (*b*) one in. Show that the protection on a single incoming feeder (other switched out) will not operate to a load of 25 MVA.

(L.C.T.) (1·2, 0·61 s)

11. A 30-MW, 37·5-MVA, 33-kV star-connected alternator is protected by a circulating current (balanced current) protective scheme using 600/1 A C.T.s and (unbiased) relays set to operate at 10% of their rated current (1 A). (*a*) If the earthing resistor is 90% based on the machine's rating, estimate the percentage of the stator winding which is not protected against an earth fault. (*b*) If it is required to protect 90% of the winding, estimate the value of the earthing resistor in ohms and percent based on machine rating.

(L.C.T.) (8·22%; 31·8 Ω, 109·3%)

12. The neutral point of a 10 000-V alternator protected by the balanced circulating-current system is earthed through a resistance of 10 ohms. The protection relay is set to operate when there is an out-of-balance current of 1 A in the pilot wires, which are connected to the secondary windings of 1000/5 A current transformers. What percentage of the winding of each stator phase is protected against a fault to earth of zero impedance, and what must be the minimum value of the earthing resistance to give protection to 90% of the winding of each phase?

(65·5%, 2·9 ohms)

13. A 30-MVA, Yd1, 132/33-kV solidly-earthed/delta-connected transformer, protected by circulating-current equipment, is supplied on the 33-kV side. If the h.v. C.T. has a primary current rating of 150 A, calculate the necessary C.T. ratios for use with relays rated at 1 A. If an earth fault of 1000 A occurs on one line terminal of the 132-kV winding (within the protected zone), determine the currents in each part of each pilot wire and in the relay coils, assuming no in-feed from the 132-kV system.

(L.C.T.) (150/0·577, 600/1 A; 3·86 A)

14. Given the data and answers in example 13, and given also that the transformer has a leakage reactance of 10% based on its rating, and that the positive-, and negative-sequence reactances of the 33-kV source are equal, calculate the fault current if a phase-to-phase fault occurs across two of the 132-kV terminals (within the protected zone). Calculate the currents in the pilot wires and relays.

(L.C.T.) (775, 2·98, 5·96 A)

15. A 50-MVA, 66/33-kV, delta/resistance-earthed star-connected transformer is protected by an overall circulating-current-protection system. The h.v. and l.v. C.T. ratios are 600/5 A and 1200/2·89 A. The unbiased relays are set at 20% of their rated current (5 A). For each of the following faults, 3-phase, phase-to-phase, phase-to-earth, at (*a*) the 33-kV terminals, (*b*) the middle of the 33 kV winding, calculate the fault current assuming a relay is on the point of pick-up.

(L.C.T.) ((*a*) 240, 208, 416; (*b*) 480, 416, 832 A)

16. For the same data as in example 15, estimate the percentage of the 33-kV winding not protected against internal earth faults assuming that the earthing resistor has a value of 100% based on the rating of the transformer, and that there is no infeed from the 33-kV system. To improve the protective scheme a restricted earth fault protective system is added to the 33-kV winding, i.e. three C.T.s residually connected in the 33-kV lines and one C.T. in the neutral, all four C.T.s having a ratio of 800/5 A, and connected as a circulating-current system using one relay set at 20% of its rated current of 5 A. Estimate the percentage of 33-kV winding not protected against earth faults.

(L.C.T.) (68·8, 18·3%)

17. A 25-MVA, 132/33-kV, solidly-earthed star/delta connected

transformer is protected overall by an unbiased circulating-current system. The C.T. ratio on the l.v. side is 600/1 A and the relays have a setting of 40% of their rated current of 1 A. If the connection of the star point into true earth has a resistance of 5 Ω, estimate the percentage of the h.v. winding protected against internal earth faults: assume that there is no infeed from the 132-kV system and that only the earth resistance limits the fault current.

(L.C.T.) (91·7%)

18. It is proposed to apply overall circulating-current protection to a transformer rated at 1000 kVA, 11 000/415 V, delta/solidly-earthed star-connected, using relays rated at 1 A. Current transformers having the following primary rated currents are available: 50, 60, 75, 1600, 2000 A. Choose the most suitable C.T. ratios on the basis of minimum error when expressed as a fraction of the nominal C.T. primary rated current. If x is the fraction of the 415-V winding unprotected against earth faults, S is the setting of the (unbiased) relay as a percentage of its rated current and R is the resistance in ohms between the star point and true earth, show that $x^2/RS =$ 0·115. Assuming that at least half the 415-V winding must be protected and that the smallest relay setting available is 20%, show that R must be less than 0·1087 Ω.

(L.C.T.) (75/1, 2000/0·577 A)

19. A summation transformer has primary terminals marked $P1$, $P2$, $P3$ and $P4$. The number of turns between pairs of terminals $P1$–$P2$, $P2$–$P3$, $P3$–$P4$ are 4, 4, 8 respectively, and the secondary has 10 turns. A 3-phase feeder with phases R, Y, B is fitted with star-connected, 400/5 A current transformers and $P1$, $P2$, $P3$ are connected respectively to the R, Y, B current transformers and $P4$ to their star point. If the currents in the R, Y, B phases, flowing from the supply, are respectively 0, $1000\underline{/100°}$, $900\underline{/230°}$ A and the summation transformer is short-circuited, calculate the current in the short-circuit.

(I.E.E.) (11·52 A)

20. If two sets of summation transformers and current transformers of the type given in example 19 comprise the two ends of a feeder unit-protective system and if the setting for an internal earth-fault on the B phase is 60% (of 5 A), calculate the settings for earth

faults on the *R* and *Y* phases, for phase faults across the *RY*, *YB*, *BR* phases and for a 3-phase fault.

(L.C.T.) (30, 40, 120, 120, 60, 277%)

21. Four 1000/5 A current transformers are connected as a restricted earth-fault protective system for a star-connected transformer winding (as described in example 16). The secondary impedance of each C.T. is $2+j\,0\,\Omega$. The three C.T.s on the circuit breaker are so close together that the resistance of the wiring, connecting them in parallel, can be neglected. The loop resistance of the pilot connecting these residually-connected C.T.s to the neutral C.T. is 1 Ω. The relay connected across the centre of the pilot loop has a setting current of 20 mA and a variable, internal stabilising resistor calibrated in volts at the relay terminals at setting current. When this resistor is all out, it reads 15 V. Calculate the resistance of the relay. If this protective system is to be stable up to a through-earth fault of 12·5 kA, estimate the voltage setting of the relay (assuming no factor of safety), and the ohmic value of the stabilising resistor. If with this setting, the relay is on the point of operating to an internal earth fault (fed from the power transformer primary only) estimate this minimum fault current as a percentage of the C.T. primary rated current, assuming that each C.T. takes an exciting current of 60 mA for a secondary voltage of 156·2 V.

(L.C.T.) (750 Ω, 156·2 V, 7·06 kΩ, 5·2%)

Appendix 1

THE PER-UNIT SYSTEM

The per-unit value of any quantity is defined as the ratio of that quantity to an arbitrarily chosen value having the same dimensions and called the base value. Thus a per-unit value is dimensionless. Base values will be indicated by a subscript *b*, and per-unit values by a subscript *pu*.

If a single-phase supply of voltage $\mathbf{V}$ feeds a current $\mathbf{I}$ amperes to a load of impedance $\mathbf{Z}\,\Omega$ and a base of V_b volts is chosen, the per-unit value of the supply voltage is $\mathbf{V}_{pu} = \mathbf{V}/V_b$. Similarly the supply current could be given as $\mathbf{I}_{pu} = \mathbf{I}/I_b$. There is a base impedance $Z_b = V_b/I_b\,\Omega$. Applying the per-unit system to impedances,

$$\mathbf{Z}_{pu} = \frac{\mathbf{Z}}{Z_b} = \frac{\mathbf{V}/\mathbf{I}}{V_b/I_b} = \frac{\mathbf{V}/V_b}{\mathbf{I}/I_b} = \frac{\mathbf{V}_{pu}}{\mathbf{I}_{pu}}. \qquad [A1.1]$$

Thus the per-unit system obeys Ohm's law (and hence also all the other network laws).

1·0 per-unit value of any quantity equals the base value of that quantity, so that per-unit values are relative to 1. Multiplying per-unit values by 100 gives the percentage (%) values which are relative to 100. The student should now show that Ohm's law is not true in the percentage system of units. Hence any power system data given in the percentage system is best changed to the per-unit system before being used in any calculations.

The choice of base values is arbitrary, so clearly the choice cannot affect the ultimate answer to any given problem. Since $Z_b = V_b/I_b\,\Omega$, only two of these three can be arbitrarily chosen. Further, the choice also fixes the base volt-amperes $S_b = V_b I_b$. Volt-amperes (and hence also power and reactive power) can be manipulated in the per-unit system since

$$\mathbf{S}_{pu} = \frac{\mathbf{S}}{S_b} = \frac{\mathbf{VI}^*}{V_b I_b} = \mathbf{V}_{pu}\mathbf{I}^*_{pu}. \qquad [A1.2]$$

The reason for using the conjugate of current in [*A1.2*] is given in Appendix 4.

The original quantities **V**, **I**, **Z**, **S** can all be represented by complex numbers having phase-angles. It is essential that the per-unit system should preserve these phase-angles. Hence it must be clearly understood that all base values are scalars having no phase-angles. Thus the phase-angle of a per-unit quantity is the phase-angle of the original quantity in its numerator. The per-unit system obeys the rules of complex algebra. If $\mathbf{Z} = R \pm \mathrm{j}\, X\,\Omega$ (equivalent series circuit) then $\mathbf{Z}_{pu} = R_{pu} \pm \mathrm{j}\, X_{pu}$. If $\mathbf{Y} = G \mp \mathrm{j}\, B$ mhos (equivalent parallel circuit) then $\mathbf{Y}_{pu} = G_{pu} \mp \mathrm{j}\, B_{pu}$. Also $Y_b = 1/Z_b$, and $\mathbf{Y}_{pu} = \mathbf{Y}/Y_b$.

There is nothing strange about the per-unit system: it is in everyday use and makes certain numerical data much more meaningful than it otherwise would be. If the current in a circuit is given as 30 A, that information by itself is not very informative unless the rated current is also given, as only then do we know whether the circuit is lightly loaded or overloaded. If the rated current were given as 40 A, then the circuit is on $\frac{3}{4}$ rated load and the current is 75% or 0·75 per-unit of rated current. Similarly, the student is quite familiar with efficiency given as a percentage value, power factor as a per-unit value and slip as either a percentage or per-unit value. Conversely, the per-unit value of a current is meaningless unless the base current is also given. The distinction between base values and rated values is dealt with in the discussion which follows. The (thermal) rated current of any circuit is that current which the circuit can carry continuously without exceeding the temperature limits specified in the appropriate British Standard Specification (B.S.S.).

Formulae which are true for the usual system of units, are also true in the per-unit system: e.g. take a series circuit of total impedance $\mathbf{Z}_t$ and components $\mathbf{Z}_1$, $\mathbf{Z}_2$... then $\mathbf{Z}_t = \mathbf{Z}_1 + \mathbf{Z}_2 + \ldots\,\Omega$. Dividing through by Z_b changes the formula to the per-unit system while leaving it still true. Thus

$$\mathbf{Z}_{t\,.\,pu} = \mathbf{Z}_{1\,.\,pu} + \mathbf{Z}_{2\,.\,pu} \ldots$$

Clearly the initial statement above must be qualified by adding 'provided that only one (common) base value is used'. This condition may become clearer by substituting $Z_b = V_b/I_b$, giving

$$I_b\mathbf{Z}_t/V_b = I_b\mathbf{Z}_1/V_b + I_b\mathbf{Z}_2/V_b + \ldots \qquad [A1.3]$$

This equation shows that ohmic impedances must be given at a common voltage level, V_b (this point is important in connection with transformers and will be dealt with later in this Appendix), and also that a common base current I_b must be used.

In general, different items of plant in a power system will have different rated currents and voltages, so it is necessary, when using the per-unit system, to change to a common or base value of current and of voltage. Hence the use of base value rather than rated value.

It follows from [A1.3] that the formula relating the ohmic value of an impedance $\mathbf{Z}$ to its per-unit value $\mathbf{Z}_{pu}$ is

$$\mathbf{Z}_{pu} = \frac{I_b \mathbf{Z}}{V_b}. \qquad [A1.4]$$

Thus the per-unit impedance of equipment at a certain MVA, i.e. a certain current and at rated voltage, is the fraction given by dividing the phase impedance voltage, when that current is flowing, by the normal rated phase voltage.

Worked example A.1

Two impedances, 1+j 2 and 2+j 2 Ω, are connected in series across a 250-V supply. Calculate the total impedance and load current using the per-unit system.

SOLUTION

Base 200 V, 20 A (the choice is arbitrary)

$$1 \text{ p.u. impedance} = 200/20 = 10\,\Omega$$
$$\mathbf{Z}_{1.pu} = (1+\text{j}\,2)/10 = 0{\cdot}1+\text{j}\,0{\cdot}2 \text{ p.u.}$$
$$\mathbf{Z}_{2.pu} = (2+\text{j}\,2)/10 = 0{\cdot}2+\text{j}\,0{\cdot}2 \text{ p.u.}$$
$$\mathbf{Z}_{t.pu} = 0{\cdot}3+\text{j}\,0{\cdot}4 = 0{\cdot}5\underline{/53{\cdot}14^\circ} \text{ p.u.}$$
$$\mathbf{V}_{pu} = 250/200\underline{/0^\circ} = 1{\cdot}25\underline{/0^\circ}$$
$$\mathbf{I}_{pu} = (1{\cdot}25\underline{/0^\circ})/(0{\cdot}5\underline{/53{\cdot}14^\circ}) = 2{\cdot}5\underline{/-53{\cdot}14^\circ} \text{ p.u.}$$
$$I = 2{\cdot}5\times 20 \text{ A} = 50 \text{ A}.$$

The impedance voltage (IZ)/phase of a machine is defined as the product of its rated phase current in amperes and its impedance in ohms/phase. If the impedance voltage, in volts/phase, is divided by the rated phase voltage, it will be seen from [A1.4] that the quotient is an impedance in per-unit. Thus if an alternator has a (synchronous) impedance of 200% or 2 p.u., then on full-load current, its impedance voltage IZ_s is twice its rated phase voltage.

A point which often causes confusion is whether a machine's

per-unit impedance refers to its rated current or to some arbitrarily chosen base current. Either can be used, but which it is should be clearly stated: if it is not stated, rated current will usually be assumed since this is the known quantity.

An advantage of the per-unit system is that if a particular type of machine is taken—say, a non-salient, 2-pole alternator—its impedance in Ω/phase varies over a wide range depending on its rating. But if this same impedance is expressed in per-unit based on its thermal rating, it varies over a much smaller range and is often a fixed (standardised) value.

The per-unit system of units can also be applied to a 3-phase circuit. The simplest procedure, for balanced 3-phase circuits, is to use equivalent per-phase-star (line-to-neutral) values for all original and base quantities. When dealing with balanced 3-phase circuits, it will always be assumed, unless otherwise stated, that the circuit is star-connected: i.e. the impedances take line current at line-to-neutral voltage.

The per-unit value of the line-to-line voltage equals the per-unit value of the line-to-neutral voltage and the per-unit value of the 3-phase volt-amperes equals the per-unit value of the volt-amperes/phase, provided always that the base values form a balanced 3-phase system, i.e. base line voltage equals $\sqrt{3} \times$ base voltage/phase and base 3-phase volt-amperes equals $3 \times$ base volt-amperes/phase. In the per-unit system, the volt-amperes (whether 3-phase or per phase) equal the product of the voltage (whether line or phase) and the current and there are no factors such as 3 or $\sqrt{3}$.

The per-unit system is a completely consistent system of units which satisfies all the usual electrical circuit formulae. Thus any two (but only two) base quantities can be arbitrarily chosen, as all the other base quantities can be derived from the two specified. 3-phase MVA and line kV are the two quantities specified (as these are usually given on the rating plate of any machine), but the base quantities chosen are the corresponding phase values, or 3-phase MVA and phase voltage.

Thus if the base values are taken as 100 MVA (3-phase) at 132 kV (line) then

$$1 \text{ p.u. voltage} = 132/\sqrt{3} = 76{\cdot}21 \text{ kV/phase}$$
$$1 \text{ p.u. current} = 100/(\sqrt{3} \times 132) = 0{\cdot}4373 \text{ kA}$$
$$1 \text{ p.u. impedance} = 76{\cdot}21/0{\cdot}4373 = 174{\cdot}3\ \Omega\text{/phase (star).}$$

Equation [*A1.4*] may be re-written as

$$\mathbf{Z}_{pu} = \frac{I_b \mathbf{Z}}{V_b} = \frac{(3V_b I_b \times 10^{-6})\mathbf{Z}}{(\sqrt{3}\,V_b \times 10^{-3})^2}$$

$$= \frac{(\text{base 3-phase MVA})\,\mathbf{Z}}{(\text{base line kV})^2}. \qquad [A1.5]$$

This formula relates an ohmic impedance to its per-unit value, in terms of the two normal base values. A given alternator, line or cable (transformers will be dealt with later) has a fixed ohmic impedance. Hence it follows from [*A1.5*] that

$$Z_{pu} \propto \frac{\text{base 3-phase MVA}}{(\text{base line kV})^2}. \qquad [A1.6]$$

Thus if a machine's per-unit impedance is given relative to its thermal rating, it can be changed to any base value using the above proportionality.

Worked example A.2

A 3-phase alternator rated at 50 MVA, 12·8 kV has a (synchronous) reactance phase of 2·0 p.u. on its thermal rating. It supplies an 11-kV overhead line 5 km long each phase of which has an impedance of 0·22+j 0·51 Ω km. Calculate the total impedance on a base of 100 MVA (3-phase) and 11 kV.

SOLUTION

$$\text{Base current} = \frac{100 \times 10^6}{\sqrt{3} \times 11\,000} = 5260 \text{ A}.$$

$$\text{The line per-unit impedance} = \frac{5260 \times 5(0{\cdot}22 + \text{j}\,0{\cdot}51)}{11\,000/\sqrt{3}}$$

$$= 0{\cdot}91 + \text{j}\,2{\cdot}11 \text{ p.u.}$$

From [*A1.4*] the alternator reactance

$$= \text{j}\,\frac{12\,800}{\sqrt{3}} \times 2{\cdot}0 \Big/ \left(\frac{50 \times 10^6}{\sqrt{3} \times 12\,800}\right) = \text{j}\,12\,800^2 \times 4 \times 10^{-8}\ \Omega.$$

The alternator per-unit reactance at 100 MVA, 11 kV

$$= \text{j}\,\frac{12\,800^2 \times 4 \times 10^{-8} \times 5260}{11\,000/\sqrt{3}} = \text{j}\,5{\cdot}42 \text{ p.u.}$$

The total impedance at 100 MVA, 11 kV = 0·91+j 7·53 p.u.

The student is strongly recommended to avoid memorising equations other than the minimum of basic relationships, (which tends also to reduce the risk of error) and for this reason [*A1.4*] has been used in Example A.2 rather than [*A1.5*]. It should be noted however that engineers accustomed to making particular calculations, are able to reduce the amount of work by using other relationships. An example of this is that [*A1.5*] gives the alternator reactance $j\,2{\cdot}0\times(100/50)\times(12{\cdot}8/11)^2 = j\,5{\cdot}42$ p.u. From [*A1.4*] the student should be able to demonstrate that if a machine has a resistance of 1% on its thermal rating, then its full-load I^2R loss is 1% of its VA rating.

A major advantage of using the per-unit method in power system analysis, arises when the system contains transformers. An analysis using ohmic impedances would involve transferring impedances through the transformers, using the square of the turns-ratio, until all impedances had been referred to a common voltage level. Having calculated the current at that voltage level (for any given conditions) that current would then need to be transferred through the transformers, using the inverse turns-ratio, to find the current at any other voltage level.

It will now be shown that the per-unit value of an impedance is the same on both sides of a two-winding transformer, and similarly so is the per-unit value of a current, provided that the base values on both sides of the transformer are properly related by transformer action. If subscript p refers to the primary side and subscript s to the secondary side, then given the base values V_{bp} and I_{bp} on the primary side, the base values on the secondary side are (assuming for simplicity a single-phase transformer)

$$\left.\begin{aligned} V_{bs} &= V_{bp}\times(N_s/N_p) \\ I_{bs} &= I_{bp}\times(N_p/N_s) \end{aligned}\right\}. \qquad [A1.7]$$

If $\mathbf{Z}_p$ is the ohmic impedance of the primary winding, its per-unit value is

$$\mathbf{Z}_{p\cdot pu} = I_{bp}\mathbf{Z}_p/V_{bp}.$$

Similarly $\mathbf{Z}_{s\cdot pu} = I_{bs}\mathbf{Z}_s/V_{bs}$.

When the secondary impedance is referred to the primary winding, the total transformer ohmic impedance is $\mathbf{Z}_p+\mathbf{Z}_s(N_p/N_s)^2\,\Omega$, and the per-unit impedance of the whole transformer on the primary side

$$= \frac{I_{bp}}{V_{bp}}(\mathbf{Z}_p + \mathbf{Z}_s(N_p/N_s)^2) \text{ p.u.}$$

$$= \frac{I_{bp}}{V_{bp}}\mathbf{Z}_p + I_{bp}\frac{N_p}{N_s} \times \frac{N_p}{N_s V_{bp}} \times \mathbf{Z}_s \text{ p.u.}$$

$$= \frac{I_{bp}\mathbf{Z}_p}{V_{bp}} + \frac{I_{bs}\mathbf{Z}_s}{V_{bs}} \text{ p.u.}$$

$$= \mathbf{Z}_{p\,.\,pu} + \mathbf{Z}_{s\,.\,pu} \text{ p.u.}$$

Thus the total per-unit impedance of the transformer, for any base MVA and voltage, is the sum of the per-unit impedances of the primary and secondary windings, and is therefore the same on both sides of the transformer. Clearly the impedance of any equipment connected to the primary and at the same voltage may be included with $\mathbf{Z}_p$, and the impedance of any equipment connected to the secondary may be included with $\mathbf{Z}_s$. The base values could equally well have been in volt-amperes and volts. Clearly base volt-amperes is the same on both sides of a two-winding transformer. The problem of 3-winding transformers, where the ratings of the three windings are not necessarily all equal, is dealt with in section 6.9.

Thus in power system analysis using the per-unit method, a transformer can be regarded as equivalent to a series impedance and the transformation of current and voltage can be disregarded. The ultimate answers must be given in the usual system of units and are obtained from the per-unit values using the base values corresponding to the voltage level at which they are required.

The above analysis is valid for a single-phase transformer or for a star/star transformer where the symbols refer to the per-phase-star (line-to-neutral) values. It will now be shown that the per-unit value of a phase impedance is the same in star or in delta. Consider a delta winding having an impedance of $\mathbf{Z}$ Ω/phase, connected to a 3-phase supply of base line voltage V_{bL} with a corresponding base phase voltage $V_b = V_{bL}/\sqrt{3}$. The base current in one phase of the winding is I_b with a corresponding base line current $I_{bL} = \sqrt{3}I_b$ and

$$\mathbf{Z}_{pu\,.\,\text{delta}} = \frac{I_b\mathbf{Z}}{V_{bL}}.$$

The equivalent star impedance is $\mathbf{Z}/3\,\Omega$ (see Appendix 2) and its per-unit value is

$$\mathbf{Z}_{pu\,.\,\text{star}} = \frac{I_{bL}(\mathbf{Z}/3)}{V_b} = \frac{\sqrt{3}I_b(\mathbf{Z}/3)}{V_{bL}/\sqrt{3}} = \mathbf{Z}_{pu\,.\,\text{delta}}.$$

Thus if a transformer impedance is specified in per-unit, it is not necessary to know whether the windings are in star or delta, except in the final analysis when the current distribution within the transformer is required. In the case of a delta/star transformer, the 30° phase shift must also be accounted for (see sections 6.3 and 8.3.5).

When using [*A1.4*] or [*A1.5*] it is essential that all quantities on the right-hand side are specified on the same side of the transformer.

Appendix 2

STAR-DELTA TRANSFORMATION

Given three (and only three) terminals (Fig. A2.1) in *any* circuit, a star-connection of three impedances can be replaced by a delta-

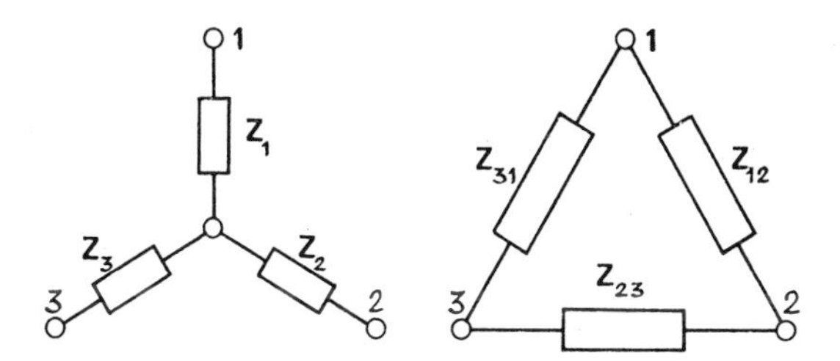

FIG. A.2.1. *Star-delta transformation.*

connection of three impedances, without affecting the remainder of the circuit, and the converse is true. Such impedances are referred to as equivalent impedances. As with most network theorems this refers to linear impedances.

The usual proof calculates the impedances between the three pairs of terminals in both circuits and equates them: this gives three simultaneous equations for three unknowns. Thus, given the three delta impedances, it can be shown that

$$\mathbf{Z}_1 = \frac{\mathbf{Z}_{12}\mathbf{Z}_{31}}{\mathbf{Z}_{12}+\mathbf{Z}_{23}+\mathbf{Z}_{31}}$$

By visualising the two circuits superimposed, this formula can be generalised as

$$\mathbf{Z}_{\text{star}} = \frac{\text{product of delta impedances at that terminal}}{\text{sum of impedances round the delta}} \qquad [A2.1]$$

Similarly, given the star impedances

$$\mathbf{Z}_{12} = \mathbf{Z}_1 + \mathbf{Z}_2 + \frac{\mathbf{Z}_1\mathbf{Z}_2}{\mathbf{Z}_3} = \frac{(\mathbf{Z}_1\mathbf{Z}_2)+(\mathbf{Z}_2\mathbf{Z}_3)+(\mathbf{Z}_3\mathbf{Z}_1)}{\mathbf{Z}_3}$$

$$\mathbf{Z}_{\text{delta}} = \frac{\text{sum of products of pairs of star impedances}}{\text{impedance opposite to the two terminals}} \qquad [A2.2]$$

If the three impedances are identical the formulae reduce to

$$\mathbf{Z}_{\text{star}} = \mathbf{Z}_{\text{delta}}/3. \qquad [A2.3]$$

It has been shown in Appendix 1 that the formulae above are true for per-unit impedances; but the per-unit impedance of *3-phase* equipment is the same whether it is star or delta connected.

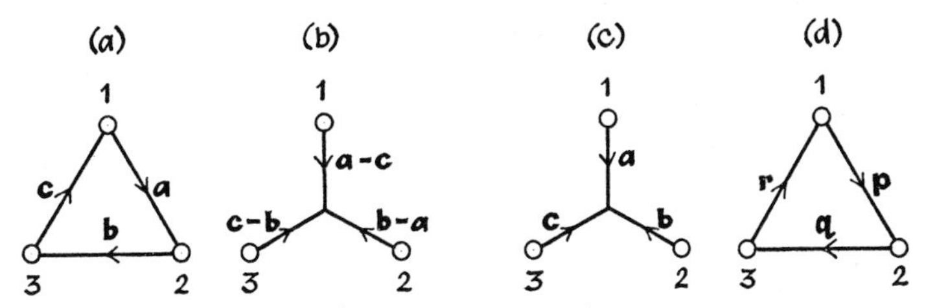

FIG. A2.2. *Current distribution.*

The star-delta transformation is used to simplify complex networks, and often the current distribution in the circuit is required. Given the delta currents in Fig. A2.2(a), the current distribution in the equivalent star (b) is obvious. Given the current distribution in the star circuit (c), the delta currents can be obtained from relations such as

$$\mathbf{p}\mathbf{Z}_{12} = \mathbf{a}\mathbf{Z}_1 - \mathbf{b}\mathbf{Z}_2.$$

Appendix 3

h-OPERATOR

It will be assumed that the student is familiar with the j-operator which is defined as an operator which multiplies a phasor (vector) so as to rotate it through 90° in the positive (anticlockwise) direction, without any change in magnitude. Thus in the conventional polar form we write $j = 1\underline{/90°}$: in the rectangular cartesian form we could write $j = 0+j\,1$. Multiplying the unit phasor $1\underline{/0°}$, Fig. A3.1, moves it from the $+x$ axis to the $+y$ axis to give $j\times 1 = 1\underline{/90°}$: multiplying this latter phasor by a further j gives $j\times(j\times 1) = 1\underline{/180°} = -1$. Thus j is obeying the usual rules of the algebra of complex

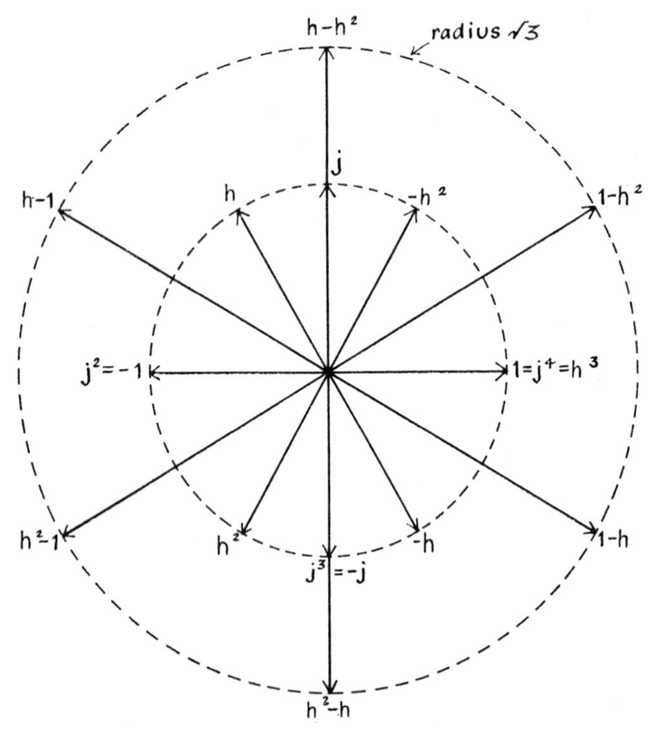

FIG. A3.1. *Operators at intervals of* 30°.

numbers, viz. to multiply two complex numbers, multiply their moduli and add their arguments. Hence $j = \sqrt{-1}$, the so-called imaginary number, in the sense that there is no real positive (or negative) number (along the x-axis) which multiplied by itself will give -1.

Since electrical power engineers are mainly concerned with 3-phase circuits, it is natural to extend the above idea to a 120°-operator.

We thus define h $= 1\underline{/120^\circ}$, an operator which multiplies a phasor to rotate it 120° in the positive (anticlockwise) direction, without any change in magnitude. Thus, $h^2 = 1\underline{/240^\circ} = 1\underline{/-120^\circ}$ and $h^3 = 1\underline{/360^\circ} = 1\underline{/0^\circ}$. Hence $h = \sqrt[3]{1}$: the three cube roots of unity are 1, h and h^2. Changing to rectangular cartesian co-ordinates we have, since $\cos 60^\circ = 1/2$ and $\sin 60^\circ = \sqrt{3}/2$,

$$h = -1/2 + j\sqrt{3}/2 \quad \text{and} \quad h^2 = -1/2 - j\sqrt{3}/2$$

Since $(1+h+h^2) = 0$, it is called the null operator—it multiplies a phasor to reduce it to zero. If **V** is any phasor then

$$(1+h+h^2)\mathbf{V} = \mathbf{V} + h\mathbf{V} + h^2\mathbf{V} = 0.$$

The geometrical interpretation of this is an equilateral triangle, the vector sum of whose sides is zero. By considering functions of j and h, a sequence of operators at intervals of 30° can be obtained as in Fig. A3.1. Such operators are important when considering symmetrical components and asymmetrical fault calculations (Chapter 8).

The above ideas can be generalised. We can define an operator $\mathbf{A}(\theta)$ as one which multiplies the magnitude of a phasor by A and increases its phase by θ in the positive sense. Thus

$$\mathbf{A}(\theta) = A\underline{/\theta}$$
$$= A(\cos\theta + j\sin\theta) = A\,e^{j\theta} = A\exp(j\,\theta).$$

The exponential method of indicating a phasor having magnitude and direction is the basis of the treatment in mathematical texts.

Appendix 4

VOLT-AMPERES, POWER AND REACTIVE POWER

In an a.c. circuit, if $v = \sqrt{2}\,V\sin\omega t$ and $i = \sqrt{2}\,I\sin(\omega t - \phi)$, the instantaneous power is given by

$$p = vi = VI[\cos\phi - \cos(2\omega t - \phi)] \text{ watts}$$

where V and I are r.m.s. values. This shows that power contains a double-frequency component which must change sign (direction of flow) in alternate half-cycles (of the double-frequency) and which must be zero when averaged over an integral number of cycles. Thus the average power is $P = VI \cos \phi$.

The above equation can be rewritten as

$$p = VI \cos \phi - (VI \cos \phi) \cos 2\omega t - (VI \sin \phi) \sin 2\omega t \text{ watts.} \quad [A4.1]$$

The first term gives the average power P, flowing from the source to the load, which when multiplied by time will give the energy consumption. The second term represents a double-frequency oscillating power flow of average value zero, whose amplitude, being equal to P, will be a maximum at unity power factor and zero at zero power factor. The third term is also a double-frequency oscillating power flow, of average value zero, and amplitude $Q = VI \sin \phi$, i.e. zero at unity power factor and maximum at zero power factor. These two oscillating powers are orthogonal to each other in time.

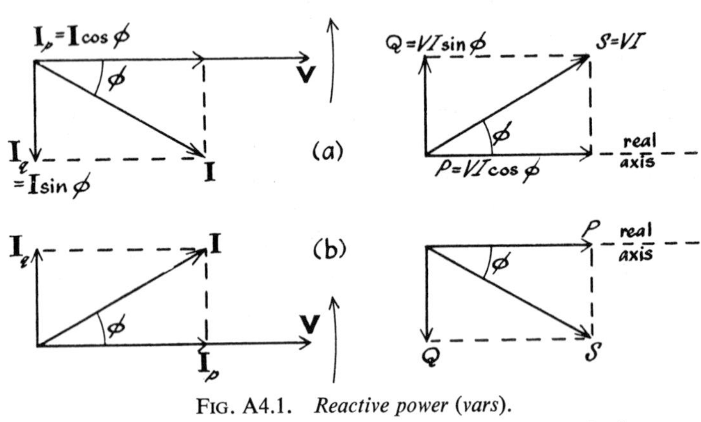

FIG. A4.1. *Reactive power (vars).*
(*a*) *Inductive load* (*b*) *Capacitive load*

If the load is regarded as a parallel RL circuit, Fig. A4.1(a), the R branch will take an active current $I_p = I \cos \phi$, and consume a power P, and a power $P \cos 2\omega t$ will oscillate between the supply and the R branch, while the L branch will take a reactive lagging current $I_q = I \sin \phi$ and a power $Q \sin 2\omega t$ will oscillate between the supply and the L branch. $Q(= VI \sin \phi)$ is called the reactive power (volt-amperes reactive or vars) of the circuit. The corresponding unit is written as VAr. The RL load is a sink of power P and is said to be a sink of lagging vars Q. It should be noted from [*A4.1*] that Q is the amplitude of a power oscillating at twice supply frequency between the supply and the electromagnetic field of the inductive load. It follows that the supply alternator is a source of power and of lagging vars.

By international agreement, the lagging vars which are taken by an inductive load, and are therefore flowing in the same direction as the power, are taken to be positive.

If now the load is assumed to be a parallel *RC* circuit (Fig. A4.1(b)), it is clear that the active (power) component of current I_p is still in phase with the corresponding current for the inductive circuit, since for both circuits the direction of power flow is the same. But the two reactive components of current are in antiphase relative to each other, so that the corresponding double-frequency oscillating powers $Q \sin 2\omega t$ are also in antiphase. The corresponding vars are therefore also in antiphase. Leading vars can thus be regarded as negative vars flowing from source to load or as positive vars flowing from load to source. A capacitor is therefore a sink of leading (negative) vars or a source of lagging (positive) vars. Thus the action of a power-factor-improving capacitor connected across an inductive load can be explained by saying that the capacitor generates lagging vars which it supplies to the load, thereby reducing the demand for lagging vars from the supply.

When dealing with synchronous motors, the direction of reference is usually taken as into the motor, i.e. the motor takes current and power. If the motor is operating at a lagging power factor, then it is a sink of lagging vars: if it is operating at a leading power factor, it is a sink of leading vars or a source of lagging vars.

In the case of a transmission line, the series inductance is a sink of lagging vars equal to I^2X_L where $X_L = 2\pi fL$, while the shunt capacitance is a sink of leading vars (source of lagging vars) equal to V^2/X_C where $X_C = 1/2\pi fC$.

In power systems it is usual to refer always to lagging vars (and omit any reference to leading vars). Thus if a power-station is generating at a lagging power factor, it is said to export watts and (lagging) vars, while if it is generating at a leading power factor it is said to export watts and import vars (the import of (positive) lagging vars being equivalent to the export of (negative) leading vars).

Fig. A4.2 shows a voltage across a load, and the current through it, drawn in arbitrary positions relative to a set of rectangular co-ordinate axes, except that **I** lags **V** by angle ϕ.

$$P = VI\cos\phi = VI\cos(\alpha-\beta)$$
$$= VI[(\cos\alpha)(\cos\beta)+(\sin\alpha)(\sin\beta)]$$
$$= ac+bd$$

and

$$Q = VI\sin\phi$$
$$= VI[(\sin\alpha)(\cos\beta)-(\cos\alpha)(\sin\beta)]$$
$$= bc-ad.$$

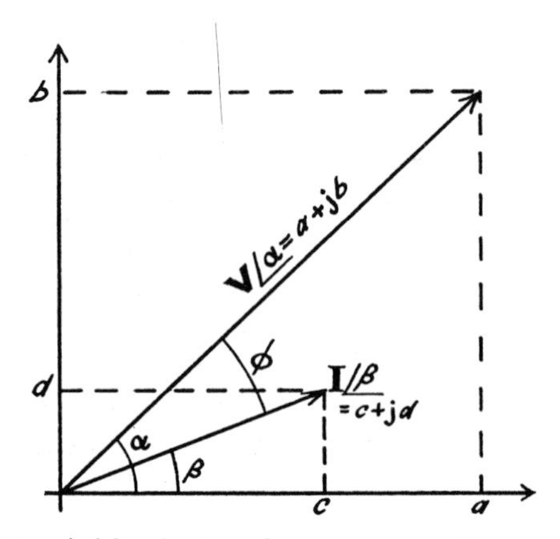

FIG. A.4.2. *Rectangular components of* **V** *and* **I**.

The interpretation of these equations is that ac represents power due to a voltage component and a current component both along the x axis, and similarly bd along the y axis, and that these add arithmetically: also that bc represents the vars due to a current component lagging a voltage component while ad represents the vars due to a current component leading a voltage component.

It would seem reasonable to attempt to obtain the volt-amperes from the multiplication of voltage and current as complex numbers. Thus

$$(a+\mathrm{j}\,b)(c+\mathrm{j}\,d) = (ac-bd)+\mathrm{j}\,(bc+ad).$$

This result is clearly incorrect. The correct result can be obtained by using the conjugate of current, $\mathbf{I}^*$, in place of the current where

$$\mathbf{I}^* = I\underline{/-\beta} = c-\mathrm{j}\,d$$

Thus

$$\mathbf{S} = \mathbf{VI}^* = (a+\mathrm{j}\,b)(c-\mathrm{j}\,d)$$
$$= (ac+bd)+\mathrm{j}\,(bc-ad)$$
$$= P+\mathrm{j}\,Q.$$

The use of conjugate current agrees with the convention of lagging vars being taken as positive when flowing in the same direction as power into an inductive load.

Volt-amperes, power and reactive power, although not phasors, are vectors in the sense that they have directions relative to each other (Fig. A4.1). Thus for an inductive load, the input volt-amperes are

$$\mathbf{S} = P + \mathrm{j}\, Q.$$

With the load voltage as the reference phasor, the corresponding load current is

$$\mathbf{I} = I_p - \mathrm{j}\, I_q.$$

The engineers' instinct to treat the most commonly occurring quantity (lagging reactive power) as positive, has resulted in this apparent contradiction in sign.

According to B.S. 1991, Part 6, volt-amperes should be called apparent power.

INDEX

Page numbers shown in **bold**-face indicate key entries

EPS 1 aa